Mathematics Classrooms in Twelve Countries

Mathematics Classrooms in Twelve Countries
The Insider's Perspective

Edited by

David Clarke
University of Melbourne
Australia

Christine Keitel
Freie Universität Berlin
Germany

Yoshinori Shimizu
University of Tsukuba
Japan

SENSE PUBLISHERS
ROTTERDAM / TAIPEI

A C.I.P. record for this book is available from the Library of Congress.

ISBN 90-77874-95-X (paperback)
ISBN 90-77874-99-2 (hardback)

Published by: Sense Publishers,
P.O. Box 21858, 3001 AW Rotterdam, The Netherlands
http://www.sensepublishers.com
Printed on acid-free paper

−

Cover design: Cameron Mitchell, ICCR, Melbourne, Australia

Cover Photo: Simulated data generation at Flinders Peak Secondary College, Corio, Victoria, Australia, Used with Permission

SERIES PREFACE

The Learner's Perspective Study provides a vehicle for the work of an international community of classroom researchers. The work of this community will be reported in a series of books of which this is the first. The documentation of the practices of classrooms in other countries causes us to question and revise our assumptions about our own practice and the theories on which that practice is based. International comparative and cross-cultural research has the capacity to inform practice, shape policy and develop theory at a level commensurate with regional, national or global priorities. International comparative research offers us more than insights into the novel, interesting and adaptable practices employed in other school systems. It also offers us insights into the strange, invisible, and unquestioned routines and rituals of our own school system and our own classrooms. In addition, a cross-cultural perspective on classrooms can help us identify common values and shared assumptions, encouraging the adaptation of practices from one classroom for use in a different cultural setting. As these findings become more widely available, they will be increasingly utilised in the professional development of teachers and in the development of new theory.

David Clarke
Series Editor

TABLE OF CONTENTS

ACKNOWLEDGEMENTS

The Editors would like to express their gratitude to Carmel Mesiti for her meticulous work in formatting the majority of the chapters in this book and in constructing both the Subject Index and Author Index. Mary Barnes' careful work in formatting the remaining chapters is also acknowledged with gratitude.

The research reported in this book benefited substantially from funding awarded by the following agencies, centres and universities:

The Australian Research Council
Ben Gurion University of the Negev *(Israel)*
Bank of Sweden Tercentenary Foundation
Centre for Research in Pedagogy and Practice, National Institute of Education, Nanyang Technological University *(Singapore)*
The Collier Charitable Trust *(Australia)*
Committee for Research and Conference Grants, University of Hong Kong, *(Hong Kong SAR, China)*
Czech Science Foundation *(Czech Republic)*
Department of Curriculum and Instruction, Purdue University *(USA)*
The Fohs Foundation *(USA)*
Global Development Network (GDN), World Bank
Japan Society for the Promotion of Science
Korean Educational Development Institute *(AP-EPRI/KEDI)*
Mathematics Association of Victoria *(Australia)*
Ministry of Education, Science, Sports and Culture *(Japan)*
Ministry of Education, Youth and Sports *(Czech Republic)*
National Research Foundation *(South Africa)*
The Potter Foundation *(Australia)*
Pundasyon sa Pagpapaunlad ng Kaalaman sa Pagtuturo ng Agham, Ink. *(The Philippines)*
Research Commission, Freie Universität Berlin *(Germany)*
Research Grants Council *(Hong Kong SAR, China)*
The Sacta-Rashi Foundation *(Israel)*
The Spencer Foundation *(USA)*
Swedish Research Council
University of KwaZulu-Natal *(South Africa)*
University of the Witwatersrand *(South Africa)*
The University of Macau, Academic Community *(China)*
The University of Melbourne *(Australia)*

All editors and authors would like to thank the teachers and students, whose cooperation and generous participation made this international study possible. Publication of this work was assisted by a publication grant from the University of Melbourne.

DAVID CLARKE, CHRISTINE KEITEL AND YOSHINORI SHIMIZU

CHAPTER ONE

The Learner's Perspective Study

INTRODUCTION

The Insider's Perspective

It is an essential thesis of the Learner's Perspective Study (LPS) that international comparative research offers unique opportunities to interrogate established practice, existing theories and entrenched assumptions. In this book, we offer you a variety of images of classrooms from twelve of the countries participating in the Learner's Perspective Study. These various portraits of classroom practice are open to at least two readings: firstly, as characterisations of salient features of practice as judged by the members of the local research group carrying out the analysis; secondly, as indicative of the diversity of practice evident in mathematics classrooms internationally. It is the first reading that prompts the title of this book: *Mathematics classrooms in twelve countries: The insider's perspective.* The term "insiders" is used in two senses: The authors of each chapter are insiders in their own cultures and school systems and carry out their analyses from that position; also, the voices that constitute the data of this research are the voices of the insiders in the classrooms studied – the students and their teachers. The resultant accounts of mathematics classrooms in twelve countries should carry a consequent weight of credibility because of their insider status. We leave it to you, the reader, to decide whether or not the accounts resonate with the practices of classrooms with which you are familiar.

The Learner's Perspective Study

The Learner's Perspective Study was designed to examine the practices of eighth grade mathematics classrooms in a more integrated and comprehensive fashion than had been attempted in previous international studies. The project was originally designed to complement research studies reporting national norms of student achievement and teaching practices with an in-depth analysis of mathematics classrooms in Australia, Germany, Japan and the USA. Since its inception, research teams from other countries have continued to join the Learner's Perspective Study. The title of the project (The Learner's Perspective Study) was

D. J. Clarke, C. Keitel & Y. Shimizu (Eds.), Mathematics Classrooms In Twelve Countries: The Insider's Perspective. 1–14. © 2006 Sense Publishers. All rights reserved.

intended to complement teacher-focused studies by foregrounding the learner's perspective. As the project grew, its purpose was progressively reinterpreted and expanded. In the same way that students, teachers and researchers are all considered to be insiders for the purposes of this book, so all can be considered learners: partners in an international collaboration to develop new knowledge and to understand and improve the practices and outcomes of our classrooms. We hope the reader will also identify with this Learner's Perspective.

The twelve research teams whose research is reported in this book are situated in universities in Australia, China, the Czech Republic, Germany, Israel, Japan, Korea, the Philippines, Singapore, South Africa, Sweden and the USA. This combination of countries gives good representation to different European and Asian educational traditions, affluent and less affluent school systems, and mono-cultural and multi-cultural societies.

A significant distinguishing characteristic of this study is its documentation of the teaching of *sequences* of lessons, rather than just single lessons. The importance of this cannot be overestimated. Analyses of classroom practice that do not take into account the situation of the lesson within the enfolding topic, ignore one of the major influences on the teacher's purposeful selection of instructional strategies. In addition, unlike previous international studies, this project has the capacity to relate identified teacher practices to antecedent student behaviours and to consequent student outcomes. The documentation of these chains of association, both within a single lesson and across several lessons, and the analysis of their cultural-specificity will contribute significantly to the improvement of both teacher and learner practices in mathematics classrooms in all participant countries. Finally, the use of post-lesson video-stimulated interviews provided an opportunity for the classroom participants' voices to be heard, in particular in relation to the meanings that each classroom activity and situation held for that participant.

Another distinguishing feature of this project is the exploration of learner practices. Previous cross-national studies have identified coherent sets of actions, and associated attitudes, beliefs and knowledge, that appear to constitute culturally-specific teacher practices. It was hypothesised in the formulation of the Learner's Perspective Study that there might be sets of actions and associated attitudes, beliefs, and knowledge of students that might constitute culturally-specific, coherent learner practices. The use of three video-cameras in the classroom, supplemented by post-lesson video-stimulated interviews, provided a data base sufficiently complex to support analysis of both individual learners' constructed meanings and their perspectives on classroom practice, as well as documenting those corporate behaviours common to the class conglomerate. It was also possible to study the consistency or variability of teacher use of particular practices, the students' construal of those teacher actions and the mathematical and social meanings constructed by students. In particular, this project facilitated the comparison of 'quality' mathematics teaching across a wide variety of school systems situated in different countries, by identifying similarities and differences in both teaching practice and in the associated student perceptions and behaviours.

2

It was of interest in this study whether the learner practices observed in the classrooms from one country showed consistency of form and purpose, sufficiently different from other classrooms, such as to suggest a culturally-specific character. Because of the highly selective nature of the classrooms studied in each country, no claims can be made about national typification of practice, however any regularities of practices sustained across thirty lessons demand some consideration of the possible causes of such consistency. Whether or not such identifiable learner characteristics exist as cultural traits, this study was predicated on a belief that international comparative studies are likely to reveal patterns of practice less evident in studies limited to a single country or community.

The findings of this study include rich descriptions of the practices of participants in eighth grade mathematics classrooms in twelve countries, predominantly from the perspective of the learner, supplemented by the perspectives of the teacher and the researcher. While the basic data collection tool was videotape supplemented by reconstructive interviews (see Clarke, 1998, 2001), the learner's perspective was constructed not only through the reporting of visual detail per se, as might be expected in a video study, but also through what the learner reported as seeing, and from an analysis of those practices in which the learner chose or did not choose to participate, together with the meanings that the student (and the teacher) reported as being associated with those practices.

This project cannot make statements of a general nature about national characteristics. It does, however, aim to situate its findings in relation to documented common teaching practices and levels of school achievement. In studying the diversity of ways in which students perceive and respond to a particular teacher action, it is helpful to know how prevalent that action is within the body of teaching practice commonly evident in the eighth grade mathematics classrooms of that country. This is where the two Third International Mathematics and Science video Studies (TIMSS) (Hiebert et al., 2003; Stigler & Hiebert, 1999) are of particular relevance to this study. The emphasis of this study, however, is not on teaching practice in isolation, but rather the manner in which this anticipates, arises from and impacts upon learner practice. It was an hypothesis to be addressed in this study that teacher and learner practices are mutually accommodating and mutually sustaining and evolve symbiotically to the mutual benefit of classroom participants, who co-construct these practices through their participation in classroom activity.

A Practice-Oriented Approach

A significant component of the Learner's Perspective Study is the utilisation of a practice-oriented analysis of learning. This approach characterises key aspects of the LPS project because it situates mathematical activity in relation to the social settings with which the project is fundamentally concerned, and also because it allows us to interrogate those settings with respect to the practices they afford and constrain. Analyses focusing upon the practices of a system (or setting) offer our best hope of accommodating the complexity of the phenomena we are interested in, but doing so in a manageable fashion. We distinguish the practice of individuals (a teacher's practice or a single learner's practice) from 'professional practice' in the sense of established 'legal practice' or 'medical practice.' In this regard, we posit the notion of an individual having constructed a body of practice in which s/he engages regularly, but which is subject to refinement, modification, rejection, and replacement over time. Such individual practice will be a subset of the practices of the various communities of which each individual has membership and will conform to the affordances and constraints of the settings and situations in which those individuals find themselves.

There are differences between this view of learning as emergent individual practice and the social theory of learning articulated by Wenger (1998), for example. These differences relate to the degree of agency accorded to the individual (i) to choose the nature of their participation in community practice, and (ii) to contribute to and change that practice. Such differences are largely ones of emphasis, with Wenger foregrounding the community into whose practice the learner is being initiated, while, by taking the "Learner's Perspective," we are more interested in the acts of interpretive affiliation, whereby the learners align themselves with various communities of practice and construct their participation and ultimately *their* practice through a customising process in which their inclinations and capabilities are expressed within the constraints and affordances of the social situation and the overlapping communities that compete for the learner's allegiance and participation. Wenger also stresses the multiplicity and overlapping character of communities of practice and the role of the individual in contributing to the practice of a community. In another respect, we are also in sympathy with Wenger's perspective.

> The kind of social theory of learning I propose is not a replacement for other theories of learning that address different aspects of the problem. But it does have its own set of assumptions and its own focus. Within this context, it does constitute a coherent level of analysis; it does yield a conceptual framework from which to derive a consistent set of general principles and recommendations for understanding and enabling learning (Wenger, 1998, p. 4).

This theoretical position, as stated by Wenger, accords a legitimate complementarity to theories of learning. The criteria for legitimacy are coherence, a domain of applicability, an implicit consistency with empirical evidence within that domain, and the potential to inform our understanding of learning and our promotion of learning in that domain. The legitimacy and utility of complementary analyses follow directly from this position.

Complementarity and Voice

Complementarity is fundamental to the approach adopted in the Learner's Perspective Study. This applies to complementarity of participants' accounts, where both the students and the teacher are offered the opportunity to provide retrospective reconstructive accounts of classroom events, through video-stimulated post-lesson interviews. It also applies to the complementarity of the accounts provided by members of the research team, where different researchers analyse a common body of data using different theoretical frameworks. This approach proved successful in a previous study (Clarke, 2001) and is evident in the chapters of this book and its companion volume (Clarke, Emanuelsson, Jablonka, & Mok, 2006). For example, Chapters 4, 12, and 20 were all written by members of the Japanese research group participating in the LPS project. Although each chapter reports an analysis of the Japanese LPS data, the emphasis and the method of analysis employed in each of these three chapters is quite different: Shimizu contrasts teacher and student perceptions of significant classroom events; Sekiguchi examines mathematical norms in the Japanese mathematics lessons; and Hino looks at teachers' support for students during seatwork.

Within the LPS project and the LPS research community, complementarity is aligned with the recognition that researchers bring to any project a set of values that reflect each researcher's cultural background and theoretical orientation. Essential to this form of complementarity is the acknowledgement that the perspective that a South African researcher brings to the analysis of classroom data is inevitably different from that of a researcher from China, the Czech Republic or the Philippines. Different priorities drive the research agenda of the researchers participating in the Learner's Perspective Study. These differences enriched the project, the community and this book.

Another agenda also contributes here: the agenda of voice. Basic limitations on affluence and available resources restrict some countries from participating in international studies, except as the objects of the research of other, more affluent, research communities. An attempt was made in the LPS project to accord all participating countries the status of research partners, rather than research objects. The diverse authorship of the chapters in this book reflects the extent to which this attempt was successful. The resulting research reports do not represent an international consensus on how classrooms should be viewed. They reflect important differences in what the researchers from each community saw as salient. It was not our intention to privilege one voice over another.

Classroom researchers around the world have constructed theories of pedagogy, instructional practice, classroom interaction and learning. Friends and colleagues, such as Ball (2000), Bauersfeld (1988), Bromme and Steinbring (1994), Cobb, Wood and Yackel (1993), Gu, Huang and Marton (2004), Voigt (1998) and their co-workers have constructed theories with varying claims to generalisability. Such theories are largely "within-culture" theories and any claims to generalisability should be seen as bounded by the cultural situatedness of the classroom contexts from which (and for which) those theories were constructed. We make no such claims of generalisability. Indeed, the legitimacy of any cross-cultural generalisation with regard to a phenomenon so socially-situated as classroom practice must be suspect. In this book, in particular, no attempt at cross-cultural comparison is made *within* any one chapter. The book itself, however, should sustain any comparisons the reader might like to make *between* chapters. Issues most salient and practices most prominent in Singapore may not resonate at all with the situation in Sweden. On the other hand, there may be unexpected echoes of the reader's classrooms in the descriptions of South African attempts to thematically connect and situate the mathematics curriculum (Chapter 8, Sethole, Goba, Adler and Vithal), or in Sekiguchi's discussion of mathematical norms in the Japanese lessons (Chapter 20). We think that the reader will benefit from the recognition of such similarities and also from the confronting differences that we have all found in the practices of classrooms overseas. Ultimately, however, the significance of any particular researcher's account is a matter for each reader to determine.

RESEARCH QUESTIONS

The Third International Mathematics and Science Study (TIMSS) (Beaton & Robitaille, 1999) established national profiles of student achievement and teacher and student beliefs regarding classroom practice. National norms for teaching practice were reported from the analysis of a statistically representative sample of videotaped eighth-grade mathematics classes in Japan, Germany and the USA (see Stigler & Hiebert, 1999). However, this research into mathematics classrooms collected only single lessons from each teacher and did not address learner practices. The LPS research design aimed to construct rich, detailed portrayals of the practices of individual well-taught mathematics classrooms over sequences of ten lessons.

It is an important feature of this project that it examined sequences of ten lessons, taught by teachers identified as competent by the local education community. Most significantly, this project adopted the position that research into classrooms, and into learning in classrooms, in particular, must address the interactive and mutually dependent character of teaching and learning. Such an approach requires the simultaneous documentation of the practices of both teacher and learners and the identification of the meanings each constructs for (and from) the practices of the other.

A series of research questions structured data collection in this project. Each of the original research questions is stated below, along with a brief summary of how each question was addressed. While each question was explored by research groups in each country through each set of local data separately, the power of the project is greatly enhanced by the access provided to matching data from other countries. A companion volume (Clarke et al., 2006) reports analyses that sought to make specific comparison between the practices of classrooms situated in different countries. This book focuses on analyses of classrooms within a single country. As a consequence, some research questions are addressed in one book and some in the other.

The first six research questions are sequenced according to the extent to which both teacher and learner practices and outcomes are integrated within the question. The issue of the practical implications of this study is addressed in question 7.

1. Within the classrooms studied in each country, is there evidence of a coherent body of student practice(s) (and to what extent might these practices be culturally-specific)?

Regularities within the practices of particular classrooms, particularly across ten lessons and at the level of detail documented in this research, can provide evidence of coherent bodies of practice. The cultural specificity of any such classroom practices requires comparison of substantial bodies of data drawn from different cultures, however diversity of practice within the classrooms of a particular country may suggest aspects of practice that are not culturally specific. It is also possible that teacher and learner practices are so interrelated as to be only meaningful when considered as aspects of classroom practice (see Chapter 3 (Keitel) or Chapter 4 (Hino) for example).

2. What are the antecedent and consequent conditions and actions (particularly learner actions) associated with teacher practices identified in earlier studies as culturally specific and nationally characteristic?

The research design had the capacity to identify particular teacher actions associated with the cultural teaching scripts identified in earlier research and relate these to the student practices that preceded or followed their use, with the potential to identify both influences and outcomes with respect to both sets of practices or, alternatively, to view such interactive connections as evidence of the extent to which teachers and learners were both complicit in the collaborative enactment of classroom practice. Some of these chains of association or interactive connections are reported in the chapters in this book, such as Chapters 11 (Begehr), 15 (Williams) and 18 (Huang, Mok & Leung). Such "patterns of participation" (Greeno, 1997, p. 9) can be more evident through the comparison of classrooms situated in very different educational cultures and, consequently, feature prominently in several chapters of the companion volume to this book (Clarke et al., 2006).

3. To what extent does an individual teacher employ a variety of pedagogical approaches (and/or lesson scripts) in the course of teaching a lesson sequence?

This study provided a validation check for earlier international studies of mathematics teaching by examining the teaching of sequences of lessons, rather than just single lessons. Variation and consistency in teacher use of pedagogical practices, including the reported lesson scripts (Stigler & Hiebert, 1997), could be examined over sequences of ten or more consecutive lessons. Evidence of the value of studying lesson sequences can be found throughout this book, but Chapters 10 (Kaur, Seah and Low), 17 (Park and Leung), and 21 (Emanuelsson and Sahlström) take particular advantage of this attribute of the research design.

4. What degree of similarity or difference (both locally and internationally) can be found in the learner (and teacher) practices occurring in classrooms identified by the local education community as constituting sites of competent teaching practice?

This study did not prescribe a common set of criteria for competent practice, but instead delegated the responsibility for the identification of competent teachers to the local research group in each country. As a consequence, the practices documented in the classrooms selected for this project constitute an illustration of teaching competence as it is conceived and practiced in each community. Not unexpectedly, competent practice in Shanghai has different characteristics from competent practice in Prague, as evidenced in Chapters 6 (Mok) and 19 (Binterová, Hošpesová, and Novotná). The availability of data from two cities within China offered the opportunity for a fine-grained within-country comparison of practice in Chapter 16 (Mok and Lopez-Real).

5. To what extent are teacher and learner practices in a mutually supportive relationship?

Because of the combination of video and interview data, it was possible to identify the ways in which the practices of learners both afford and constrain specific teacher practices (including the realisation of the teachers' goals). Equally, the same assessment could be carried out of the extent to which teacher practices represent affordances and constraints on the students' practices and goals. Conclusions are drawn in a variety of contexts as to whether teacher and learner practices are best seen as conflicting or as mutually sustaining. This should significantly inform theorising on classroom practice. Interestingly, this question can be answered at both local and international levels. In this book, local answers are prioritised, and in Chapter 12 (Shimizu) juxtaposes the teacher's and student's perspectives on Japanese mathematics classrooms very powerfully. In Chapter 14 (Fried and Amit), tensions between collaboration and authority and between public and private domains are explored from both the teacher's and the students' perspectives. The capacity to contrast public and private discourse is a direct consequence of the research design and is exploited particularly effectively in Chapter 13 (Gallos).

6. To what extent are particular documented teacher and learner practices associated with student construction of valued social and mathematical meanings?

Since learners' constructed meanings and perceptions were accessed within this study, it was possible to assess the effectiveness of teacher and learner actions in promoting particular forms of student learning. This can be seen particularly clearly in Chapter 15 (Williams).

7. What are the implications for teacher education and the organisation of schools of the identification of those teacher and learner practices that appear to be consistent with the realisation of local goals (and those which are not)?

The evaluation of teacher and learner practices against local curricular goals provides a critique of teacher education and the resourcing and organisation of schools. Chapter 9 (Ulep) examines the consequences of an idiosyncratic instructional strategy prevalent in mathematics classrooms in the Philippines. Chapter 7 (Kaur, Low and Seah) looks at textbook and homework use in Singapore. In Chapter 5 (Wood), the US reform agenda becomes the lens through which classrooms are viewed. In this way, findings have the capacity to inform local practice in each participating country and, given sufficient international commonality of educational goals, to identify generic teacher and learner practices worthy of more widespread emulation.

Taking these research questions in combination, the LPS project sought to document both the practices of eighth-grade mathematics classrooms and the meanings, mathematical and social, associated with those practices and to utilise the data collected to draw conclusions, both locally and internationally situated, concerning those practices most likely to lead to the optimisation of learning. The chapters of this book represent the separate attempts of the participating local research groups to interpret the above research questions in locally relevant terms. Researchers within each local group focused their analyses of the mathematics classroom data they had collected on those aspects of practice and meaning most closely aligned with the concerns of both the local school system and the researcher. In order that readers might better understand the local educational context in each country, an appendix is provided in which salient features of the school system and mathematics curriculum are summarised for each of the countries whose classrooms feature in this book.

INTERNATIONAL STUDIES IN EDUCATION: WHAT CAN WE HOPE TO LEARN?

Alternative Approaches

There is a contemporary enthusiasm for studying the mathematics classrooms of other countries. Such international studies are expensive to conduct and those of us who are doing them must justify what it is that they offer as return for our efforts. This brief discussion addresses both the justification of international studies in general and the potential value of this project.

Among the studies undertaken to address the challenge of legitimate international comparison of curriculum as policy and practice, the two studies by Schmidt exemplify these two contrasting challenges. The Survey of Mathematics and Science Opportunities Study (Schmidt et al., 1996) involved over 120 classroom observations in mathematics and science classrooms in six countries, in an attempt to characterise 'a typical mathematics or science lesson' for 9-year-olds or 13-year-olds in the countries being studied. The Curriculum Analysis Study

(Schmidt, McKnight, Valverde, Houang & Wiley, 1997) involved the analysis of curricular guidelines, programs and textbooks from about 60 different countries. While both these studies are indicative of the current activity in international comparative research in education, they employ quite different methodologies and were intended to address different questions from those pertinent to the LPS project. In combination with the work of Stigler and Hiebert (1999), the first of these studies informed the LPS project by offering a documentation of practices identified as nationally characteristic. One message arising from the Curriculum Analysis Study is that the results of the LPS project must be situated in relation to local curriculum priorities in each of the countries in which the classrooms are situated.

One of the best known international comparative studies is the videotape study carried out by Stigler and Hiebert (1999). This study collected samples of classroom instruction from 231 eighth grade mathematics classrooms in Germany, Japan and the USA. Given the success of Japanese students on the Third International Mathematics and Science Study (TIMSS) of student achievement, it would be a simple but naïve approach to identify what it is that Japanese teachers are doing differently from everyone else, and then explore ways to emulate them. Stigler and Hiebert have consistently challenged any such reading of their findings. Instead, they have drawn attention to the essentially cultural nature of teaching. Their research has linked the practices of the mathematics classroom to deeply held values and beliefs about teaching, learning, mathematics and the role of schools.

A couple of simple examples can illustrate the cultural character of mathematics classrooms:
- Japanese lessons are never interrupted from the outside – not by announcements from the public address system, not by lunch monitors, not by anyone. The lesson as a unit is the central element in the culture of the Japanese school, and each lesson must tell a coherent story;
- By comparison, US lessons are conceived as combinations of smaller units, which have their own integrity. From such a perspective, it is not so important if something interrupts the lesson; the number of constitutive activities, such as the practising of procedures, may be reduced, but learning is not irrevocably disrupted.

Classrooms around the world also differ with respect to what can be reasonably asked of students. The public solution of a problem by one student in front of the class is routine in some countries but rare in others. Such time-honoured practices and the values and beliefs that they embody are deeply "cultural" in character. The power of studies such as Stigler and Hiebert's is that they offer us images of possibilities for practice that we might not otherwise have considered, and they ask us to question the assumptions on which our present practice is based.

Not everyone is unquestioningly enthusiastic about such international studies. Bracey (1997) has questioned the legitimacy of comparing the practices and the products of classrooms embedded in very different cultures seeking to implement very different curricula. As Keitel and Kilpatrick (1999) have argued, these concerns apply equally to studies of student achievement and of teacher practice.

For example, the priority and significance attached to the teaching of mathematical proof in Europe is almost completely absent in Australia. What then is the significance of comparing the practices and outcomes of educational programs that have such different goals? One simple answer is, of course, that it causes us to review those goals, as well as the practices by which we hope to achieve them. Despite the legitimate concerns regarding lack of attention to curricular difference in some research studies, the international education community continues to find many common goals and issues, suggesting sufficient points of common interest to justify comparative investigation.

One of the more intriguing outcomes of recent international comparative research is the diversity of classroom practice that characterises even those countries with similar levels of student achievement. Students in Japan, the Netherlands, and the Czech Republic have performed consistently well on international tests of mathematics performance, yet the pictures that are emerging from contemporary studies of mathematics classrooms in these countries are very different. Such findings do not encourage us to seek the 'ideal' mathematics classroom, but they do suggest that "good practice" is a culturally-determined entity. There are practices employed routinely in classrooms in the Czech Republic that would probably never translate to an American setting, for example, and the same comment would apply to the translation of some Hong Kong classroom practices to Australia or Sweden. However, even these untranslatable practices have power as catalysts for discussion and reflection on the practices of our classrooms and the values that underlie them. Other classrooms, in Singapore for example, might offer models of practice that might be successfully emulated by American teachers, and *vice versa*. One principle will hold true in any such emulation: teachers seldom adopt, they almost always adapt. It is a reflection of teacher professionalism that this is the case. One criterion for the effectiveness of the adapted activity will be its congruence with the existing culture.

The inclusion of the learner's perspective must enrich our portrayal of mathematics classrooms. Indeed, its absence from international comparative research calls into question the adequacy of previous research to do more than describe teacher practice, lacking either associative or explanatory potential. It seems reasonable to suppose that the meanings which students ascribe to the actions of their teachers and their classmates are as culturally-specific, and as significant for our understanding of classrooms, as the actions themselves. Any portrayal of mathematics classrooms in different countries can hardly avoid the obligation to document both the practices extant in those classrooms and the meanings that participants ascribe to those practices. Further, any evaluation of the relative merits of the practices of such classrooms can only be made through the accumulation of data on the meanings constructed by learners in these classrooms.

As will become evident in Chapter Two, we cannot characterise the teaching of a country or a culture on the basis of our selective sampling of teachers and lesson sequences and this was never our intention. Nor do we claim to compare teaching in one country with teaching in another. The research design was developed to support analyses intended to portray, to compare and to contrast teachers and their classrooms, not cultures. Of course, the original choice of school systems (Germany, Japan, and the USA) was not accidental. It was intended to complement any general claims of national typicality made by the TIMSS video study (Stigler & Hiebert, 1999) by situating identified prevalent practice in relation to the antecedent conditions and consequent outcomes that might transform description into explanation.

The documentation of the practices of mathematics classrooms in other countries causes us to question our assumptions about our own practice. As the findings of these studies become more widely available, they will be increasingly utilised in the professional development of mathematics teachers. If our goal is continual improvement, then we must explore the practices of others whose goals resemble ours, and we must seize every opportunity for a fresh perspective on our own practice. Participating research teams (and you, the reader) have the opportunity to compare and contrast the practices occurring in some of the better mathematics classrooms in their country with those occurring in the classrooms of teachers adjudged competent by the education communities of other countries. Such comparisons should be universally beneficial.

The essential characteristic of this study of mathematics classrooms is the commitment to an integrative approach. Adaptation of a research design developed for an Australian study for use in researching classrooms in twelve different countries anticipated a much more global examination of practice than that possible with data grounded in a single country. The commitment to examining the interdependence of teaching and learning as related activities within an integrated body of classroom practice accepts an obligation to document (and analyse) relationships between participants' practices as well as the occurrence of the individual practices themselves. The importance attached to the meanings that participants attribute to their actions and the actions of others and to the mathematical and social meanings that are the major products of the classroom requires a methodology able to access participants' accounts of those meanings and to integrate these within a coherent picture of the classroom.

Whereas the companion volume to this book (Clarke et al., 2006) reports the results of analyses that sought to make explicit comparisons of practice between classrooms internationally, this book adopts a more introspective approach. What follows are accounts of researchers investigating the classrooms of competent teachers in their "local" community. The issues addressed are those most salient to the researcher and the community. This book reflects the eclectic spirit of the LPS community. Rather than subjecting classroom practice in every country to a single constraining analytical framework, it celebrates diversity. The Learner's Perspective Study was guided by a belief that we need to learn from each other, rather than decide who is doing what better: inclusion rather than evaluation. The resulting chapters offer you deeply situated insights into the practices of mathematics classrooms in twelve countries: an insider's perspective.

REFERENCES

Ball, D. L. (2000). Bridging practices: Intertwining content and pedagogy in teaching and learning to teach. *Journal of Teacher Education, 51*, 241-247.

Bauersfeld, H. (1988). Interaction, construction and knowledge: Alternative perspectives for mathematics education. In D. Grouws, T. J. Cooney & D. Jones (Eds.), *Effective Mathematics Teaching* (pp. 27-46). Reston, VA: NCTM & Lawrence Erlbaum.

Beaton, A. E. & Robitaille, D. F. (1999). An overview of the Third International Mathematics and Science Study. Chapter 3 in G. Kaiser, E. Luna & I. Huntley (Eds.), *International comparisons in mathematics education* (pp. 19-29). London: Falmer Press.

Bracey, G. W. (1997). On comparing the incomparable: A response to Baker and Stedman. *Educational Researcher, 26*(4), 19-26.

Bromme, R. & Steinbring, H. (1994) Interactive development of subject matter in the mathematics classroom. *Educational Studies in Mathematics, 27*(3), 217-248.

Clarke, D. J. (1998). Studying the classroom negotiation of meaning: Complementary accounts methodology, Chapter 7 in A. Teppo (Ed.), *Qualitative research methods in mathematics education*, monograph number 9 of the *Journal for Research in Mathematics Education*, Reston, VA: NCTM, 98-111.

Clarke, D. J. (Ed.). (2001). *Perspectives on practice and meaning in mathematics and science classrooms.* Dordrecht, Netherlands: Kluwer Academic Press.

Clarke, D. J., Emanuelsson, J., Jablonka, E., & Mok, I. A. C. (Eds.). (2006). *Making connections: Comparing mathematics classrooms around the world.* Rotterdam: Sense Publishers.

Cobb, P., Wood, T., & Yackel, E. (1993). Discourse, mathematical thinking, and classroom practice. In E. Forman, N. Minick, & A. Stone (Eds.), *Contexts for learning: Social cultural dynamics in children's development* (pp. 91-119). Oxford, England: Oxford University Press.

Greeno, J. (997). On claims that answer the wrong questions. *Educational Researcher, 26*(1), 5-17.

Gu, L., Huang, R., & Marton, F. (2004). Teaching with variation: A Chinese way of promoting effective mathematics learning. Chapter 12 in L. Fan, N. Y. Wong, J. Cai, & S. Li (Eds.), *How Chinese learn mathematics: Perspectives from insiders* (pp. 309-347). New Jersey: World Scientific.

Hiebert, J., Gallimore, R., Garnier, H., Givvin, K., Hollingsworth, H., Jacobs, J., Chui, A., Wearne, D., Smith, M., Kersting, N., Manaster, A., Tseng, E., Etterbeck, W., Manaster, C., Gonzales, P., & Stigler, J. (2003). *Teaching mathematics in seven countries: Results from the TIMSS 1999 video study.* Washington, DC: U.S. Department of Education, National Center for Education Statistics.

Keitel, C., & Kilpatrick, J. (1999). The rationality and irrationality of international comparative studies. Chapter 16 in G. Kaiser, E. Luna, & I. Huntley (Eds.). *International comparisons in mathematics education* (pp.241-256). London: Falmer Press.

Schmidt, W. H., Jorde, D., Cogan, L. S., Barrier, E., Gonzalo, I., Moser, U., Shimizu, K., Sawada, T., Valverde, G. A., McKnight, C., Prawat, R. S., Wiley, D. E., Raizen, S. A., Britton, E. D., & Wolfe, R. G. (1996). *Characterizing pedagogical flow: An investigation of mathematics and science teaching in six countries.* Dordrecht: Kluwer.

Schmidt, W. H., McKnight, C. C., Valverde, G. A., Houang, R. T., & Wiley, D. E. (1997). *Many visions, many aims volume 1: A cross-national investigation of curricular intentions in school mathematics.* Dordrecht: Kluwer.

Stigler, J., & Hiebert, J. (1997). Understanding and improving classroom mathematics instruction: An overview of the TIMSS video study. *Phi Delta Kappan, 79*(1), 14-21.

Stigler, J., & Hiebert, J. (1999). *The teaching gap.* New York: Free Press.

Voigt, J. (1998). The culture of the mathematics classroom: Negotiating the mathematical meaning of empirical phenomena. In F. Seeger, J. Voigt, & U. Waschescio (Eds.), *The culture of the mathematics classroom* (pp. 191-220). Cambridge, UK: CUP.

Wenger, E. (1998). *Communities of practice: Learning, meaning, and identity.* Cambridge: Cambridge University Press.

DAVID CLARKE, CHRISTINE KEITEL AND YOSHINORI SHIMIZU

David Clarke
International Centre for Classroom Research
Faculty of Education
University of Melbourne
Australia

Christine Keitel
Fachbereich Erziehungswissenschaft und Psychologie
Freie Universität Berlin
Germany

Yoshinori Shimizu
Graduate School of Comprehensive Human Sciences
University of Tsukuba
Japan

DAVID CLARKE

CHAPTER TWO

The LPS Research Design

INTRODUCTION

The initiation of the Learner's Perspective Study (LPS) was motivated to a significant extent by a perceived need to complement the survey-style approach characteristic of the research of Stigler and his co-workers with a more in-depth approach that accorded more prominent voice to the perspective of the learner. The originators of the LPS project, Clarke, Keitel and Shimizu, felt that the methodology developed by Clarke and known as complementary accounts (Clarke, 1998), which had already demonstrated its efficacy in a large-scale classroom study (subsequently reported in Clarke, 2001) could be adapted to meet the needs of the Learner's Perspective Study. These needs centred on three key requirements: (i) the recording of interpersonal conversations between focus students during the lesson; (ii) the documentation of sequences of lessons, ideally of an entire mathematics topic; and, (iii) the identification of the intentions and interpretations underlying the participants' statements and actions during the lesson. The methodology developed by Clarke (1998, 2001) met requirements (i) and (iii) directly, and required only minor adaptation to address requirement (ii). This chapter sets out the theoretical basis and technical details of this approach.

Methodological Position

Recent classroom research (Alton-Lee, Nuthall & Patrick, 1993; Clarke, 2001; Sahlström & Lindblad, 1998), backed by more sophisticated ways of collecting and analysing data, has shown that some of the findings of the classroom research classics such as Bellack, Kliebard, Hyman and Smith (1966), Sinclair and Coulthard (1975) and Mehan (1979) are seriously skewed because of technological issues in data collection. In particular, this has concerned the ability to simultaneously record both student and teacher interactions, and the ability to facilitate ways of working with these data within educational research project funding constraints.

Clarke (2001) argued that since a classroom takes on a different aspect according to how you are positioned within it or in relation to it, our research

D. J. Clarke, C. Keitel & Y. Shimizu (Eds.), Mathematics Classrooms In Twelve Countries: The Insider's Perspective. 15–36. © *2006 Sense Publishers. All rights reserved.*

methodology must be sufficiently sophisticated to accommodate and represent the multiple perspectives of the many participants in complex social settings such as classrooms. Only by seeing classroom situations from the perspectives of all participants can we come to an understanding of the motivations and meanings that underlie their participation. Our capacity to improve classroom learning depends on such understanding. The methodological challenge is how to document and analyse the fundamental differences in how each participant experiences any particular social (classroom) situation. Lindblad and Sahlström (1999, 2002) have argued that if early researchers had access to the same tools for data generation and analysis as are available today, the general view of classroom interaction would be quite different.

The most striking of these differences, and a very important one from an education point of view, concerns the role of students in classrooms. Single-camera and single-microphone approaches, with a focus on the teacher, embody a view of the passive, silent student, which is at odds with contemporary learning theory and classroom experience. Research done with technologically more sophisticated approaches has described a quite different classroom, where different students are active in different ways, contributing significantly to their own learning (cf. Clarke, 2001; Sahlström & Lindblad, 1998).

Research as the Enactment of the Researcher's Epistemology

In discussing the emergence of mathematical meaning in a second-grade classroom, Krummheuer (1995) invoked Goffman's (1959) notion of a "working consensus" as the immediate goal of classroom argumentation. Goffman's conception of a working consensus as a transient convergence on a locally viable interpretation is a particularly apt characterisation of the goal of the consensus process operating in many interpretive research teams (for example, Cobb & Bauersfeld, 1995; Stigler & Hiebert, 1999). Our research (Clarke, 2001, for example) problematises such consensus and attempts to synthesise portrayals of practice from 'complementary accounts.' In both approaches, the alignment of methodology with theory and the reflexive relationship between them reflects an inevitable symbiosis.

The Learner's Perspective Study research design embodies the inevitable existence of multiple reflexivities between theory, research into practice, and the practice of research. This argument is predicated on three basic premises:
- The discourse of the classroom acts to position participants in ways that afford and constrain certain practices.
- The discourse of educational research acts to position participants in ways that afford and constrain certain interpretations.
- The adoption of a theory of learning in social situations will inevitably find its reflection in the manner in which those situations are researched.

These fundamental reflexivities are seldom acknowledged.

In her paper "Psychometricians' beliefs about learning", Shepard (1991) contended that the disputes of the testing community can be explained in terms of differences in the beliefs about learning held by the various educational measurement specialists. In particular, Shepard argued that the beliefs of many psychometricians derive from an implicit behaviourist learning theory in flagrant contradiction with evidence from cognitive psychology.

What Shepard did to good effect in her paper was to reverse engineer psychometricians' learning theories on the basis of their test instruments. Reverse engineering consists of analysing an artifact (a procedure, a tool or a test) from the perspective of the purpose it was intended to serve (see Dawkins, 1995, p. 120). Dawkins combined the notion of reverse engineering with the idea of 'utility function' (that which is maximised in a system) to identify the biological mechanisms underlying the survival of species. The fruitfulness of this approach is fully evident in Shepard's provocative question, "But what if learning is not linear and is not acquired by assembling bits of simpler learning" (Shepard, 1991, p. 7).

Miles and Huberman's text on qualitative data analysis (Miles & Huberman, 2004) focused attention on 'data reduction.'

> Even before data are collected . . . anticipatory data reduction is occurring as the researcher decides (often without full awareness) which conceptual framework, which cases, which research questions, and which data approaches to use. As data collection proceeds, further episodes of data reduction occur (p. 10).

This process of data reduction pervades any classroom video study. The choice of classroom, the number of cameras used, who is kept in view continuously and who appears only given particular circumstances, all contribute to a process that might better be called 'data construction' or 'data generation' than 'data reduction.' Every decision to zoom in for a closer shot or to pull back for a wide angle view represents a purposeful act by the researcher to selectively construct a data set optimally amenable to the type of analysis anticipated and maximally aligned with the particular research questions of interest to the researcher. The process of data construction does not stop with the video record, since which statements (or whose voices) are transcribed, and which actions, objects or statements are coded, all constitute further decisions made by the researcher, more or less explicitly justified in terms of the project's conceptual framework or the focus of the researcher's interest. The researcher is the principle agent in this process of data construction. As such, the researcher must accept responsibility for decisions made and data constructed, and place on public record a transparent account of the decisions made in the process of data generation and analysis. In case the bases for some of these decisions might go unrecognised, it is essential that we interrogate our actions as researchers and reverse engineer our study designs in order to deconstruct the epistemology and the learning theory on which our research is predicated. Inevitably, such deconstruction should extend to our use of technology and the epistemologies implicit in that use.

In the case of the Learner's Perspective Study: Research guided by a theory of learning that accords significance to both individual subjectivities and to the constraints of setting and community practice must frame its conclusions (and collect its data) accordingly. Such a theory must accommodate complementarity rather than require convergence and accord both subjectivity and agency to individuals not just to participate in social practice but to shape that practice. The assumption that each social situation is constituted through (and in) the multiple lived realities of the participants in that situation aligns the Learner's Perspective Study with the broad field of interpretivist research. In a research project predicated on acknowledging, documenting and studying agency, both individual and collective, the challenge of constructing a corresponding methodology was considerable. Among the research reports in this book and its companion volume (Clarke, Emanuelsson, Jablonka, & Mok, 2006) are analyses that apply variations on the social constructivism of Cobb and Bauersfeld (1995), the discursive psychologies of Vygotsky (1962) and Bruner (1990), notions of situated cognition (Lave & Wenger, 1991) and communities of practice (Wenger, 1998), social positioning theory (Harre & Langenhove, 1999), and variation theory (Gu & Marton, 2004). This theoretical eclecticism is an immediate and pragmatic consequence of the manner in which the various analytical approaches reflect the different research foci of the international researchers that make up the research community of the Learner's Perspective Study.

Research that aims to support the application of such theories must construct its methodologies accordingly and draw from available technologies in ways that afford rather than constrain the methodological and theoretical ambitions of the researcher. The LPS research design was developed to address the particular research interests and theoretical orientations of the Australian, German and Japanese research groups that initiated the project. On joining the LPS research community, each research group had to evaluate the capacity of the research design to generate data relevant to their area of research interest and amenable to analysis from their particular theoretical perspective. As has already been demonstrated (Clarke, 2001), the complementary accounts provided by such alternative analytical approaches offer in their combination a rich and insightful portrayal of the contemporary classroom.

If contemporary research is to generate data commensurate with the sophistication of contemporary theory, it is imperative that educational research makes optimal use of available technology. International comparative classroom research, in particular, poses methodological and technical challenges that are only now being adequately addressed through advances in:
– techniques and equipment for the collection of audio-visual data in classrooms;
– tools for the compression, editing and storage of digitised video and other data;
– storage facilities that support networked access to large complex databases; and
– analytical tools capable of supporting sophisticated analyses of such complex databases.

In the course of the Learner's Perspective Study, the international research team have collaborated in the utilisation of all possible available technologies and the progressive refinement of the technical aspects of the original research design. This chapter sets out the further refinement of the complementary accounts methodology as it has been adapted and developed in the course of the Learner's Perspective Study.

DATA GENERATION IN THE LEARNER'S PERSPECTIVE STUDY

Data generation in the Learner's Perspective Study (LPS) (see Figure 1) used a three-camera approach (Teacher camera, Student camera, Whole Class camera) that included the onsite mixing of the Teacher and Student camera images into a picture-in-picture video record (see Figure 2, teacher in top right-hand corner) that was then used in post-lesson interviews to stimulate participant reconstructive accounts of classroom events. These data were generated for sequences of at least ten consecutive lessons occurring in the "well-taught" eighth grade mathematics classrooms of teachers in Australia, the Czech Republic, Germany, Hong Kong and mainland China, Israel, Japan, Korea, The Philippines, Singapore, South Africa, Sweden and the USA. As noted earlier this combination of countries gives good representation to European and Asian educational traditions, affluent and less affluent school systems, and mono-cultural and multi-cultural societies.

Each participating country used the same research design to generate videotaped classroom data for at least ten consecutive mathematics lessons and post-lesson video-stimulated interviews with at least twenty students in each of three participating eighth grade classrooms. The three mathematics teachers in each country were identified for their locally-defined 'teaching competence' and for their situation in demographically diverse government schools in major urban settings. Rather than attempt to apply the same definition of teaching competence across a dozen countries, which would have required teachers in Uppsala and Shanghai, for instance, to meet the same eligibility criteria, teacher selection was made by each local research group according to local criteria. These local criteria included such things as status within the profession, respect of peers or the school community, or visibility in presenting at teacher conferences or contributing to teacher professional development programs. As a result, the diverse enactment of teaching competence is one of the most interesting aspects of the project.

In most countries, the three lesson sequences were spread across the academic year in order to gain maximum diversity within local curricular content. In Sweden, China and Korea, it was decided to focus specifically on algebra, reflecting the anticipated analytical emphases of those three research groups. Algebra forms a significant part of the eighth grade mathematics curriculum in most participating LPS countries, with some variation regarding the sophistication of the content dealt with at eighth grade. As a result, the data set from most of the LPS countries included at least one algebra lesson sequence.

In the key element of the post-lesson student interviews, in which a picture-in-picture video record was used as stimulus for student reconstructions of classroom events (see Figure 2), students were given control of the video replay and asked to identify and comment upon classroom events of personal importance (see Appendix B of this chapter). The post-lesson student interviews were conducted as individual interviews in all countries except Germany, Israel and South Africa, where student preference for group interviews was sufficiently strong to make that approach essential. Each teacher was interviewed at least three times using a similar protocol.

Figure 1. Classroom data generation

With regard to both classroom videotaping and the post-lesson interviews, the principles governing data generation were the minimisation of atypical classroom activity (caused by the data generation activity) and the maximisation of respondent control in the interview context. To achieve this, each videotaped lesson sequence was preceded by a one-week familiarisation period in which all aspects of data generation were conducted until the teacher indicated that the class was functioning as normally as might reasonably be expected. A detailed set of in-class data generation guidelines were developed (Appendix A of this chapter). In interviews, the location of control of the video player with the student ensured that the reconstructive accounts focused primarily on the student's parsing of the lesson. Only after the student's selection of significant events had been exhausted did the interviewer ask for reconstructive accounts of other events of interest to the research team. Documentation of the participant's perspective (learner or teacher) remained the priority.

In every facet of this data generation, technical quality was a priority. The technical capacity to visually juxtapose the teacher's actions with the physical and oral responses of the children was matched by the capacity to replay both the public statements by teacher or student and the private conversations of students as they struggled to construct meaning. Students could be confronted, immediately after the lesson, with a video record of their actions and the actions of their classmates.

In the picture-in-picture video record generated on-site in the classroom (Figure 2), students could see both their actions and the actions of those students around them, and, in the inset (top right-hand corner), the actions of the teacher at that time. This combined video record captured the classroom world of the student. The video record captured through the whole-class camera allowed the actions of the focus students to be seen in relation to the actions of the rest of the class.

Figure 2. Picture-in-picture video display

To reiterate: In this study, students were interviewed after each lesson using the video record as stimulus for their reconstructions of classroom events. It is a feature of this study that students were given control of the video replay and asked to identify and comment upon classroom events of personal importance. Because of the significance of interviews within the study, the validity of students' and teachers' verbal reconstructions of their motivations, feelings and thoughts was given significant thought. The circumstances under which such verbal accounts may provide legitimate data have been detailed in two seminal papers (Ericsson & Simon, 1980; Nisbett & Wilson, 1977).

It is our contention that videotapes of classroom interactions constitute salient stimuli for interviewing purposes, and that individuals' verbal reports of their thoughts and feelings during classroom interactions, when prompted by videos of the particular associated events, can provide useful insights into those individuals' learning behaviour. Videotapes provide a specific and immediate stimulus that optimises the conditions for effective recall of associated feelings and thoughts. Nonetheless, an individual's video-stimulated account will be prone to the same potential for unintentional misrepresentation and deliberate distortion that apply in any social situation in which individuals are obliged to explain their actions. A significant part of the power of video-stimulated recall resides in the juxtaposition of the interviewee's account and the video record to which it is related. Any apparent discrepancies revealed by such a comparison warrant particular scrutiny and careful interpretation by the researcher. Having relinquished the positivist commitment to identifying 'what really happened,' both correspondence and contradiction can be exploited. The interview protocols for student and teacher interviews were prescribed in the LPS Research Design and are reproduced as Appendix B of this chapter.

Inevitably, some variation in interviewer prompt occurred, where each interviewer exercised their own discretion regarding how best to probe the interviewee's replies to the various prompts set out in Appendix B. This variation was a consequence of the deliberate devolution to the interviewee of control over which classroom episodes were discussed. This was true of both individual and group interviews. As noted earlier, if sufficient interview time was available, the interviewee might be asked to comment on a classroom episode of interest to the researcher, but only once the interviewee's significant moments had been discussed fully.

Other Sources of Data

Teacher questionnaires were used to establish teacher beliefs and purposes related to the lesson sequence studied. Student tests were used to situate each student group and each student in relation to student performance on eighth-grade mathematics tasks. Student mathematics achievement was assessed in three ways:

Student written work in class. Analyses of student written work were undertaken both during and after the period of videotaping. For this purpose, the written work of all "focus students" in each lesson was photocopied, clearly labelled with the student's name, the class, and the date, and filed. Additional data on student achievement was also collected, where this was available. In particular, student scores were obtained on any topic tests administered by the teacher, in relation to mathematical content dealt with in the videotaped lesson sequence.

Student performance to place the class in relation to the national eighth grade population. In Australia, Japan, Korea, China and the USA, this was done by using the International Benchmark Test for Mathematics (administered immediately after the completion of videotaping). The International Benchmark Test (IBT) was developed by the Australian Council for Educational Research (ACER) by combining a selection of items from the TIMSS Student Achievement test. In the case of this project, the test for Population Two was used, since this was in closest correspondence with the grade level of the students taking part in the LPS project. In administering the IBT, the local research group in each country constructed an equivalent test using the corresponding version of each of the TIMSS items, as administered in that country. In some countries, where this was not possible (Germany, for example), the typical school performance was characterised in relation to other schools by comparison of the senior secondary mathematics performance with national norms.

Student performance in relation to other students in that class. Since student-student interactions may be influenced by perceptions of peer competence, it was advantageous to collect recent performance data on all students in the class. Two forms of student mathematics achievement at class level were accessed, where available: (a) student scores from recent mathematics tests administered by the teacher, and (b) brief annotated comments by the teacher on a list of all students in the class – commenting on the mathematics achievement and competence of each student.

Teacher Goals and Perceptions

Three questionnaires were administered to each participating teacher:
− A *preliminary* teacher questionnaire about each teacher's goals in the teaching of mathematics (TQ1);
− A *post-lesson* questionnaire (TQ2 − either the short TQ2S or the long TQ2L version − if the short version was used, the researcher's field notes provided as much as possible of the additional detail sought in the long version);
− A *post-videotaping* questionnaire (TQ3) (also employed by some research groups as the basis of a final teacher interview).

The Integrated Data Set

In summary, the LPS Research Design generated the following data set in relation to any particular eighth-grade mathematics classroom:

Table 1. The Integrated Data Set for any particular lesson

Initial Data Set	Videotape from Teacher Camera
	Videotape from Student Camera
	Videotape of composite Image from Student Camera and Teacher Camera (The "Learner's" Perspective Composite Image)
	Videotape from Whole Class Camera (The Whole Class Image)
	Audiotapes or videotapes of interviews with at least two students (and possibly the Teacher)
	Photocopies of written work produced by all focus students
	Photocopies of textbook pages, worksheets or other written materials as appropriate
	Post-lesson teacher questionnaire
Additional General Data Set	Student test (only administered once, after completion of videotaping)
	Other student achievement data (class tests, teacher comments)
	Teacher questionnaire data on teacher goals and beliefs
	Teacher interview data

DATA CONFIGURATION AND STORAGE

Transcription and Translation

A detailed Technical Guide was developed to provide guidelines for the transcription and translation of classroom and interview, video and audiotape data. Also included in the Technical Guide were the specifications of Lesson Tables. These Lesson Tables served as a navigational aid and summative description of each lesson. Lesson tables included such details as the time and duration of each distinct lesson component (event), the type of social organisation involved, a description of the activity, and a description of the mathematical content that constituted the focus of the activity. A sample Lesson Table has been included as Appendix C of this chapter.

It was essential that all research groups transcribe their own data. Local language variants (eg. the Berliner dialect) required a "local ear" for accurate transcription. Translation into English was also the responsibility of the local research group. The Technical Guide specified both transcription conventions, such as how to represent pauses or overlapping statements, and translation conventions, such as how to represent colloquialisms. In the case of local colloquial expressions in a language other than English, the translator was presented with a major challenge. A literal English translation of the colloquialism may convey no meaning at all to a reader from another country, while the replacement of the colloquialism by a similar English colloquialism may capture the essence and spirit of the expression, but sacrifice the semantic connotations of the particular words used. And there is a third problem: If no precise English equivalent can be found, then the translation inevitably misrepresents the communicative exchange. In such instances, the original language, as transcribed, was included together with its literal English translation. Any researcher experiencing difficulties of interpretation in analysing the data could contact a member of the research group responsible for the generation of those data and request additional detail.

Data Storage

To carry out serious systematic empirical work in classroom research, there is a need for both close and detailed analysis of selected event sequences, and for more general descriptions of the material from within which the analysed sample has been chosen. To be able to perform this work with good-quality multiple-source video and audio data, video and audio materials have to be compressed and stored in a form accessible by desktop computers. Software tools such as *Final Cut Pro* are essential for the efficient and economical storage of the very large video data files. Compression decisions are dictated by current storage and back-up alternatives and change as these change. For example, when the Learner's Perspective Study was established in 1999, it was anticipated that data would be exchanged between research teams by CD-ROM and compression ratios were set at 20:1 in order to get maximum data quality within a file size that would allow one video record of one lesson to be stored on a single CD. As a result, the complete US data set in 2001 took the form of a set of over fifty separate CDs. Later, it was possible to store all the data related to a single lesson (including four compressed video records) on a single DVD. The contemporary availability of pocket drives with capacities of 60 gigabytes and higher, has made data sharing both more efficient and cheaper. It is possible to store all the data from a single school in compressed form on such a pocket drive, making secure data transfer between international research groups much more cost-effective.

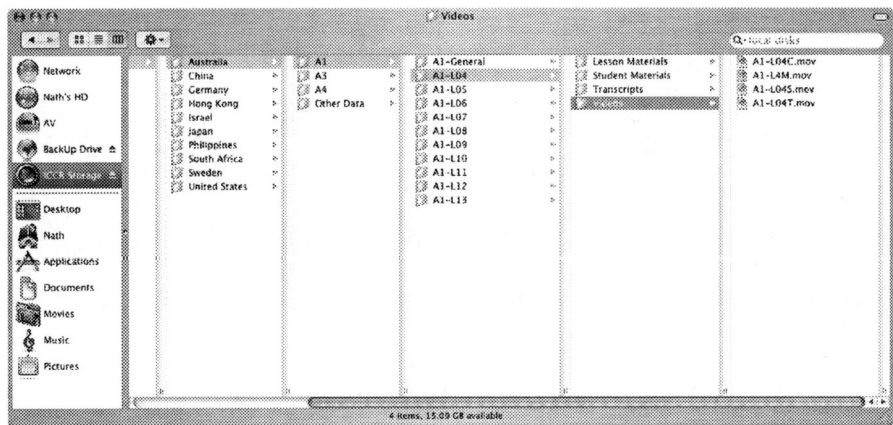

Figure 3. Structure of the LPS database at the ICCR circa 2004

The materials on the database have to be represented in a searchable fashion. In Figure 3, the configuration of the LPS database is displayed as a stratified hierarchy of: Country (column 1), school (column 2), lesson (column 3), data source (column 4), specific file (column 5). Any particular file, such as the teacher camera view of lesson 4 at school 2 in Japan, can then be uniquely located.

Setting up data in this way enables researchers to move between different layers of data, without losing sight of the way they are related to each other. Further, data can be made accessible to other researchers. This is a sharp contrast to more traditional ways of storing video data on tapes, with little or no searchable record available, and with data access limited to very small numbers of people. At the International Centre for Classroom Research (ICCR) at the University of Melbourne, for example, several researchers can simultaneously access the full range of classroom data. This capacity for the simultaneous analysis of a common body of classroom data is the technical realisation of the methodological and theoretical commitment to complementary analyses proposed by Clarke (2001, 1998) as essential to any research attempting to characterise social phenomena as complex as those found in classrooms.

ANALYTICAL TOOLS CAPABLE OF SUPPORTING SOPHISTICATED ANALYSES
OF SUCH COMPLEX DATABASES

Research along the lines argued for above requires the development of software tools for analysing video efficiently. The reasons for this are, in short, that video editing software (such as *Final Cut Pro*) is not analytically resourceful enough, whereas qualitative analysis software (such as *Nudist* or *nVivo*) is not well enough adapted to video and audio work. Early examples of video analysis software (such as *vPrism*) have been hampered by problems arising from their project-specific origins, leading to a lack of flexibility in customising the analysis to the demands of each particular project or research focus.

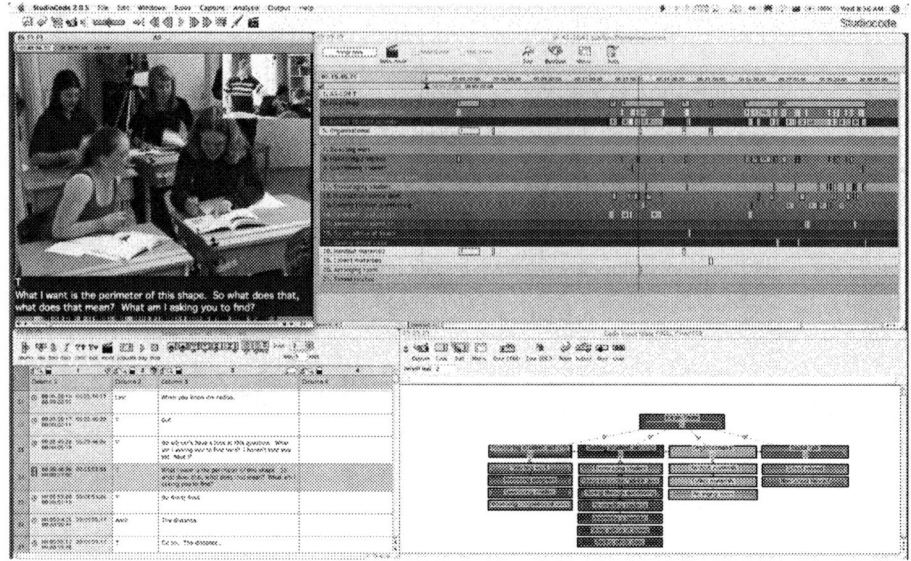

Figure 4. Sample analytical display (Studiocode) – video window (top left), time-line (top right), transcript window (bottom-left) and coding facility (bottom-right)

Collaboration with the Australian software company, Sportstec, was carried out to adapt the video analysis software *Studiocode* for use with classroom video data. These adaptations were driven by specific methodological, theoretical and practical needs. For example, the commitment to the capturing and juxtaposition of multiple perspectives on classroom events was partially addressed with the onsite capture of the picture-in-picture display shown in Figure 2, but the need to 'calibrate' the actions of the focus students against the actions of the rest of the class required multiple viewing windows.

Figure 4 displays the key analytical elements provided within *Studiocode*: video window, time-line, transcript window, and coding scheme. The researcher has the option of analysing and coding the events shown in the video window, or the utterances shown in the transcript window, or both. The resultant codes can be displayed in timelines (as shown in Figure 4) or in frequency tables. Once coded, single lessons, events within single lessons, or combinations of lessons can be merged into a single analysis.

The continual addition of new countries to the Learner's Perspective Study community required that video data already coded should not need to be recoded when additional data (eg. from a different country) were incrementally added to the database. Only the new data should require coding and the newly-coded data should be accessible for analysis as part of the growing pool of classroom data.

This flexibility is ideally suited to a project such as the Learner's Perspective Study, with many collaborating researchers adopting a wide range of different analytical approaches to a commonly held body of classroom data.

APPROACHES TO ANALYSIS

The *Studiocode* software described above is only one of the many analytical tools available to the classroom researcher. Increasingly sophisticated public access software tools are being developed continually. Most of the chapters in this book and in the companion volume (Clarke et al., 2006) report specific analyses of different subsets of the large body of LPS classroom data. Each analysis is distinctive and interrogates and interprets the data consistent with the purpose of the authoring researcher(s). Analytical tools such as *nVivo* and *Studiocode* can support the researcher's analysis but ideally should not constrain the consequent interpretation of the data. In reality, all such tools, including statistical procedures, constrain the researcher's possible interpretations by limiting the type of data compatible with the analytical tool being used, by restricting the variety of codes, categories or values that can be managed, and by constraining the range of possible results able to be generated by the particular analytical tool.

The methodological stance of the LPS project is fundamentally interpretive, but the means by which each analysis carries out the process of interpretation is distinctive. The author(s) of each chapter accept the responsibility to render the interpretive process as transparent as possible, in order that the reader might be best positioned to accord credibility to each resultant account. Such transparency of process is also important if the reader is to synthesise the findings of two or more accounts into a composite portrayal of classroom practice. It is entirely possible that two accounts might employ sufficiently divergent analytical frameworks as to defy synthesis. For example, an analysis of the contrasting perspectives of teachers and students in Japanese mathematics classrooms (Chapter 12) may not appear to share many common elements with an analysis of the role of the textbook and homework in mathematics classrooms in Singapore (Chapter 7). However, the first analysis may alert us to the possibility that even such familiar elements as textbooks and homework may be perceived and used differently by teachers and students. And the second analysis may highlight the significance of these familiar elements in one context and prompt us to question the significance they are accorded in another context and with what difference in meaning if the perspective considered is the teacher's or the learner's.

The most important act of synthesis is the one undertaken by the reader. The chapters that follow describe eighth grade mathematics classrooms in twelve countries through a kaleidoscopic array of perspectives. No-one can predict or prescribe which combination of perspectives will appear most relevant to each reader. It is to be hoped that each chapter provides sufficient detail and transparency of process to sustain the reader's integration of the various accounts, and their interest.

DAVID CLARKE

REFERENCES

Alton-Lee, A., Nuthall, G., & Patrick, J. (1993). Reframing classroom research: A lesson from the private world of children, *Harvard Educational Review, 63*(1), 50-84.

Bellack, A. A., Kliebard, H. M., Hyman, R. T., & Smith, F. L. (1966). *The language of the classroom.* New York: Teachers College Press.

Bruner, J. (1990). *Acts of Meaning.* Cambridge, Mass: Harvard University Press.

Clarke, D. J. (1998). Studying the classroom negotiation of meaning: Complementary accounts methodology, (chapter 7). In A. Teppo (Ed.), *Qualitative research methods in mathematics education*, monograph number 9 of the *Journal for Research in Mathematics Education*, Reston, VA: NCTM, 98-111.

Clarke, D. J. (Ed.). (2001). *Perspectives on practice and meaning in mathematics and science classrooms.* Dordrecht, Netherlands: Kluwer Academic Press.

Clarke, D. J., Emanuelsson, J., Jablonka, E., & Mok, I. A. C. (Eds.). (2006). *Making connections: Comparing mathematics classrooms around the world.* Rotterdam: Sense Publishers.

Cobb, P., & Bauersfeld, H. (Eds.). (1995). *The emergence of mathematical meaning: Interaction in classroom cultures.* Hillsdale, NJ: Lawrence Erlbaum.

Dawkins, R. (1995). *River out of Eden.* London: Orion.

Ericsson, K. A., & Simon, H. A. (1980). Verbal reports as data. *Psychological Review, 87*(3), 215-251.

Goffman, E. (1959). *The presentation of self in everyday* (New York: Doubleday, 1959), cited in Krummheuer (1995).

Gu, L., Huang, R., & Marton, F. (2004). Teaching with variation: A Chinese way of promoting effective mathematics learning. In L. Fan, N. Y. Wong, J. Cai & S. Li (Eds.), *How Chinese learn mathematics: Perspectives from insiders* (pp. 309-347). Singapore: World Scientific.

Harre, R., & Langenhove, V. (1999). The dynamics of social episodes. In R. Harre, L. V. Lagenhove & L. Berman (Eds.), *Positioning theory* (pp. 1-13). Oxford: Blackwell Publishers Ltd.

Krummheuer, G. (1995). The ethnography of argumentation. In P. Cobb & H. Bauersfeld (Eds.), *The emergence of mathematical meaning: Interaction in classroom cultures* (pp. 229-269). Hillsdale, NJ: Lawrence Erlbaum.

Lave, J., & Wenger, E. (1991). Situated learning: Legitimate peripheral participation. New York: Cambridge University Press.

Lindblad, S., & Sahlström, F. (1999). Ramfaktorteori och klassrumsinteraktion. Gamla mönster och nya gränser [Frame factor theory and classroom interaction. Old patterns and new borders]. *Pedagogisk Forskning i Sverige, 4*(1), 73-92, English summary available at http://www.ped.gu.se/biorn/journal/pedfo/eng.html.

Lindblad, S., & Sahlström, F. (2002, May). *From teaching to interaction: On recent changes in the perspectives and approaches to classroom research.* Invited Plenary Lecture, Current Issues in Classroom Research: Practices, Praises and Perspectives Conference, Oslo, May 22-24.

Mehan, H. (1979). *Learning lessons: Social organization in the classroom.* Cambridge, MA: Harvard University Press.

Miles, M. B., & Huberman, A. M. (2004). *Qualitative data analysis* (2nd edition). Thousand Oaks, CA: Sage Publications.

Nisbett, R. E. & Wilson, T. D. (1977). Telling more than we can know: Verbal reports on mental processes. *Psychological Review, 84*(3), 231-259.

Sahlström, F., & Lindblad, S. (1998). Subtexts in the science classroom - an exploration of the social construction of science lessons and school careers, *Learning and Instruction, 8*(3), 195-214.

Shepard, L. A. (1991). Psychometrician's Beliefs about Learning. *Educational Researcher, 20*(6), 2-16.

Sinclair, J., & Coulthard, M. (1975). *Towards an analysis of discourse.* London: Oxford University press.

Stigler, J., & Hiebert, J. (1999). *The teaching gap.* New York: Free Press.

Vygotsky, L. S. (1962). *Thought and language.* Cambridge, Mass: MIT Press.

Wenger, E. (1998). Communities of practice: Learning, meaning, and identity. Cambridge: Cambridge University Press.

David Clarke
International Centre for Classroom Research
Faculty of Education
University of Melbourne
Australia

APPENDIX A: CLASSROOM DATA GENERATION

Camera Configuration

Data generation employed three cameras in the classroom – a "Teacher Camera", a "Student Camera" and a "Whole Class Camera". The protocol below was written primarily for a single research assistant/videographer, but brief notes were provided suggesting variations possible if a second videographer was available. In order to ensure consistency of data generation across all schools in several countries, the protocol was written as a low inference protocol, requiring as few decisions by the videographer as possible. One or two possible anomalous cases were specifically discussed – such as when a student presents to the entire class. However, the general principles were constant for each camera: The Teacher Camera maintained a continuous record of the teacher's statements and actions. The Student Camera maintained a continuous record of the statements and actions of a group of students. The Whole Class Camera was set up in the front of the classroom to capture, as far as was possible, the actions of every student – that is, of the "Whole Class." The Whole Class Camera can also be thought of as the "Teacher View Camera." While no teacher can see exactly what every individual student is doing, the teacher will have a sense of the general level of activity and types of behaviors of the whole class at any time – this is what was intended to be captured on the Whole Class Camera.

Camera One: The Teacher Camera

The "Teacher Camera" maintained the teacher in centre screen as large as possible *provided that all gestures and all tools or equipment used could be seen* – if overhead transparencies or boardwork or other visual aids were used then these had to be captured fully at the point at which they were generated or employed in the first instance or subsequently amended – but did not need to be kept in view at the expense of keeping the teacher in frame (provided at least one full image was recorded, this could be retrieved for later analysis – the priority was to keep the teacher in view). The *sole exception* to this protocol occurred when a student worked at the board or presented to the whole class. In this case, the Teacher Camera focused on the "student as teacher." The actions of the Teacher during such occasions should have been recorded by the Whole Class Camera. If the teacher was positioned out of view of the Whole Class Camera (eg. front of classroom, at the side), then the Teacher Camera might "zoom out" to keep both the student and teacher on view, but documentation of the gestures, statements, and any written or drawn work by the student at the board should be kept clearly visible. Note: Although the teacher was radio-miked, in the simulated situations we trialled it was not necessary for the teacher to hand the lapel microphone to the student. The student's public statements to the class could be adequately captured on the student microphone connected to the Student Camera. The first few lessons in a particular classroom (during the familiarisation period) provided an opportunity to learn to "read" the teacher's teaching style, level of mobility, types of whole class discussion employed, and so on. A variety of practical decisions about the optimal camera locations could be made during the familiarisation period and as events dictated during videotaping.

Camera Two: The Student Camera

Where only a single videographer was used, the "Student Camera" was set up prior to the commencement of the lesson to include at least two adjacent students and was re-focussed in the first two minutes of the lesson during the teacher's introductory comments – during this time the Teacher Camera could be set up to record a sufficiently wide image to include most likely positions of the teacher during these opening minutes. Once the Student Camera was adequately focused on the focus students for that lesson, it remained fixed unless student movement necessitated its realignment. After aligning the Student Camera, the videographer returned to the Teacher Camera and maintained focus on the teacher, subject to the above guidelines.

If two research assistants ("videographers") were available (and this was frequently the case), then it became possible for the Student Camera to "zoom in" on each student's written work every five minutes or so, to maintain an on-going record of the student's progress on any written tasks. This "zooming in" was done sufficiently briefly to provide visual cues as to the progress of the student's written work, but any such zooming in had to be done without losing the continuity of the video record of all focus students, since that would be needed for the subsequent interviews. Since it was Learner Practices that were the priority in this study, the continuous documentation of the actions of the focus students and their interactions (including non-verbal interactions) was most important. A copy of the students' written work was obtained at the end of the lesson. The video record generated by this camera served to display each student's activities in relation to the teacher's actions, the tasks assigned, and the activities of their nearby classmates.

Camera Three: The Whole Class Camera

The "Whole Class (or Teacher-View) Camera" was set up to one side of whichever part of the room the teacher spoke from (typically, to one side at the "front" of the classroom). All students should be within the field of view of this camera (it is necessary to use a wide-angle lens). Apart from capturing the "corporate" behavior of the class, this camera provided an approximation to a "teacher's-eye view" of the class. It was also this camera that documented teacher actions during any periods when a student was working at the board or making a presentation to the entire class.

Microphone Position

The teacher was radio-miked to the Teacher Camera. The focus student group was recorded with a microphone placed as centrally as possible in relation to the focus students and recorded through the Student Camera (use of a radio microphone minimised intrusive cables). The Whole Class Camera audio was recorded through that camera's internal microphone.

Fieldnotes

Depending on the available research personnel, field notes were maintained to record the time and type of all *changes* in instructional activity. Such field notes could be very simple, for example:
    00:00 Teacher Introduction
    09:50 Students do Chalkboard Problem
    17:45 Whole Class Discussion
    24:30 Individual Textbook Work
    41:45 Teacher Summation
Specific events of interest to the researcher could be included as annotations to such field notes.

Where a third researcher was available, in addition to the operators of the Teacher and Student cameras, this person was able to take more detailed field notes, including detail of possible moments of significance for the progress of the lesson (eg. public or private negotiations of meaning). In such cases, the field notes became a useful aid in the post-lesson interview, and the interviewee could be asked to comment on particular events, if these had not been already identified by the interviewee earlier in the interview.

Student Written Work

All written work produced by the focus students "in camera" during any lesson was photocopied together with any text materials or handouts used during the lesson. Students brought with them to the interview their textbook and all written material produced in class. This material (textbook pages, worksheets, and student written work) was photocopied immediately after the interview and returned to the student.

APPENDIX B: INTERVIEW PROTOCOLS

Individual student interviews

Prompt One: Please tell me what you think that lesson was about (lesson content/lesson purpose).

Prompt Two: How, do you think, you best learn something like that?

Prompt Three: What were your personal goals for that lesson? What did you hope to achieve? Do you have similar goals for every lesson?

Prompt Four: Here is the remote control for the videoplayer. Do you understand how it works? (Allow time for a short familiarisation with the control). I would like you to comment on the videotape for me. You do not need to comment on all of the lesson. Fast forward the videotape until you find sections of the lesson that you think were important. Play these sections at normal speed and describe for me what you were doing, thinking and feeling during each of these videotape sequences. You can comment while the videotape is playing, but pause the tape if there is something that you want to talk about in detail.

Prompt Five: After watching the videotape, is there anything you would like to add to your description of what the lesson was about?

Prompt Six: What did you learn during that lesson?

[Whenever a claim is made to new mathematical knowledge, this should be probed. Suitable probing cues would be a request for examples of tasks or methods of solution that are now understood or the posing by the interviewer of succinct probing questions related to common misconceptions in the content domain.]

Prompt Seven: Would you describe that lesson as a good* one for you? What has to happen for you to feel that a lesson was a "good" lesson? Did you achieve your goals? What are the important things you should learn in a mathematics lesson?

[*"Good" may be not be a sufficiently neutral prompt in some countries – the specific term used should be chosen to be as neutral as possible in order to obtain data on those outcomes of the lesson which the student values. It is possible that these valued outcomes may have little connection to "knowing", "learning" or "understanding", and that students may have very localised or personal ways to describe lesson outcomes. These personalised and possibly culturally-specific conceptions of lesson outcomes constitute important data.]

Prompt Eight: Was this lesson a typical [geometry, algebra, etc] lesson? What was not typical about it?

Prompt Nine: How would you generally assess your own achievement in mathematics?

Prompt Ten: Do you enjoy mathematics and mathematics classes?

Prompt Eleven: Why do you think you are good [or not so good] at mathematics?

Prompt Twelve: Do you do very much mathematical work at home? Have you ever had private tutoring in mathematics or attended additional mathematics classes outside normal school hours?

Prompts 9 through 12 could be covered in a student questionnaire – the choice of method may be made locally, provided the data is collected.

Student Group Interviews

Prompt One: Please tell me what you think that lesson was about (lesson content/lesson purpose) (Discuss with the group – identify points of agreement and disagreement – there is NO need to achieve consensus).

Prompt Two: Here is the remote control for the videoplayer. I would like you to comment on the videotape for me. You do not need to comment on all of the lesson. I will fast forward the videotape until anyone tells me to stop. I want you to find sections of the lesson that you think were important. We will play these sections at normal speed and I would like each of you to describe for me what you were doing, thinking and feeling during each of these videotape sequences. You can comment while the videotape is playing, but tell me to pause the tape if there is something that you want to talk about in detail.

Prompt Three: After watching the videotape, is there anything anyone would like to add to the description of what the lesson was about?

Prompt Four: What did you learn during that lesson? (Discuss)
[As for the individual interview protocol, all claims to new mathematical knowledge should be probed. BUT, before probing an individual's responses directly, the interviewer should ask other members of the group to comment.]

Prompt Five: Would you describe that lesson as a good* one for you? (Discuss) What has to happen for you to feel that a lesson was a "good" lesson? (Discuss) What are the important things you should learn in a mathematics lesson?
[*As for the student individual interviews, "good" may be not be a sufficiently neutral prompt in some countries – the specific term used should be chosen to be as neutral as possible in order to obtain data on those outcomes of the lesson which the student values].

Prompt Six: Was this lesson a typical [geometry, algebra, etc] lesson? What was not typical about it?

The Teacher Interview

The goal was to complete one interview per week, according to teacher availability. The Whole Class Camera image was used as the stimulus. In selecting the lesson about which to seek teacher comment, choose either (1) the lesson with the greatest diversity of classroom activities, or (2) the lesson with the most evident student interactions. Should the teacher express a strong preference to discuss a particular lesson, then this lesson should take priority. Tapes of the other lessons should be available in the interview, in case the teacher should indicate an interest in any aspect of a particular lesson.

Prompt One: Please tell me what were your goals in that lesson (lesson content/lesson purpose).

Prompt Two: In relation to your content goal(s), why do you think this content is important for students to learn?
What do you think your students might have answered to this question?

Prompt Three: Here is the remote control for the videoplayer. Do you understand how it works? (Allow time for a short familiarisation with the control). I would like you to comment on the videotape for me. You do not need to comment on all of the lesson. Fast forward the videotape until you find sections of the lesson that you think were important. Play these sections at normal speed and describe for me what you were doing, thinking and feeling during each of these videotape sequences. You can comment while the videotape is playing, but pause the tape if there is something that you want to talk about in detail.
In particular, I would like you to comment on:
(a) Why you said or did a particular thing (for example, conducting a particular activity, using a particular example, asking a question, or making a statement).
(b) What you were thinking at key points during each video excerpt (for example, I was confused, I was wondering what to do next, I was trying to think of a good example).
(c) How you were feeling? (for example, I was worried that we would not cover all the content)
(d) Students' actions or statements that you consider to be significant and explain why you feel the action or statement was significant.
(e) How typical was that lesson of the sort of lesson you would normally teach? What do you see as the features of that lesson that are most typical of the way you teach? Were there any aspects of your behavior or the students' behavior that were unusual?

Prompt Four: Would you describe that lesson as a good lesson for you? What has to happen for you to feel that a lesson is a "good" lesson?

Prompt Five: Do your students work a lot at home? Do they have private tutors?

APPENDIX C: EXAMPLE OF PART OF A LESSON TABLE

School: School 1
Date: 24/8/2000
Teacher's Goal: 1) **Realisation of the relationship between circumference and diameter of a circle (ie. π)**
2) **Knowledge about π: (a) irrational number and approximate value of π; (b) is used in calculating circumference.**
Number of Lesson in Unit: 4

Time	Organisation of Interaction		Description of Activity	Description of Content
	Framing	Focus Group (Sarah, Jaira)		
03:07	Classwork		**Setting up** Getting materials ready to continue yesterday's activity. Revision: Concepts.	Measuring the circumference (C), diameter (D) and radius (R) of circles printed on textbook p174
07:29		Both responding to teacher question.	**Sharing** Revision: see 1. Discussing 2,3. [T-Ss]	1. Definition of circumference, perimeter, diameter and radius including symbols (eg. D=diameter). 2. Relationship of diameter and radius in maths formula (eg. r=d÷2, d=2r). 3. Differences in calculating d÷2 and d+1/2 (algebra).
12:48			**Teacher** Explaining investigative task.	1. Textbook p174 (a-f circles). 2. Additional tasks (measurement of circles on board. 3. Accuracy of measurement.
13:20			**Discipline**	
14:13			**Sharing: Others (non-maths)**	

CHRISTINE KEITEL

CHAPTER THREE

'Setting a Task' in German Schools:
Different Frames for Different Ambitions

INTRODUCTION

What insight into mathematics and its social role in reality is conveyed by tasks set in school mathematics? And what image of mathematics or interest in it is created by which kind of tasks in a specific setting of a task? This chapter focuses on the most typical representation of school mathematics in Germany: tasks either formulated by the teacher or chosen from a textbook to be practised by students. For many students at the university as well as many adults, doing tasks is not only the predominant activity encountered in school mathematics, but a synonym for school mathematics in general. And among students in other school systems it also seems to be taken for granted that working on mathematical tasks in school has no meaning beyond the classroom. Goodchild refers to such a perception of students' goals in mathematics classroom practice as the most significant and shared when he summarises his investigations:

> The main outcome of experience in mathematics classrooms is to learn how to do mathematics classroom practice. In the classroom, mathematics is set within a specialised 'classroom' discourse, that allows students to locate and follow cues and signals, skip over peripheral text and apply a variety of resources to bring the highly stylised tasks to some form of resolution. Activity within the classroom is not mathematics and for all its pretence it is not about the students' current or future experience of the world outside the classroom. Success in classroom practice does not prepare a student for the practice of mathematics or any other activity outside the classroom. (Goodchild, 2001, p. 227)

In analysing data of the Learner's Perspective Study (LPS) from Germany we want to look closer at mathematics classroom practice and find out what kind of tasks are set, what are the differences in the ways teachers set tasks and if, and how, these differences may affect students' learning, understanding and their relationship to mathematics as a school subject and an important scientific and social enterprise.

D. J. Clarke, C. Keitel & Y. Shimizu (Eds.), Mathematics Classrooms In Twelve Countries: The Insider's Perspective. 37–57. © 2006 Sense Publishers. All rights reserved.

CHRISTINE KEITEL

CHARACTERISTICS OF THE THREE GERMAN SCHOOLS IN LPS

The selection of teachers, schools and classrooms and the data collecting followed the general guidelines of the Learner's Perspective Study as reported elsewhere in this book. The German educational system is especially characterised by a strong and early segregation of students by performance levels based on assessment results. This also forces teachers and parents to decide rather early where to place students within the tripartite system: College-bound high school (Gymnasium), commercial and technical middle school (Realschule or comprehensive school with an inner segregation) and the lower secondary or common school (Hauptschule – schools of this type often do not have qualified mathematics teachers). Among the three German schools which participated in the LPS, G1 and G2 represent the Gymnasium type of school, while G3 is a Comprehensive School with courses on two achievement levels in Grade 8.

G1 is situated in the former centre of East Berlin and is an old secondary school which was transformed into a Gymnasium after the reunification of Germany (Keitel, 1993a); the teacher is a qualified and well experienced mathematics and physics teacher who acts as leader of the group of mathematics teachers in the school responsible for common planning of mathematics classroom issues in general. There were 27 students in the class. Their parents belong mainly to the middle class, mostly tradespeople and small business owners or employees of larger enterprises, with only few academics among them.

G2 is a traditional Gymnasium situated in a rather wealthy residential area of former West Berlin; the teacher is a very experienced and well-qualified teacher of mathematics and geography, who also serves as principal headmaster of the school. The 30 students in G2 also have a middle-class background, with more parents from middle-level academics like barristers, teachers or small business owners. Parents in both groups very much care about a good education and want high achievement of their children.

In contrast to G1 and G2, the third school G3 is a Comprehensive School located in an area with a very high percentage of larger immigrant families, mostly of Turkish origin, and only very few German students, mostly lower or middle-class. G3 uses streaming according to achievement or ability into two courses: a basic course for a majority (Grundkurs) and a high ability or extension course (Leistungs- or Erweiterungskurs), generally considered as equivalent to a Gymnasium class of the same grade. The group in the extension course comprises only 15 students. The teacher reported a high drop-out rate from the extension course during the school year, because students have a free choice of where to attend. The teacher is a very experienced and specially engaged mathematics and physics teacher. She also cares very much about all the students – she is the class teacher responsible for most of the students in her course, so she knows them very well. For the purposes of this investigation of 'setting a task', however, we concentrated in a first analysis on the two Gymnasium-type schools, because G3 showed so many different features and circumstances that any comparison would widen our analysis substantially beyond this theme.

METHODOLOGY USED FOR THE ANALYSIS OF 'SETTING A TASK'

In mathematics classroom practice, teachers can choose to set tasks for students in various ways. We have decided to differentiate the following possible or assumed characteristics of these settings:

1. What type of task might be presented? It was anticipated that by setting a task the teacher might explicitly or implicitly refer to a special function of the task given, which might be a) for practising a routine, or a new skill or method; or b) a problem that requires systematic reorganisation of students' knowledge or acquiring new methods or creating new insight.
2. What kind of language or representation is used for presenting a task? What are the consequences of choosing particular language terms or representational forms? It was anticipated that there might be: a) symbolic representations with formal language only; b) context-related representation with a mixture of daily life and formal language; or c) presentation of an authentic context problem to be differentiated.
3. What kind of explicit justification does the teacher offer when setting the task? A task might be accompanied by a) explanations or declarations about the meaning of the task or problem; b) explanations of the importance of the task or its particular role within the teaching sequence or lesson; or c) a teacher may make claims regarding students' special interests for this task, for example, for training or special use of an algorithm, as an exploratory means for providing understanding, for clarification of a special case, for generalisation or transfer to another field or domain, for offering a concrete application, for supporting an argument or leading to the proof of a statement.
4. Does the teacher explicitly focus only on working instructions? Or does he refer to various conceptual aspects? The setting can a) aim at general methods or techniques, leading to rules or 'recipes'; b) emphasise theoretical importance or connections; c) focus on specific working instructions to be followed explicitly and correctly as a pattern; or d) emphasise technical importance in references to usages of the rules embedded in the task.

From a preliminary analysis, we added a specific question:

5. Which explicit value is attributed to the tasks by the teacher in setting them, and by the students when working on the tasks and presenting their solutions?

By applying a frame of analysis that includes the above described differences in form and function of tasks, in language, context and content of tasks, in the activities and settings that are closely related to working on tasks, we used interpretative methods of observation, hermeneutic and discourse analysis of coded transcripts of lessons, and of the discussions with students during lessons and in the interviews following the lessons (Huber & Mandl, 1994; Clarke, 2000, 2002, 2003). In this chapter, analysis focused on data from the teacher and student video records and from copies of text materials, student written materials, and any mathematical representations employed by the teacher. Some use was made of teacher or student post-lesson interviews to identify the meanings that participants attributed to particular classroom activities and to identify the importance they attached to particular types of task. The observations and first analyses by the individual researchers were complemented by in-depth reflections among the German group of researchers, in particular when discussing the unstructured post-lesson conversations with students about their activities. On the basis of the data analysis we tried to differentiate what aim, function or method was provided when the teacher began to set a task. A crucial aspect was the time within the lesson for setting a particular task, because this might determine a specific role and also a specific form to embody the task. It was important to interpret the specific characteristics attributed to the tasks when presented, and the different students' activities that were either directly demanded by the teacher or came up in students' interpretation of the task and the teacher's setting of the task.

Qualitative analyses of the differences in setting a task and students' reactions are used to differentiate among school types, among the teachers and student groups, while some quantitative analyses offer information about the regularity and frequency of certain settings and framings in terms of aims and importance, justification or necessity. Preliminary analyses had shown that a schools' general organisational frame might have a major influence on the setting of a task and also the type and function of tasks. The differences might relate to types of school, but also to the social environment of the students. Thus the theoretical background of our analyses necessarily draws on socio-cultural research in mathematics education and sociology of education (Begehr, 2002, 2004; Jablonka, 2002, 2003a, 2003b 2003c, 2003d, 2004; Keitel, 1986a, 1986b, 2004; Krummheuer & Brandt, 2001; Seeger, Voigt & Waschescio, 1998; Voigt, 1995; Yackel & Cobb, 1996) with the additional focus on subject matter analyses (Keitel, 1986b, 1993b; Lenné, 1969) of why and for what purpose tasks and problems are chosen, from the perspective of both teacher and students.

EXCURSION: WHY DO WE FOCUS ON TASKS?

Mathematical Texts as Presentations of Tasks

The close relationship between mathematics and tasks and the crucial role of tasks in representing mathematics is manifested very early on by ancient texts. The oldest available and decoded mathematical texts from Uruk (3000 BC) only represent mathematics in form of tasks and their solution; how to solve daily practice problems, like those of measuring in agriculture, calculating quantities harvested, supply of food and slave workers or the like. Mathematics was implicit in rules telling 'how to do' and what to gain. The later Babylonian texts (2000 BC) show more formal tasks which, although still embedded in a context, are more detached from a concrete problem and seem to have functioned for teaching purposes in schools of scribes (Damerow & Lefèvre, 1981; Nissen, Damerow & Englund, 1990). The representation of mathematics as tasks and problems is a cross-cultural phenomenon, which supports the assumption that mathematics came into being as a device for solving social problems that has been theorised much later into a systematic body of connected and more abstracted knowledge for teaching or research purposes. The systematic collection by the Greeks of the available geometrical knowledge of the time (Euclid 500 BC) is an example of a more theoretical and unified approach, and led to the acknowledgement of Greek mathematics as a first scientific system – mathematics as a science (Kline, 1985). But here also collections of tasks follow descriptions and arguments, this time with purely mathematical intentions. Connections are to be established, analysed and proved, detached from any direct concrete use in reality. The collections of tasks and problems that originate from the early times of mankind and have gradually been enriched and refined, are huge and come from many cultures and traditions, with slightly different contexts but astonishingly similar mathematical devices at the core (Keitel, Schubring & Stowasser, 1985). They had different concrete functions and importance during the course of history, but since institutions for teaching and training mathematical knowledge and techniques were established, the accompanying materials used, such as clay tablets as school books, followed this early pattern. They offered collections of descriptions of tasks and explanations of their solutions, with some general hints for commonalities and patterns. Such collections were used for early general exams giving access to privileged positions, such as Chinese exams for state administrators (see Leung, 2004), for certain trades and crafts, and finally for public schools. Patterns became rules and methods for doing tasks, the focus of teaching was often only for students 'to get tasks right', a major aim was fluency and reliability in coping with the underlying rules and techniques. In social reality, deep understanding or reflexive insight was only for a few.

The Role of Tasks in German School Mathematics and Teacher Education

Mathematics became an important and indispensable school subject for the first time in the 19th century as a result of the institutionalisation of public secondary state-determined schools leading to universities (Keitel, Schubring, & Stowasser, 1985; Keitel, 1993a, 1996). The Prussian state Gymnasium served for the education of an elite, with an orientation towards general education: 'Bildung' has been considered as a general and formal intellectual or mental training. Mathematics was perceived as a universal mental tool, as the purest manifestation of the rational endowment of mankind, with the utmost formalisation of thinking. It was given a teleological perspective, with the most formalised mathematics seen as closest to the absolute idea of 'Bildung'. As a consequence, theoretical or formalised mathematics was credited with a value per se, and a hierarchy was established from this summit downward to lower school types. Accordingly, mathematics which, for whatever reason, was not freed from the trifles of concrete reality had a lower ranking. The invention of tasks and certain routines in setting a task followed this philosophy easily. It allowed teachers to disregard any concrete meaning or context for a task, but to 'train the brain' by formal algorithms and intensive attention to memorising rules, techniques and general methods.

In German educational philosophy, mathematical knowledge for teaching and subject matter didactics for teacher education early on focussed strongly and solely on designing and teaching tasks for various formal purposes. Lenné created the term 'Aufgabendidaktik' (didactics of tasks) to describe the main and predominant ingredients of teacher education and mathematics education activities for schools (Lenné, 1969). Formalised and purely mathematical tasks focussing on algebraic rules for solving equations and equivalent transformations or geometric constructions served as means to general intellectual training, disregarding any concrete importance or aims. In elementary schools for students of lower social class, arithmetical techniques and rules were taught by using stylised tasks together with strict working behaviour and disciplinary rules. When tasks were set in a context, this context was reduced to a very stylised form and completely framed by the underlying rules: Learn to behave in a very much disciplined way and to follow rules strictly.

There was another important obligation that teachers had to learn in teacher education – they had to cope with the burden of a demanding teacher-based assessment system. Tasks therefore had a double function: They were the most important means of teaching and assessing at the same time, and they had to secure objective standards of evaluation. A strange amalgam of setting up various tasks for conflicting purposes developed. The teaching and assessing on one hand means that tasks are designed to develop performance abilities, to prepare for assessment, and to serve as means of assessment that evaluates these abilities (Keitel, 1992, 1993a; Kaiser & Keitel, 1994).

Several reform movements in the late 20th century, which finally led to the recent call for 'mathematical literacy', confronted school mathematics in the German high school with socially defined claims for the knowledge conveyed and for equal access to a broader perception of mathematics education (Damerow, Elwitz, Keitel & Zimmer, 1974; Gellert, Jablonka, & Keitel, 2001; Jablonka, 2003a; Jablonka & Keitel, 2005; Keitel, 1986a, 1996, 1997; Keitel & Kilpatrick, 2005). However, the persistence of the tripartite system did not support a balanced and comprehensive determination of school mathematics that mediated both theoretical and scientific knowledge and socially-defined practical knowledge. In school books there is still an abundance of formal and technically oriented tasks; problem-solving and proof are rarely touched on or only dealt with technically (Knipping, 2003); contexts for mathematical modelling or application are reduced to 'dressing up' formal algorithms and technical skills (Keitel, 1993b). It is left to the teacher to make a difference in changing the tasks, to invent a particular way of selecting and setting the task.

What insight into mathematics and its social role in reality is conveyed by mere technical or formal tasks? What image of or interest in mathematics is created by which way of setting the tasks? Our cultural mathematical techniques and tools have two crucial characteristics:

(1) Instruments or rules are constructed to replace an explicit use of mathematics and to facilitate working with them in daily life, but it is not always easy to directly see or discover the mathematics behind or within the instrument or technique. In emphasising this aspect, Freudenthal has taken mathematical instruments or explicit mathematical models in reality as starting points for a 'reinvention' of the

mathematics behind them, as in his view not only the most promising, but also the most challenging and successful approach to mathematics learning (Freudenthal, 1991, pp. 45-66).

(2) The application of mathematics in social life has two essential aspects: the first is to establish a proved, reliable solution for a social problem; the second consists in operationalising this procedure in a way that it eventually is encapsulated instrumentally. As a result, the problem-solving operation can henceforward be executed with reliable security even without 'knowing' the particular mathematics behind the instrument. Technical instruments, formulas, tables, and technical rules originate from this process; mathematics at this stage means the purposeful use of mathematical devices in suitable problem contexts. The high standard of mathematical ingenuity acquired for the development of such devices is encapsulated within the rule or instrument, which by machines and computers today might be shrunk to the simplest stimulus-response bond: one has to push a button (Jablonka, 2003a; Keitel, 1986a, 1993b, 1997). To handle technical devices does not require consciousness of the mathematics inside them. However, to learn mathematics and to evaluate critically the use of technical devices, instruments or social rules demands the reconstruction of mathematics and therefore a much broader knowledge than simply knowing the rules, 're-calculating' or 'checking the numbers' (Keitel, 1989). If the application is to be understood and critically evaluated, translation and interpretation are required. Taking the reinvention seriously means creating specific conditions for setting a series of well-defined tasks in classroom practice (Freudenthal, 1991, pp. 66-86).

The difficulties that our students traditionally have with mathematical applications in word problems, or with attributing meaning to parts of school mathematics are notorious and well known, recently recalled by comparative studies like the Third International Mathematics and Science Study (TIMSS) and the OECD Programme for International Student Assessment (PISA) (Baumert et al., 2002; Keitel & Kilpatrick, 1999). It can be assumed that many of these difficulties are rooted in the conception of school mathematics that does not address the constructive interplay and interaction within the development of mathematics of both mathematical theory and applications. When teaching focuses on mathematical theoretical or formal knowledge only on one side, or replacing theoretical knowledge by an application pattern on the other side, mathematics is taught in unconnected halves. This analysis should contribute to the explication and investigation of assumptions about the function and impact of tasks and the possible ways of setting tasks that foster or hinder a reasonable insight into mathematics and its social role.

PATTERNS OF SETTING TASKS IN CLASSROOM PRACTICE

The Lesson Structures in G1 and G2

In G1 the teacher had established a very regular structure for his lessons. They mostly had three distinct parts:

- A 10- to 15-minute repetition or training phase of activities and tasks practised in variations as individual written work of students, sometimes complemented by publicly comparing homework and collecting feedback from students about their achievement or technical problems;
- A more extended phase (30 to 40 minutes) of teacher-guided question-and-answer debate, either introducing new concepts or techniques, or concentrating on special rules from a new singular aspect;
- A short round-up lasting 10 minutes that included assigning homework and preparation for the next phase of the work, or announcement of future activities, for example, indicating the regular class test or written classwork.

This pattern, which was clearly visible in the lesson plan and mentioned in the daily questionnaire of the teacher collected in the LPS-data, served mainly for developing routines in solving formalised algebraic tasks – in this case equivalent algebraic transformations – and was maintained throughout the video-recording, with very few exceptions. Three of these were identified explicitly by students during interviews: G1-L06 had a larger session of unusual group work, in G1-L07 three groups of students were out at front working and presenting collectively their results on an 'unusual' activity on the blackboard, and G1-L10 contained an activity or game like dominoes, using equivalent transformation patterns of algebraic terms as rules for playing the game.

In G2, the teacher structured the lessons less formally and regularly, although he also followed a general lesson plan. However this plan seemed to allow more improvisation and immediate reaction to actual needs formulated by students; for example, regarding what is still not clear in 'how to do a task or procedure' and has to be worked on again. Although he used formal routine tasks for equivalent transformations of algebraic terms or expressions in a similar way as in G1, the teacher in G2 was more supportive of students' self-initiated activities on tasks from the textbook. In particular, he explicitly asked students, more often than the teacher in G1, to work and discuss in groups, to present their results on the board in front of the whole class and to lead a class discussion. However, his duties as headmaster often forced him to change plans and leave the classroom, forcing him to ask the students to work independently, in groups or alone. This also required him to improvise tasks or assignments for student activities because he was absent or late. However, he trusted his students to work collaboratively and successfully on problems and tasks that he set up. In the interviews, students did not see major changes in his teaching style or in the tasks set up for them, with only very few exceptions of new kinds of problems. Unusual lesson events happened in G2-L04 when an improvised task on equivalent transformations caused special and unforeseen problems for students. These had to be discussed in detail with the class. The teacher justified this different activity by claiming that students learnt more from the discussion than by doing the usual well-organised textbook tasks. In G2-L05, the teacher set up an unusual activity with group work on 'brain sport'. In G2-L13, the whole class spent quite a long time discussing a special homework assignment, namely to design a 'recipe-like' or 'automatically working' set of instructions for those rules that apply to algebraic equations and include algebraic terms as fractions. The students were to probe these instructions with their parents or friends.

Forms and Functions of Setting a Task

In the following first and preliminary analysis of setting a task, we concentrate on the two Gymnasium types of secondary school, which have rather similar characteristics although the teachers came from different biographical background and education.

The G1 teacher regularly started with a series of tasks for practising and memorising specific small algebraic rules for expanding brackets or simplifying algebraic expressions or becoming accustomed to a special sort of task, here equivalent transformations of algebraic terms. He wrote a couple of formal algebraic terms, usually a mixture of formal expressions with letters and numbers, on the board and asked each student to individually work on simplifying by transforming them: either changing products of algebraic terms into equivalent sums, transforming sums into products, or simplifying expressions by combining or adding similar terms. In setting these tasks, the teacher made clear by repeating, that he considered practice as a very necessary and regular part of each lesson. The students followed him because he also told them repeatedly that it served to prepare for regular intermediate tests and the more demanding written class work as part of the teacher-based mode of assessment. Therefore small assessment tests were given frequently, at least every two weeks. Regular practice should especially prepare for the prescribed series of six more serious and comprehensive written classwork papers during the school year that counted for final marks in the assessment at the end of the year. These types of assessment were specially announced, while smaller tests could happen more irregularly. The pupils were accustomed to this kind of routine task; mostly they worked individually, although a few tried to collaborate. The teacher did not explicitly encourage cooperation, but left this decision to the students. This activity regularly lasted for about one third of each day's lesson and was introduced by quite similar wording:

Transcript 1. Teacher (G1-L05)

T: Now get on with a few problems that played an *important part in the test* … let's try please to do each following task … expand the brackets and work them out … I've made a *simple start again* here so that we start first *with numbers*. You have exactly 10 minutes.

Key to symbols used in transcripts in this chapter:
...   A pause of three seconds or less
( )   Indecipherable utterance
[text]   Descriptions of non-verbal action, or comments and annotations by the author.
text   Italicised text indicates emphasis added by the author.

By setting the tasks in this way the teacher emphasised the specific function of the tasks: because the last informal test some days ago had not been so successful, students have to practise again for the upcoming more serious test. The teacher tried to make use of so-called 'easier' tasks by starting with terms that only contain numbers and look like arithmetical tasks, before coming to algebraic expressions with letters. Students did not see this necessarily as easier, nor did they ever check an equivalent algebraic transformation by simply using numbers for letters. Students accepted the tasks and did not question them, because the passing of the test or written class work was the most important aim of the whole enterprise – one has to collect good marks for whatever purpose. The content of tasks was also clearly described by the teacher as necessary routine and involved the practising of rules and algorithms only, irrespective of their meaning or importance beyond the classroom. The most frequent expression describing the transformation of terms was 'calculation' (reckoning), an explicit reduction of the formal character of the tasks which was repeated nearly the same way each day at the beginning of a lesson:

Transcript 2. Teacher. (G1-L06)

T: Who was able to calculate everything in the corrections? …
       Well … right then. Or I'll put that another way … who …
       still had, in the corrections, questions which he couldn't
       solve … or isn't sure whether they're right. Put your hands
       up. Oh right, there is still some uncertainty there. That's
       why … I'd like us to practice these things again in this
       lesson. Thus to begin with, I have four small questions for
       everyday practice. So … we can start again with a small
       arithmetical problem with numbers … but not so difficult. So
       from question B onwards the brackets have to be expanded and
       then summarized … which we also had done the last time.

Transcript 3. Teacher. (G1-L07)

T: Well for a start we begin the *daily exercise orally* … we go
       through the homework and then we might revise again the
       multiplying part … so that tomorrow, *when we do the test
       again so that no difficulties come up.*

Students knew the pattern of the lessons very well. Often they referred to practising as the most important part of the lesson. They even seemed to be rather grateful to the teacher for his intensive caring to 'get them doing it right'. The formal character of the tasks did not make any sense other than passing the assessment. Understanding meant knowing how, not what it means, and mathematics is not useful outside the classroom except perhaps for (maybe?) mathematicians or banking people – but even this was not clear to them. What does a mathematician need? No idea.

In the interviews, the teacher explicitly stated a quite different intention of his mathematics classroom practice: He wanted to develop generic thinking skills and conceptual understanding. He insisted that the purpose of mathematics is not about procedures, and wanted the students learn to appreciate algebra as a wonderful system and toolbox. He sought to create connections between different mathematical domains like algebra and geometry. However, he strongly believed that understanding only follows after rules and algorithms have been mastered, and that there is no better way than to pay extensive attention to mastering operations, rules and techniques – this is indispensable.

The overwhelming emphasis on mastering routines and algorithms seemed to be extended to all parts of mathematics classroom practice in G1 and overshadowed even those events that actually broke with routines and tried to offer a new kind of insight and enlightenment. Disappointed by the fact that dealing with the binominal formula, although practised intensively, had not yet been mastered sufficiently by the students, the teacher used an unusual approach to practising these formulas. He divided the class into three groups, chosen according to ability – that is, some groups were called high achievers, others middle level performers and further others as low performance students – and assigned three different tasks to represent the first, second and third binominal formula by geometrical means:

Formula 1: $(a + b)^2 = a^2 + 2ab + b^2$;
Formula 2: $(a - b)^2 = a^2 - 2ab + b^2$;
Formula 3: $(a + b)(a - b) = a^2 - b^2$).

The formulas had already been developed by calculation or multiplying out; the new task was to find a geometrical representation for each of them, on the basis of a picture for such a representation of Formula 1 in the schoolbook (Figure 1).

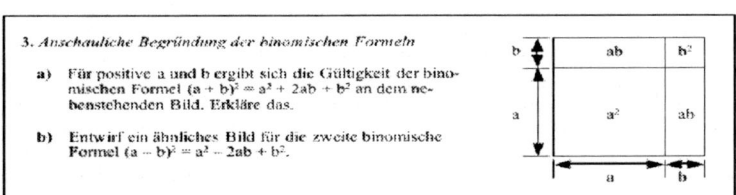

Figure 1

This activity was to be extended to the other two formulas as well and also offer a different explanation and new insight for these formulas as he explicitly wrote in the questionnaire: "To relate algebra to geometry". The teacher began setting the new task by the following:

Transcript 4. Teacher, students: Thorben, Albert, Karolina as well as unidentified student (S) and students (Ss). (G1-L06)

T: Right … in the lesson today I want us to start *something new*
          … *forming groups based on achievement levels* … that means a
          … in groups of four to try … uh to somehow try and represent
          these binomial formulas as areas … similar to what we did
          when we multiplied out. To that end … I'd just like us now
          to repeat … how we can depict or construe the product of A
          times B as a geometrical form … the product A times B …
          Thorben?
Thorben: A rectangle.
T: Yeah … what part of the rectangle is A times B?
Thorben: … the area formed.

T: Right … the area of a rectangle with the length of the sides A and B is the product of A times B. How can I depict A squared geometrically … Alexis?

Albert: A square.

T: With the sides of what length?

Alexis: A

T: And what part of the square … with the length of the sides A … is A squared?

Albert: The area formed.

T: Right. You can do it with areas then … and … I'd like us then to depict these three binomial formulas as areas. There's an introduction to that in the book … which you can open now to page … one moment

T: Forty-four. And let's look please at question three.

S: Forty-four

T: Page forty-four.

T: Karolina … read question three out please.

Karolina Um … descriptive foundations of the binomial formulas. Positive A and B demonstrates itself in the validity of the binomial formula: the square of bracket on A plus B bracket off equals A squared plus two A B plus B squared in the accompanying diagram; demonstrate this.

T: Okay. I think then you have to explain that. It's necessary to think about that for a bit … but that could be given to a group that isn't quite so strong in mathematics. So now read out B.

Karolina Create a similar picture for the second binomial formula … um the square of bracket on A minus B bracket off equals A squared minus two A B plus B squared.

T: Yeah that's a good thing of course … you have to work out a new diagram and try to illustrate it with this diagram. I think an intermediary group could manage that … and problem C … which isn't to be found here … which consequently … of course … I can work out myself … and that is the third binomial formula which is A minus B times A plus B which was so much as …

T: What is A minus B times A plus B?

Thorben A squared minus B squared

T: Yeah … which you can make clear … illustrate that … by using areas. I think that a more advanced group could uh manage that … well and how are we going to go about it. Um … I'll form three groups … three attainment … well you'll work together in groups of four … but there'll be advanced groups … intermediary groups … low-performance groups … which incidentally doesn't mean any sort of downgrading … it's just simply that the required standards uh …

Albert [to Thorben sarcastically] You'll land in the low-performance group.

T: Aren't so demanding.

Ss: [ironically] *Low-performance*.

T: Yeah perhaps the … choice of terms isn't quite so fortunate … but anyway it's a matter of us getting these results together at the end … that … from every group … that an attempt is made *to present and make it plausible* for the others. I'd just like us to try that today … and to that end I've considered the following org … arrangement … that corresponds to the performing of you.

The teacher started by announcing something new, and underlined that it was new in two respects: the chosen organisational pattern of working in performance level groups and a new content, namely representing a formula differently. However, he did not mention the new theme, namely to represent algebra by geometry, explicitly. The new setting was a break in two routines at the same time and attributed a special quality to the task assigned. But already early on he played down the changes to the normal daily classroom routine. He emphasised the necessity of drawing a distinction between the three tasks because of the differences in demands. The representation of the first formula he considered the easiest because it was already given by the textbook. This therefore demanded only a reproductive activity – students had to make sense of the already presented geometric demonstration offered by textbook authors. Then he mentioned explicitly that the second and third formulas were much more difficult to represent, because an intellectual action of transfer was demanded, a creative transformation of a given scheme to another problem. The setting of these tasks should facilitate independent work in the assigned groups that were composed according to the difficulty of the tasks into three ability groups. He also reminded the students of some 'facts' about representing length and area for rectangles and squares in using terms from the formula. While students were working in the groups, the teacher walked between the desks and talked to students, sometimes helping them or clarifying his intentions for the task, or giving some organisational hints and so on. The lesson time was over before the task was completed, so he announced the continuation and presentation of the tasks in the next lesson. The next day, the teacher began by again problematising the need to use these formulas:

Transcript 5. Teacher, Student. (G1-L07)

T:	In today's lesson we want to … try as well to evaluate this, at sometime today, but before doing so, I'd like to know again … why or which … reason is there now for these binomial formulas, why are they considered separately? Why … couldn't you just … multiply?
S:	Good question!
S:	It would simply be too strenuous and you'd have too much writing to do. That's why there are the binomial formulas that explain the special case that … uh … makes it easier for oneself.
T:	Right. … That is … a very important aspect of this, that really not each time … you want to multiply that afresh, so if you can write that down straight away, because *this multiplying is so very easy with these binomial formulas.* Right then … please think about the last lesson. So we want to present the results now. Those who found anything, well … out of these. Who had this part, the easy part A plus B in brackets to the power of two equals A squared plus two A B plus B squared. Then the group who had this and now, well, it would be nice if, well … one of each group could come out to the front.

Presenters as delegates from all three types of activity group went to the board and started to draw geometrical representations of the algebraic expressions for each

49

formula and described what they drew with some lines of algebraic equations, to show that it worked. This was one of the rare lesson events, when students worked independently on different tasks and presented their work at the board to share with others and explain to them what they had done, why it was done this way, and why this was a kind of validation or "proof for the formulas", as said by one student in the interview.

During the students' presentation, the teacher did not intervene as often as usual, which he explained later as an unusual and risky behaviour. But he helped those at the board who needed some support by calling on other members of their groups. The first formula was extensively explained by students, with some help by others sitting as well, but the task was considered as 'simple' by the teacher and the result did not get so much attention. The other two representations were reported rather quickly, although they were announced as more difficult: it was somewhat clear that these were only for the high-achievers, that is, not for the majority of students. However, there was an excited dispute among students afterwards in the interviews, whether the teacher really was right when he stated that the third formula was the hardest to represent, as in their view the second one was harder; and whether the ability grouping – using only the most recent test results – was appropriately showing the low, middle and high achievers. There was agreement in the interview with the students that actually the second formula was the hardest to present and explain, and those who had had to do this felt underestimated and under-appreciated.

While the teacher repeated that dealing with the binominal formula is mainly to make it easier to multiply algebraic terms, two students in the interview pointed to another important possible gain from this activity: They believed that it was about a kind of proof for understanding, a proof which gave reasons why these formulas work this way. This was something which the two students really valued:

Transcript 6. Interviewer, students: Clarissa, Simone (G1-int07)

```
I:        Tell me, before we turn to the video what was the most
          important in the lesson … for you? Was there such a moment?
Clarissa: Hm … well … what could have been important here? Well,
          everything that was there — the problems or — that, what we
          simply did with these squares … yes.
I:        And how would you name the topic of the lesson?
Simone:   Binomial formulas, again.
I:        How would you describe it?
Clarissa: Hm … that's difficult … that … the proofs with the binomial
          formulas is probably, well, the most important …
Clarissa: What I mean is that it is also about understanding and not
          only about knowing the formula, or something like that.
```

The students liked working collaboratively together, although many did not agree with the division into low, middle and high level ability groups. It was the first time they had worked in this way, and also unusually time-intensive for just three tasks. Although two girls agreed that this activity was more demanding than the daily exercises or the typical tasks they had and led to some kind of 'proof', the way the teacher set the task did not point to this special importance, and the highly interesting aspect of connecting geometry and algebra in a mutually supportive way became lost in the perception of the majority of students. The opportunity to develop meaning and encourage further reflection was missed, and an interesting problem was restricted to a simple routine-like task.

The G2 teacher usually started the lessons by checking homework and continued with practice using tasks chosen from the schoolbook, or by leading a whole class discussion to develop a new topic or rule. But there were a very few lesson events in which he announced the setting of a new task quite explicitly:

Transcript 7. Teacher, student (G2-L05)

```
T:      A wonderful good morning to all of you.
S:      Morning
T:      You didn't have any homework for today … and that's
        something special when there's no mathematics homework … we
        want to do something We want to do something new is the
        right cue … and today it's sport.
S:      Ah, sport.
T:      Brain sport. That means you should use as much brainpower as
        possible. You'll get really difficult tasks to work on in
        groups.
S:      Group work.
T:      There are two different types of tasks they are principally
        both as difficult as each other … the difference is that …
        eh the one assignment is based on reality that means it can
        actually happen while the other one is
T:      More or less something that mathematicians like to think up.
        I've taken them both out of your exercise books. They're not
        … they're not from me. I've made a copy. The point of the
        whole thing is that you all close your math books and put
        them away because it doesn't depend on you finding the right
        way, the usual way to deal with these tasks, rather you only
        need to find the solution somehow … .
S:      Made up.
S:      Somehow.
S:      One needs not even a fixed procedure, a recipe? [The German
        word is 'Rezept', a literal translation would be 'recipe'.
        The teacher uses this term often to describe instruction-
        like rules or just sentences students have to write down
        that explain how to proceed. Therefore the instructions can
        be mathematical rules or solution methods or might contain
        definitions.]
T:      Somehow.
T:      Not even a procedure we're nowhere near that far … So … for
        your group. You get this … task, and you get this … task.
```

The teacher asked students to go into groups between four to six according to their own wish, and then distributed to each group one little worksheet with one of the two tasks he had chosen:

Transcript 8. Teacher, student: Mihat. (G2-L05)

```
T:      So before we start I'd like to ask Mihat to read the
        introductory text out loud.
```

Mihat: *Brain sport*. Dear school children, today I'm presenting you
with an exercise out of the math book that requires a
certain way of thinking … a special way of thinking. First
of all you only have to get an answer to the exercise, the
answer. You don't have to use any special method. Work in
the group, … write down your trains of thought, agree on one
or two. Lecture end.

A time of fifteen minutes was given to get the first interim reports of each group about their way to a solution. Students talked quite loudly as they worked, discussing all possibilities that came to mind. They did not look in the mathematics book, but tried to develop and work on their own ideas.

The way in which the task was set had several special features that the teacher emphasised: It was about a humorous kind of problem, a kind of riddle or made-up story, which represents some sport for the brain, a bit of a competition. The groups were assured that they could do them even in the short time of fifteen minutes, although both tasks were unusual problems without simple or known rules or methods to be applied. The teacher emphasised that the two tasks were intellectually more demanding for all members of the groups as they are referred to their 'brain' only. The teacher rationalised the purpose of working in groups by promising that this would provide a better chance to find solutions easily, and also that solutions would be jointly defended and explained later. The explanation was considered as the most important part of the common solution. Two different forms of problem were set up: One of the tasks was the well-known context problem of movements of persons or objects from different directions approaching each other, here of two ships on a river. The other a simple word-problem with funny names (Threestone) and unrealistic features that could be described as 'dressing up' algebraic equations and relationships by introducing contextual elements. The teacher considered the two tasks to have different levels of difficulty, therefore he distributed the worksheets according to assumed task difficulty and student ability, but did not make students aware of this. So students did not argue about which of the problems assigned was easier or more demanding. They liked to work together, although they argued that this would not help them to pass tests ("Well, you can't always ask other people, you have to be able to do stuff on your own"). However they were sure that the teacher saw this task and the way in which it was posed as very important and necessary, and accepted in advance that what they learned would be related to the more general and important aims of mathematics education. In the interview after the lesson, students actually referred to this kind of more demanding and general aim of the activity in collaboration by explaining:

Transcript 9. Teacher, student: Benedikt, Nina, Cem, Maya, Julia. (G2-L05)

Benedikt:What am I supposed, oh yeah that's right, *think
mathematically*, you have to *find new ways on your own*, and
if you can't do that, … then you can't do math by yourself
either, … if you can't, um, find other things by yourself.
Nina: And with these ones, *you had to think about it, there was
nowhere you could look things up or anything*, how you're
supposed to do it.

```
Cem:      all in all, the way I see it (  ) like all in all - it's
          pretty good like this.
Maya:     You notice that others have problems as well. You don't feel
          so alone.
Julia:    You're even quicker then. Other ideas occur to you as well.
          For example, Jaqueline said that they were binomial
          formulas. That wouldn't have occurred to me so quickly.
          Working in groups isn't ideal for all questions though.
```

RESUMÉ

Setting a task is an influential part of the teaching process, it obviously is decisive for students' understanding and engagement, their valuing of mathematics, and the chances for new insight that they get out of their activities. Such opportunities for deriving new insight from their activities were rather rare in both schools.

With reference to the various characteristics of setting a task that we listed and described at the beginning of this chapter, both teachers preferred similar settings: (1) Concerning the function of setting a specific task, in nearly all lessons in both schools the tasks set served for the function of a) practising familiar routines or b) practising newly acquired skills or methods only to become familiar with them. (2) Tasks were mostly presented in formal symbolic language, authentic contexts or daily life problems were not found. (3) The predominant justification was the training for the upcoming assessment procedures like tests or written classwork. There were nearly no explanations beyond the classroom in G1, while in G2 the teacher referred to 'brain sport' as a general aim of formal education and offered some fantasy features in an exceptional situation. The focus in setting a task was working instructions and reference to rules and patterns used to gain correct results. Emphasis was merely on the technical aspect and importance of the tasks and the rules embedded. Among the 10 lessons from one class and 15 from the other, students noticed three breaks of the normal daily routine activities, which were also explicitly identified by the teachers as exceptions.

In setting the tasks, the two teachers followed slightly different ambitions and were guided by different general aims of their teaching: the overwhelming emphasis of the teacher in G1 on internalising procedures, mathematical techniques and rules might have shaped more significantly his way of setting the tasks, not only for the daily purposes of practising exercises, but even on occasions when he wanted to make a difference. His tasks predominantly served to internalise routine procedures, the purposes were the tests. In this respect, his teaching was very predictable, reliable and clear for students, and they appreciated his reliability. However, mathematics classroom practice was mostly boring and uninspiring and neither made sense nor created any meaning beyond the classroom. This was particularly visible when the teacher set the fascinating and challenging task for a self-determined student activity in G1-L05 and G1-L06 which, as he had declared in the questionnaire, could have lead to a deeper insight into relationships between geometry and algebra. This task could have developed a more concrete meaning for the three binominal formulas as comparisons of areas and sides of squares and rectangles, that is, general concepts like relationships between change of size in length and areas in relation to forms and different representations. In a more reflective and theoretical discussion afterwards, it could have led to an idea of what constitutes proving and a proof and what counts as an argument.

Students in G1 specially liked their class mates' different explanations of how they had worked and which 'reasons why' others provided for their decisions during the presentation. This was much more interesting for them than the teacher had expected. From the perspective of the teacher, the performance of individual students at the board merely counted as an additional base for assessment. He obviously felt obliged to assess students regularly and 'objectively' in order to assure their parents about the success of his teaching. He took this responsibility very seriously. However, it also shaped the whole planning and teaching, the setting of tasks and of regular small tests to an extent that counteracted other pedagogical intentions. In his teacher education he had been forced to constantly reflect on and respect the system of central exams at the end of the year designed by the National Academy of Pedagogical Sciences (Keitel, 1993a, 1998). The fact that he now had to work in a teacher-based assessment system was still a novel and special demand for him. So he may still have been influenced in his thinking by the former need to prepare for central exams, and have been following patterns of preparation for such exams, something which might also be expected by the parents. However, the teaching was uniformly standardised and restricted to assessment demands, and the clear intellectual opportunities provided by the use of more ambitious and exceptional tasks and pedagogical chances (as in L05/L06) were counteracted by organisational measures. The grouping according to ability level distracted the students from the activity, so that they not only saw these measures as unjust and inappropriate for the chosen theme and activity, but also lost any mathematical interest. It was mostly the setting that distracted them from the special opportunities to gain deeper insight, and the possibly challenging task failed to establish connections between algebra and geometry, or between argumentation and proving.

The teacher in German school G2 was not so obviously occupied with assessment demands; for him this was mainly the duty of the students and their parents. Although nearly all of the lessons also followed the same pattern of setting tasks taken from the textbook and working on similar exercises formulated by the teacher, he obviously enjoyed the rare opportunity to offer additional reasons and purposes that was provided by special tasks that interrupted the daily routine. In setting such tasks he emphasised clearly that they represented extra-challenging opportunities for students to learn more and go beyond daily rituals. He explicitly valued 'brain sport' as a challenging activity, and he used a different language to value students' activities when thinking mathematically was needed and was a goal. He called for collaboration in order to combine the intellectual forces of a group for better success and achievement in solving the problems and so on. It was clear from the data that on the specific and exceptional occasions described above, the setting of the tasks was much more ambitiously framed by an explicit justification of the tasks, particularly pointing to specific aspects of the problems that did not just ask for following rules or applying procedures. Although the somewhat artificial 'fun' in the word problem did not really persuade the students of a special quality of that task, and although the context of the realistic problem was as boring and strange to them as most other context problems they usually encounter in textbooks, they said that they did like to work on them for a while. It was the setting of the common activity, and the chance for students to be deliberately encouraged to collaborate and explain their thoughts or approaches to others that attracted them. A task that was still fairly boring was accorded a special function in its own right, enough for them to enjoy it as an 'extra' activity, and to have some fun and maybe a little bit of challenge for their brains. They honoured the extra activity and the arrangement set by the teacher, and applauded him for this excursion. It was amazing how modest students become with respect to what school life offers them.

REFERENCES

Baumert, J. Artelt, C., Klieme, E., Neubrand, M., Prenzel, M., Schiefele, U., Schneider, W., Tillmann, K.-J. & Weiss, M. (Eds.). (2002). *PISA 2000: Die Länder der Bundesrepublik Deutschland im Vergleich, Zusammenfassung zentraler Befunde*. Opladen: Leske & Budrich.

Begehr, A. (2003) "Wer nicht fragt, bleibt dumm!" - Eine Analyse von TIMSS-Video-Daten zu Qualität und Quantität verbaler Schülerpartizipation ["If you don't ask, you won't learn!" - A qualitative and quantitative analysis of students' verbal participation of the TIMSS-videotape-study]. In H.-W. Henn (Ed.), *Beitraege zum Mathematikunterricht* (pp. 93-96). Hildesheim: Verlag Franzbecker.

Begehr, A. (2004). *Teilnahme und Teilhabe am Mathematikunterricht. Eine Analyse von Schülerpartizipation*. [Participation and taking part: An analysis of students' participation in mathematics classroom practice]. PhD Thesis, Faculty for Education and Psychology, Freie University Berlin.

Clarke, D. J. (Ed.). (2001). *Perspectives on meaning in mathematics and science classrooms*. Dordrecht: Kluwer.

Clarke, D. J. (2002, April). The learner's perspective study: Exploiting the potential for complementary analyses. In D. Clarke (Chair), *Primary research, secondary research and research synthesis: Theory, value and linkage*. Symposium conducted at the Annual Conference of the American Educational Research Association, New Orleans.

Clarke, D. J. (2003). International comparative studies in mathematics education. In A.J. Bishop, M.A. Clements, C. Keitel, J. Kilpatrick, and F.K.S. Leung (Eds.), *Second International Handbook of Mathematics Education* (pp. 145-186). Dordrecht: Kluwer.

Damerow, P., Elwitz, U., Keitel, C., Zimmer, J. (1974). *Elementarmathematik: Lernen für die Praxis? Ein exemplarischer Versuch zur Bestimmung fachüberschreitender Curriculumziele* [Elementary mathematics: Learning for practice? An exemplary attempt to determine curricular goals across disciplines]. Stuttgart: Klett

Damerow, P. & Lefèvre, W. (Eds.) (1981). *Rechenstein, Experiment, Sprache. Historische Fallstudien zur Entstehung der exakten Wissenschaften* [Calculus, experiment, language. Historical case studies on the origin of the exact sciences]. Stuttgart: Klett-Cotta.

Freudenthal, H. (1991). *Revisiting mathematics education. China lectures*. Dordrecht: Kluwer

Gellert, U., Jablonka, E. & Keitel, C. (2001). Mathematical literacy and common sense in mathematics education. In B. Atweh, H. Forgasz & B. Nebres (Eds.) *Sociocultural research on mathematics education: An international perspective* (pp. 57-73). Mahwah, NJ: Lawrence Erlbaum.

Goodchild, J. (2001). *Students' goals*. Bergen: Kaspar Forlag.

Huber, G. & Mandl, H. (Eds.) (1994). *Verbale Daten. Eine Einführung in die Grundlagen und Methoden der Erhebung und Auswertung* [Verbal data. An introduction to the foundations and methods of gathering and interpretation]. Weinheim: Beltz.

Jablonka, E. (2002, October). *Values in reasoning discourses in mathematics classrooms.* Paper presented at the Invitational Conference on Values in Mathematics and Science Education, Monash University, Melbourne.

Jablonka, E. (2003a). Mathematical literacy. In A.J. Bishop, M.A. Clements, C. Keitel, J. Kilpatrick, F.K.S. Leung (Eds.), *Second International Handbook of Mathematics Education* (Vol. 1, pp. 75-102). Dordrecht, The Netherlands: Kluwer.

Jablonka, E. (2003b, April). The structure of mathematics lessons in German classrooms: Variations on a theme. In D.J. Clarke (Chair), *Mathematics lessons in Germany, Japan, the USA and Australia: Structure in diversity and diversity in structure.* Symposium conducted at the Annual Meeting of the American Educational Research Association, Chicago.

Jablonka, E. (2003c, August). Cross-national elements in lesson structure. Paper presented at the symposium *Perspectives on international comparisons of lesson structure in mathematics classrooms in Germany, Japan, the USA, and Australia,* at the 10th Conference of the European Association on Learning and Instruction (EARLI), Padova.

Jablonka, E. (2003d, August). Constraints and affordances of reasoning discourses in mathematics classrooms: Examples from Germany, Hong Kong and the United States. Paper presented at the symposium *Social interaction and learning in mathematics classrooms in Australia, Germany, Hong Kong, Japan, Sweden, and the United States* at the 10th Conference of the European Association on Learning and Instruction (EARLI), Padova.

Jablonka, E. (2004, April). Student(s) at the front: Variations of a form with different functions. In D. J. Clarke (Chair), *Lesson events as the basis for international comparisons of classroom practice.* Symposium conducted at the Annual Meeting of the American Educational Research Association, San Diego.

Jablonka, E. & Keitel, C. (2005). Funktionale Kompetenz oder mathematische Allgemeinbildung? [Functional competency or mathematical literacy?] In J. Schloemerkemper (Ed.) *Bildung und Standards. Zur Kritik der Instandardsetzung des Deutschen Bildungswesens. Die Deutsche Schule, Special Issue 8,* 135-144.

Kaiser, G. & Keitel, C. (1994). Gender differences in assessment modes: FRG. In L. Burton (Ed.), *Who counts? Assessing mathematics in Europe* (pp. 87-122). London: Trentham Books.

Keitel, C. (1986a). On the relationship between scientific knowledge and school knowledge. In Équipe de Didactique des Mathématiques et de l'Informatique (Eds.) *Sur la didactique des mathématiques en France et l'Allemagne. Un colloque bilateral* (pp. 69-75). Marseille-Luminy: CNRS.

Keitel, C. (1986b). Social needs and secondary mathematics education. *For the Learning of Mathematics 6*(3), 27-33.

Keitel, C. (1989). Mathematics education and technology. *For the Learning of Mathematics 9*(1) 7-13.

Keitel, C. (1992). Mathematician or pedagogue? On the education of mathematics teachers in Germany. *The Curriculum Journal, 3*(3), 291-309.

Keitel, C. (1993a). Notes on the unification of the two German states and the integration of two different educational systems. *The Curriculum Journal, 4*(1), 130-135.

Keitel, C. (1993b). Implicit mathematical models in social practice and explicit mathematics teaching by application. In I. Huntley, C. Keitel, J. De Lange & M. Niss (Eds.), *New trends in teaching mathematics by modelling and applications* (pp. 19-30). Chichester: Ellis & Horwood.

Keitel, C. (1996). Réformes et développements de l'enseignement mathématique en R.F.A. depuis 1950 [Reforms and developments in the teaching of mathematics in Germany since 1950]. In B. Belhoste, H. Gispert & N. Hulin (Eds.) *Les Sciences au Lycée. Un siècle de réformes des mathématiques et de la physique en France et à l'étranger* (pp. 302-310). Paris: Vuibert.

Keitel, C. (1997). Numeracy and scientific and technological literacy. In W.E. Jenkins (Ed), *Innovations in science and technology.* (Vol. VI, pp. 165-185). Paris: UNESCO.

Keitel, C. (1998). Mathematics education in Germany: Ideal and reality. In Japanese-German Center Berlin (Ed.) *Education in Japan and Germany. A comparison*, (pp. 24-37). Berlin: JGC.

Keitel, C. (2004). Establish the common in the diverse – Respect the diverse in the common! Students' struggle for sense making. In J. Gimenez G. FitzSimons & C. Hahn (Eds.), *A challenge for mathematics education: To reconcile commonalities and differences. Proceedings of* CIEAEM 54 (pp. 142-149). Barcelona: GRAO.

Keitel, C. & Kilpatrick, J. (1999). Rationality and irrationality of international comparative studies. In G. Kaiser, I. Huntley & E. Luna (Eds.), *International comparative studies in mathematics education* (pp. 241-257). London: Falmer.

Keitel, C. & Kilpatrick, J. (2005). Mathematics education and common sense. In J. Kilpatrick, C. Hoyles & O. Skovsmose (Eds.), *Meaning in mathematics education* (pp. 105-128). Dordrecht, The Netherlands: Kluwer.

Keitel, C., Schubring, G. & Stowasser, R. (1985). History of mathematics teaching. In T. Husén, & N. Postlethwaite (Eds.), *International encyclopedia of education: Research and studies*. (Vol. 6, 3236-3239). Oxford: Pergamon Press.

Kline, M. (1985). *Mathematics and the search for knowledge.* Oxford: Oxford University Press.

Knipping, C. (2003). *Beweisprozesse in der Unterrichtspraxis. Vergleichende Analysen von Mathematikunterricht in Deutschland und Frankreich* [Processes of proving and proofs in classroom practice. Comparative analyses of German and French mathematics classroom practices]. Hildesheim: Franzbecker.

Krummheuer, G., & Brandt, B. (2001). *Paraphrase und Traduktion* [Paraphrase and transduction]. Weinheim: Beltz.

Lenné, H. (1969). *Analyse der Mathematikdidaktik in Deutschland* [Analysis of mathematics didactics in Germany]. Stuttgart: Klett.

Leung, F. (2004, January). *Mathematics assessment and culture: The case of Hong Kong.* Paper presented at the APEC Summit Conference, Beijing.

Nissen, H., Damerow, P. & Englund, R. K. (1990). *Frühe Schrift und Techniken der Wirtschaftsverwaltung im Vorderen Orient: Informationsspeicherung und verarbeitung vor 5000 Jahren* [Early writing and technologies of the management and administration of economics in the Near East: Information storing and processing 5000 years ago]. Bad Salzdetfurth: Franzbecker.

Seeger, F., Voigt, J., & Waschescio, U. (Eds.). *The Culture of the mathematics classroom.* Cambridge, UK: Cambridge University Press.

Voigt, J. (1995). Thematic patterns of interaction and sociomathematical norms. In P. Cobb & H. Bauersfeld (Eds.), *The emergence of mathematical meaning: interaction in classroom cultures.* (pp. 163-201). Hillsdale, NJ: Lawrence Erlbaum.

Yackel, E., & Cobb, P. (1996). Sociomathematical norms, argumentation, and autonomy in mathematics. *Journal for Research in Mathematics Education,* 27, 458-477.

Christine Keitel
Fachbereich Erziehungswissenschaft und Psychologie
Freie Universität Berlin
Germany

KEIKO HINO

CHAPTER FOUR

The Role of Seatwork in Three Japanese Classrooms

PURPOSE OF THE STUDY

In mathematics lessons, students engage in different activities such as review of the previous lesson, listening to the teacher's explanation, discussion about solution methods, practising, and summarising. These activities usually take the form of whole-class interaction, small-group interaction or individual work. 'Seatwork' activity is defined as "a period of time during the lesson when students work independently on assigned tasks, either alone or in small groups (two students or more)" (Stigler, Gonzales, Kawanaka, Knoll, & Serrano, 1999, p. 74). A good image of Seatwork activity is where students are solving problems individually or in groups, with predominantly private talk, and the teacher is walking around their desks to assist their work.

In the Third International Mathematics and Science Study (TIMSS), Seatwork as an organisational segment is examined in terms of its frequency, length, pattern and process (Stigler et al., 1999). As processes during Seatwork, they identified different tasks and situations involved in Seatwork. However, they did not delve into the process of teaching/learning during Seatwork. The purpose of this chapter is to investigate how experienced Japanese teachers make use of Seatwork in their classrooms. In this chapter, I examine the reality of the teacher's support for individual students during Seatwork, the role of Seatwork in the later development of the lesson, and the relationship of Seatwork to students' learning.

There are two reasons for this analysis. Firstly, teacher support for students during and through Seatwork is seen not only in Japan but also in other countries and is considered as an important aspect of teaching. Therefore, it gives a common reference point for comparison. In this chapter, I will describe the case of Japanese lessons, which will enable a discussion of different mechanisms of learning that take place in the classroom across countries. Secondly, in Japan, teacher support for students during Seatwork is called 'Kikan-Shido.' According to Shimizu (1999), it means instruction at students' desks and includes a purposeful scanning by the teacher of the students' individual problem-solving processes. He further says,

D. J. Clarke, C. Keitel & Y. Shimizu (Eds.), Mathematics Classrooms In Twelve Countries: The Insider's Perspective. 59–73. © *2006 Sense Publishers. All rights reserved.*

> While the teacher moves about the classroom silently monitoring students' activities, he performs two important activities that are closely tied to the whole-class discussion that will follow the individual work. First, the teacher assesses students' problem-solving progress. .... Second, the teacher makes mental notes as to which students used the expected approaches and which students used different approaches to the problem. These students will be asked to present their solutions later (p. 110).

The interest of this chapter extends across the reality of Kikan-Shido and its role in students' learning by looking at the behaviour of experienced Japanese teachers.

Clarke (2004) proposes an analytical unit, 'Lesson Event', as a basis for comparison of mathematics classrooms internationally. It is "an event type sharing certain features common across the classrooms of the different countries studied" (p. 5). He then identifies 'Between Desks Instruction' as a Lesson Event and documents its occurrence in four countries. My analysis is also concerned with the Lesson Event of Between Desks Instruction. In this chapter, I will concentrate on the three Japanese classrooms in the Learner's Perspective Study (LPS) and deepen the analysis of its functions in these classrooms. In doing so, I will offer information on its unique feature as exemplified in the Japanese lessons and provide a basis for further comparison with lessons in other countries.

METHOD

Japanese mathematics lessons by three experienced teachers (throughout this chapter, I will use J1, J2 and J3 to refer to these three sites) are the object of this analysis. Topics that were dealt with in the three sites are as follows:
− J1: Linear functions,
− J2: Similar figures,
− J3: Simultaneous equations.

I began the analysis by watching the videos and reading the lesson transcripts and by specifying situations where students are working individually or in groups. Here, problems students are working are not necessarily the problems for the day. Still, the problems should require students to think mathematically in some way. Therefore, I excluded situations where the students were just copying from the blackboard or writing certain symbols mechanically. Analysis of the specified situations consisted of three steps:

1. Making a summary table of features of each situation and comparing the features across the three sites. Features on which I focused include (i) problem being worked, (ii) duration, (iii) tools used by the students such as worksheets, (iv) how it begins and how it ends, (v) the teacher's behaviour such as verbal interaction with individual students and with all students, and the way of naming students, and (vi) how the lesson evolved after the situation.

2. Analysis of the situation from the perspective of students. In the students' interviews, they were asked to watch the video of the lesson and to identify personally important situations in the lessons and give their reasons. I categorised the important situations identified and investigated their meaning of the Seatwork activity both quantitatively and qualitatively. Other interview data with students were also used to validate the results.

3. Inquiring into the meaning of the Seatwork activity both from the perspective of teachers and from the perspective of students across the three sites. Based on the results of 1 and 2, I investigated the similarities and differences in the meaning of the Seatwork activity across the three sites. Here, I also used part of LPS Australian data and TIMSS 1995 and 1999 video studies (Stigler & Hiebert, 1999; Stigler et al., 1999; Hiebert et al., 2003) in order to clarify what I got from the analysis.

SEATWORK ACTIVITY FROM THE TEACHER'S PERSPECTIVE

Problem and Time Used in the Seatwork Activity

I identified 26 Seatwork activities in J1, 16 Seatwork activities in J2, and 21 Seatwork activities in J3. I categorised the problems used in the activity into five types:
- Main problem or the problem for the day,
- Preparatory problem that is followed by the main problem or by explanation of the main content by the teacher,
- Problem that is related to the main problem,
- Problem that emerged from the exchange of ideas during whole-class discussion,
- Exercise problems.

Table 1. Numbers of types of problems that were used in Seatwork activity during the 10 consecutive lessons across three sites

Types of problems	J1	J2	J3
Main problem	12	6	8
Preparatory problem	3	2	3
Problem relating to main problem	3	1	0
Emerged problem	3	0	0
Exercise problem	5	7	13

Table 1 shows the frequencies of different types of problems used in the Seatwork activities. The first four types relate to the presentation of the main content in the lesson. Table 1 shows that in all three sites about half (81% in J1) of the problems in the Seatwork activity relates to the presentation of main content. It also shows that in J1 three problems emerged from the exchange of ideas among students. This was not observed in the other sites.

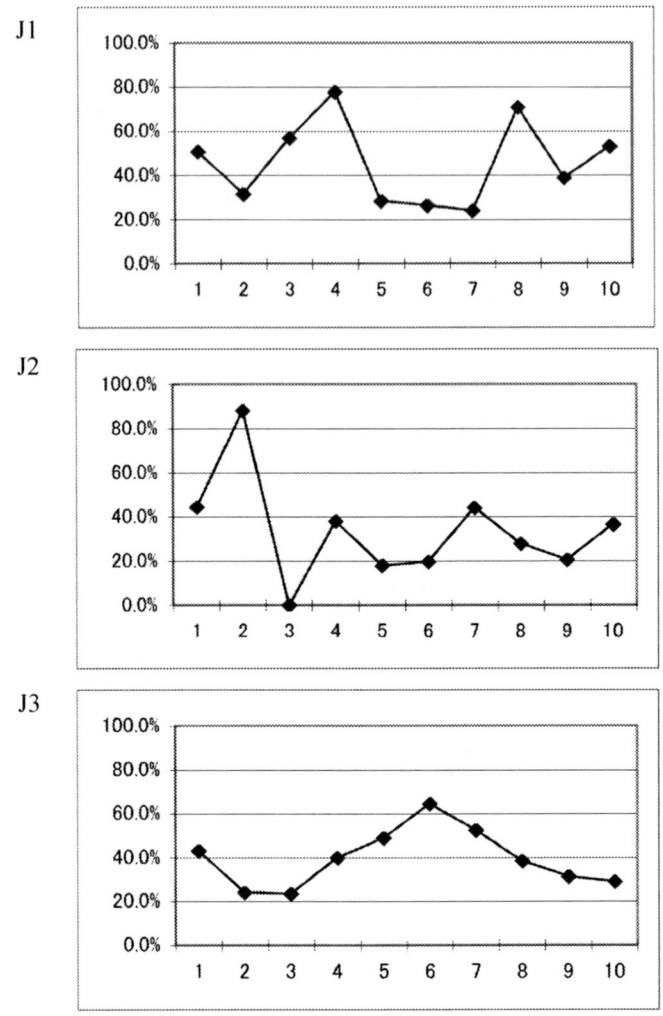

Figure 1. Percentage of time spent on Seatwork activity in each lesson across three sites

Further analysis revealed that each teacher placed the preparatory problem in the lesson in somewhat different ways. In J1, the teacher placed the preparatory problem before going into the main problem. For example, she asked the students to examine the relationship between two designated variables (number of folds of origami and circumference) before going into main problem where they choose two variables on their own and examine their relationship. The teacher emphasised a problem-solving approach in which problems are connected developmentally. In J1, most problems were in the form of worksheets prepared by the teacher. In J2, there were two preparatory problems. Both problems asked the students to construct figures which the teacher could use to present the main content of the lesson. For example, the teacher asked them to construct a 3-times-enlarged triangle. Later, he used the triangle to explain the ratio of similarity. In class, problems were prepared by the teacher and the students were not allowed to open the textbooks. In J3, a unique feature was that the same problem changed type according to the situation. For example, in order to teach verifying the answer, the teacher began by encouraging the students to verify the answer when working on exercise problems. Later, the teacher pointed out there were two ways of verifying the answer and asked them to examine the differences. Here, the exercise problem can also be described as a preparatory problem. Because of this complexity, it was not easy to identify the type of each problem. In J3, in contrast to J2, most problems came from the textbook.

The average time spent on the Seatwork activity was 8.5 minutes in J1, 16.6 minutes in J2, and 18.4 minutes in J3. However, it varied widely across the activities in all sites. Figure 1 shows the percentage of time spent on Seatwork activity in each of ten consecutive lessons. It also shows variations across lessons. In J2 and J3, the variations came mainly from the extent to which the lesson was centred on exercises. However, in J1, the wave on the graph reflected the unique role of Seatwork activity in the lessons. Namely, in learning new content the Seatwork activity is used to construct a shared base among the students, whereby discussion about the content is made possible. Therefore, the first half lessons (1 or 2 lessons) of introducing new content is spent more on Seatwork activity and the second half lessons (2 or 3 lessons) is spent more on discussion and summary.

Teacher's Support of Individual Students during the Seatwork Activity

In all three sites, it was observed that the teacher moved around the students' desks and supported and directed individual students' activities and the activity of students as a whole. In J1 and J3, the teacher also asked some students to write their solutions on the blackboard. At the beginning of the Seatwork activity, all teachers watched how the students were reacting. Here, they quickly gathered information about difficulties the students faced and ways of coping with them. They tended to use these points and responses in their subsequent support during the Seatwork activity.

There were both similarities and differences across the three sites in the teachers' verbal interaction with students. All the teachers made approaches to the students in different ways according to their states of understanding and motivation. All of them were giving behavioural directions, assisting their understanding of the problem, giving directions on procedures, leading their ways of thinking, pointing out mistakes, responding to their questions, giving directions on their expressions, and taking an interest in their thinking and solutions. On the other hand, there were also some verbal interactions that were observed characteristically in only one site. These were 'encouraging the student's own decision making' and 'emphasising peer collaboration' (J1), 'evaluating the correctness of the answer by the student' (J2), and 'encouraging to ask questions' and 'referring to the intention and meaning of the problem' (J3).

Quantitative and qualitative analysis of these teachers' support and direction for students during the Seatwork activity revealed that there were differences in the extent to which 'others' are incorporated into the learning of mathematical content. In J1, compared with the other sites, the teacher emphasised the importance of learning as a whole group, and regularly put this into practice, even when she was supporting individual students. Two characteristic teaching behaviours were seen in J1. Firstly, in the beginning of the Seatwork activity, a 'Sharing' activity was sometimes observed. It began with the teacher usually starting the Seatwork activity by calling for all students, "Now I will take some time." However, she did not just let students begin the work. A certain amount of time was spent in developing among students a common image of the work that will come. The Sharing activity was observed more often in the earlier part of the ten consecutive lessons. Secondly, during the scanning of the students' progress, the teacher sometimes looked up and talked to all students. Such talk often reflected on her previous verbal interaction with individual students. When she made a decision that taking more time on the Seatwork activity was not useful because there were so many difficulties, she also let students quit their work and went into the explanation. The observation of the teacher's emphasis of peer collaboration in her verbal interaction with the students described above is also related to this.

In contrast, the teacher in J2 paid less attention to learning as a whole group. Rather, his behaviours reflected his emphasis that each student should understand the content and solve problems correctly by him/herself. Contrary to J1, a Sharing activity was not observed. Nor was it observed that the teacher talked to all students based on his previous verbal interaction with individual students. The teacher repeated the same explanation to individual students and did not share it with all students. These behaviours were consistent with observation of the teacher's frequent verbal interaction with the students. His interactions included giving directions on procedures and pointing out mistakes as well as evaluating the correctness of the student's answer.

The teacher in J3 can be described as placed between the teacher in J1 and the teacher in J2. On one hand, by creating the time to talk, he encouraged the students to talk with their neighbours about the work, which suggested his emphasis on the aspect of mutual learning. On the other hand, like the teacher in J2, a Sharing Activity or reflective talking to all students were rarely observed. Most talk to all students was concerned with time management or information about homework. At the same time, there was also a teacher behaviour that was uniquely observed in J3, that is, the frequent observation of questions by the students and responses by the teacher. In class, the teacher verbalised his interest in the students' questions, for example, "Ask questions whenever you do not understand" and "Say it, say it, let me know what you do not know." The teacher also dealt with the students' questions in the whole-class discussion. In J3, questioning by the student had the double roles of fostering the individual student's understanding and of establishing learning from 'others.'

Development of the Lesson after the Seatwork Activity

In all sites, during the Seatwork activity the teachers supported individual students by taking into consideration the development of the lesson after the activity. However, there were more differences than similarities among the teachers in the subsequent development of the lesson. Table 2 shows variations of the development of the lesson after the Seatwork activity.

Table 2. Development of the lesson and use of student's work after Seatwork activity across three sites

J1	J2	J3
Development of the lesson		
Check the answer	Check the answer	Check the answer
Student presentation and	Teacher's	Teacher's
sharing the solution	(complementary)	(complementary)
Further development based	explanation	explanation
on the shared solutions	Relating to the next	Dealing with questions
Leading to the point based	problem	Identify intention of
on the shared solutions	Teacher's summary	problem
Relating to the next problem		Relating to the next
Discussion based on		problem
different solutions		Deepening thinking on
Teacher's summary		different solutions
Connection to textbook		Teacher's summary
Use of students' work after Seatwork activity		
Eliciting some of students'	Pointing out different	Taking up student's
work and making use of	solutions and giving	question
them	explanation	Taking up student's way
Eliciting student's mistake	Pointing out difficulty	of thinking as a better way
Eliciting student's	and giving explanation	Eliciting different
puzzlement		solutions and deepening
Eliciting opposing solutions		their thinking
and drawing students'		
attentions		

It was in J1 that the Seatwork activity played the most important role in the subsequent development. In J1, after the Seatwork activity, there was usually time for students' presentations. Based on the presentations, the teacher led students to the point of the lesson, supported further discussion or encouraged inquiry. In J3, the teacher often asked some students to write their solutions on the board during the Seatwork activity. Development of the lesson after the activity was usually based on the solutions they wrote. The teacher pointed out mistakes, supplemented the students' solutions or made supplementary explanations. Furthermore, he sometimes asked several students, whose solutions were different, to write their thinking on the board and asked the class to find out the differences by themselves.

In his lesson development, questions asked by the students during the Seatwork activity were also dealt with.

There was less variation of the development of the lesson after the Seatwork activity in J2, compared with the other sites. As was described earlier, the teacher made verbal interactions with the students by considering later development. Directions and explanations made during the Seatwork activity were actually repeated in the whole-class activity. However, they were only repeated and pointed out, not elaborated. In some situations, the teacher pointed out different ways of thinking during the Seatwork activity. However, it was the teacher who explained and summarised these. The teacher did not ask students to write their solutions on the board during the Seatwork activity. Furthermore, 7 out of 16 Seatwork activities came to an end because of the bell (the end of the class). This observation also suggests the role of Seatwork activity in this site. Namely, the Seatwork activity did not carry much weight in the subsequent development of the lesson.

SEATWORK ACTIVITY FROM THE PERSPECTIVE OF THE STUDENT

Students' Responses to Some of the Question Items

In the student interview, the target students were asked 13 questions in all. These included questions about the content they learned, their goals of the lesson and their image of a good lesson. Table 3 summarises their responses to the question of a good lesson in the three sites. As the table shows, the students' responses in all sites contained "I can understand the lesson," "I can engage in the lesson positively by talking my thinking and so on" and "I can think well." They suggest the students' common image of a good lesson. On the other hand, there were responses that were unique to each site. In J1, 7 students out of 12 who were interviewed responded with phrases about relating to 'others': for example, "Everyone gives their opinions in a lively way" and "We make up the lesson together." Their concern for 'others' is also seen in their responses to other question items.

In J2, there was just one student who referred to 'others.' Rather, the students (8 out of 12) referred to 'the teacher' in response to the questions of best way of learning the content and their goals for the lesson. For instance, they said, "I try to understand the teacher's explanation better," "I listen to the teacher," or "I try to do what the teacher explains." For them, the big concern was to understand the content the teacher teaches them. In J3, five students referred to 'others' and two students referred to 'listening to the teacher.' It is also curious that three students in J3 talked about asking questions of the teacher as a condition of good lesson.

Table 3. Samples of students' responses to the question "What is a good lesson for you?" across three sites

J1	J2	J3
I can solve the problem correctly	I can understand teacher's explanation	I can think sufficiently
Everyone say opinions and we think about them	I have the attitude to attend the class positively	Everyone propose their opinions
I speak out my thinking	I think of something by myself	I am willing to ask questions and speak out my thinking
I can understand the content	I understand what I did not understand	I listen to the teacher well
My question is resolved	I can solve the problem by myself	I concentrate
I tackle the problem well	I listen to the teacher	I approach the problem in discussion with my friends
Some manipulative is used		I can understand the content
		I learn something new
		There is enjoyable atmosphere

Concerning the question "What have you learned in today's lesson?" the students in J1 pointed out varieties of matters such as a new term and its meaning, ways of solving and writing, ways of thinking, learning with 'others,' and ways of taking notes. Their responses were not particularly concentrated on specific matters. Here again, five students talked about 'others.' In J2, the students' responses were more concentrated on ways of solving the problem and understanding the content. Moreover, they often replied by referring to the name of the topic, for example, "I learned about similarity" or "It's the condition of similar figures." The mathematical content of linear functions (J1) and similar figures (J2) are considered to be difficult in lower secondary schools, which seems to be reflected in these observations. In J3, since the topic of the lessons was simultaneous equations, it was expected that many students would refer to ways of solving the problem and this was the case. However, more students than expected also mentioned learning with 'others.' The students in J3 seem to have been learning the content by interacting with their friends in class.

Seatwork Activity Viewed from Classroom Events of Personal Importance

In the post-lesson video-stimulated interviews, the target students were given control of the video replay and asked to identify and comment upon classroom events of personal importance. In the analysis, these events were roughly classified into four categories: Seatwork activity, 'Whole-Class Discussion' activity, 'Instruction' by teacher and 'Other' (unrelated to classroom activities). Table 4 shows the percentage of these categories in the three sites. As the figure shows, Seatwork activity was one of the most important from the perspective of students. More detailed categorisation of events that belong to the category of Seatwork activity further shows that many students identified events of discussion with their neighbours as being of personal importance. Informal discussion with their neighbours during the Seatwork activity was as important as discussion with the teacher. Here, it is possible to say that the time for Seatwork was important as a time to exchange information and opinions with their neighbours and to think about the problematic points together.

Table 4. Percentages of four categories of classroom events that were identified by students as personally important

Activities mentioned by students (%)	J1	J2	J3
Seatwork (including informal talk among neighbours, sharing problems)	33	29	39
Whole-Class Discussion (including making connections to own thinking)	32	4	37
Instruction by the teacher	35	65	18
Other (unrelated to classroom activities)	0	2	6

Of interest are the results found after classifying the events (comments) belonging to 'Whole-Class Discussion' into the following two categories:
– events in which the students especially make connections to their own thinking during the previous Seatwork activity and
– events in which the students express their impressions to what their friends say without making connections to their thinking during the Seatwork activity.

The proportion of the first category in the entire events belonging to the Whole-Class Discussion varied across the sites. Still, some students commented on the important points of the work presented by their friends by comparing or connecting to their own methods and thinking, for example:

Transcript 1: J1-L10 (Obata)

O   One problem was we didn't have much time, I suppose. Well, so, I was listening to the presentation carefully. Yes, it was quite difficult for me to understand. So, I listened to the presenter carefully... Well, so, I didn't understand it perfectly, but I could get the rough idea of solving the question.

Transcript 2: J3-L1 (Kooriyama)

K   Well, first I didn't understand the meaning. But I could understand after Endou explained it.

Key to symbols used in transcripts in this chapter
...   A short pause of one second or less

From this observation, it is also possible to say that the Seatwork activity was important to the students because it serves as a basis to make sense of the development of the work that followed.

DISCUSSION AND SUMMARY

In this study, I focused on the Seatwork activity that is commonly observed in mathematics lessons in different countries, and investigated its features and functions in the case of the Japanese mathematics lessons in the LPS study, from the perspective of both teacher and student. As a result, both similarities and differences were observed in the three sites. One similarity is that the Seatwork activity is placed before the presentation of main content of the lesson. In classrooms J2 and J3, generally half of the problems in the Seatwork activity related to the presentation of main content while in classroom J1, it was found that more than half of the problems were related. This result is consistent with the TIMSS 1995 video study (Stigler & Hiebert, 1999). They identified the Japanese pattern of teaching a lesson as a sequence of five activities (p. 79):
− Reviewing the previous lesson
− Presenting the problem for the day
− Students working individually or in groups
− Discussing solution methods
− Highlighting and summarising the major points.

Here, Seatwork activity is seen in the activity of "Students working individually or in groups." In Germany and US on the other hand, Seatwork activity usually followed the presentation of main content. In fact, as the time for Seatwork, they identified the activity of 'Practicing' in the German pattern and 'Practicing' and 'Correcting Seatwork and Assigning Homework' in the US pattern. The three Japanese teachers in this study also spent the Seatwork activity as the time for exercises. However, when presenting main content, they took the time to let students think about the content beforehand. Also in the TIMSS 1999 video study (Hiebert et al., 2003), based on the data of time spent on different purposes in the lesson, it was found that in Japanese mathematics lessons, introducing new content is emphasised more than in the other six countries. One reason for this, conjectured from this study, is that the teachers spent time that was allocated to Seatwork activity on introducing new content.

The placement of Seatwork activity before presenting main content gives a foundation for interpreting the teacher's actions during the Seatwork activity. In all sites, during Seatwork activity the teachers supported individual students by taking into consideration the development of the lesson after the activity. Here, another similarity is the emphasis on different ways of thinking. In all sites, observations were made that the teacher dealt with different solutions, ideas or opinions by the students or that the students sought for and found different solutions. Especially in the Seatwork activity before the presentation of main content, it was a natural tendency that the students used somewhat naïve ideas and informal ways of thinking. This was considered to be good by the teachers. Rather than preventing these, the teachers in this study tried to elicit and make use of them when presenting the main content of the lesson.

Similarities were also observed in the students with respect to their recognition of the Seatwork activity. From the viewpoint of students, time for Seatwork had two significances. Firstly, it was the time to exchange information and opinions with their neighbours and to think about the problematic points together. Secondly, the time devoted to the Seatwork activity proved valuable with respect to their thinking and understanding later in the lesson. Students found that after the Seatwork activity, when their teacher and their peers presented and explained work, they could better reflect and connect the explanations with what they had worked on during Seatwork. The students remembered what they had thought about the problem during the Seatwork activity and listened to the later development with these thoughts in mind. These included, for instance, what they did not try, what they did not understand, or what they wanted to ask. While listening to the teacher and to their friends, the students were reconstructing their understandings. Nakamura (2002) argues, based on his analysis of LPS data, that Japanese students listen very well to what their friends have to say. A Japanese lesson is said to be like a drama that has a time of climax (Yoshida, 1992). The style of drama may impact students' ways of attending to the lesson. At the same time, the style of drama may also be supported by students' careful listening to 'others' in class.

There were also differences in the features and functions of the Seatwork activity among the three sites. This shows that even within a country, neither the role of the Seatwork activity nor the intention by the teacher in doing the activity were the same. In J1, Seatwork activity had an important role in shaping the flow of lesson. This was seen in the variety of problems dealt with in the Seatwork activity and in the differences in proportion of time devoted to Seatwork in each lesson over a ten-lesson sequence. The teacher in J1 was also practising repeatedly the importance of learning as a whole group, even when she was supporting individual students, and making use of the students' responses in the subsequent development of the lesson. Conversely, in J2, the role of Seatwork activity was secondary. It served as a time for preparation for the teacher's explanation of the main content and as a time for practising. During the Seatwork activity, the teacher evaluated individual students' states of understanding and assisted their progress mainly by giving directions on procedures. In J3, Seatwork activity had the role of eliciting students' questions and different solutions so that the teacher would be able to use them in the later development of the lesson. Transcript 3 shows the teacher's idea of the importance of students' activities in the lesson.

Transcript 3: J3 Teacher Interview 1

```
T        If we do not pay close attention to the student's activities,
         I think that students wouldn't start thinking. The responses
         from the students are of course varied, er that student think
         ... If I ask a question about the scene, (the students) start
         responding again ... I want the lessons to teach student to
         figure things out and I think that to figure things out is the
         most important activity in a lesson.
```

Here, it is worth noting that he had his conception of the mechanism of learning that contains questions and variety of thinking as major driving forces of learning.

The analysis, from the students' perspective, revealed differences among the three sites. The students in J1 were learning the content from interacting with their friends. Their learning also extended beyond learning how to solve the problem. The students in J2 considered it important to listen to the teacher's explanation in class and many of them talked about how to solve the problem and understanding of the content as their products of learning from the lesson. The students in J3 talked not only about how to solve the problem but also about thinking as the product of learning and referred to different ideas by their friends even though they were learning the procedure for solving simultaneous equations. These differences are somewhat consistent with the differences on the part of teachers as described above. Figure 2 is a sketch of locations of the three sites on the two axes: 'how much the teacher intentionally incorporates 'others' in learning the content' and 'how much variety of learning the teacher practises in his/her teaching beyond the understanding of content.'

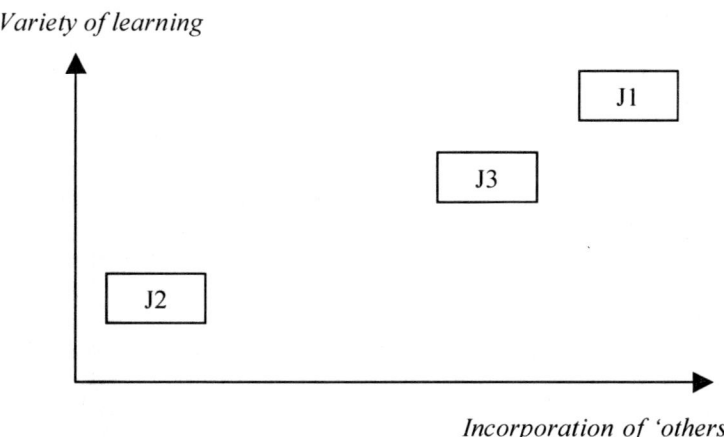

Figure 2. Three sites viewed from 'incorporation of 'others'' and 'variety of learning'

Tasks for the Future

Similarities seen in the three sites indicate that the three teachers held certain shared conceptions of how to teach mathematics effectively in the classroom. Jacobs and Morita (2002) point out that Japanese teachers have more shared conceptions of effective teaching than their US counterparts. The result of this study is partly consistent with their argument. It means that there are still significant differences in their teaching across the three sites. It is possible that these differences may reflect differences in the mathematical content they were teaching. Or it may be that these experienced teachers change their ways of teaching according to the classrooms. Although I should be careful in generalising the results obtained in this study, it is interesting to pursue the similarities seen in Japanese teachers compared to their counterparts in other countries, because they indicate the pedagogical values of mathematical content and effective teaching of mathematics. My preliminary analysis of the LPS Australian data suggests differences in teachers' decision-making between the two countries: for example, whether listening to their friends' ways of thinking enhances students' understanding of mathematics, whether students who lack self-confidence in mathematics should think by themselves first or should think with their peers first, or whether the teacher talking with other teachers who visit the classroom during the lesson should never happen or should be accepted from the views of students. In order to deepen our understanding, it would be necessary to go into deeper examination of the three teachers' intentions and furthermore to get more information about the cases of other experienced teachers.

Another task for the future is more in-depth analysis of the learning by individual students during these lessons. The results described in this chapter suggest that how Seatwork activity is placed in the lesson influences the actual learning of students. In the TIMSS 1999 video study, the organisation of mathematics lesson is carefully examined and compared across the participating countries. The authors conclude that the organisation of the lesson may constrain both the mathematics content that is taught and the way it is taught (Hiebert et al., 2003). For the purpose of deepening the analysis of lesson organisation, it would be useful to incorporate the viewpoint of the structure of students' participation in the lesson. Clarke (2004) states that the specific enactment of Between Desks Instruction as a Lesson Event in each classroom displays distinct functional features reflecting patterns of participation specific to that classroom. Here, one important mark should be put on the placement of Seatwork in the lesson because it suggests idiosyncratic patterns of participation in the three Japanese classrooms. In these classrooms, the place of the Seatwork in the lesson seems to realise two stages of students' participation: participation through Seatwork activity and participation through whole-class activity after the Seatwork. This participation structure would enrich students' learning substantially, but at the same time would give them double burdens to learn in class (Hino, 2003). Examination of the real-time learning of individual students in the lessons would reveal more about the relationship between the lesson organisation and their process of learning, and give important information about different mechanisms of learning in the classrooms of different countries.

Finally, I would like to mention the implications of this study within the Japanese community of mathematics teachers. The similarities observed among the teachers of this study are not always seen in everyday mathematics lessons in Japan. For example, there are many lessons that do not place the Seatwork activity before presenting main content (Soma, 2003; Kato & Senuma, 2005). Therefore, it would be meaningful to analyse the features of lessons conducted by experienced teachers. In doing so, we should be able to describe the strong points of these lessons by carefully collecting the information about process and product of students' learning. Moreover, we should demonstrate the difficulties of conducting these lessons both from on the part of the teacher and on the part of the students.

REFERENCES

Clarke, D. (2004). *Kikan-Shido - between desks instruction.* Paper presented as part of the symposium "Lesson events as the basis for international comparisons of classroom practice" at the Annual Meeting of the American Educational Research Association, San Diego, April 12-16, 2004.

Hiebert, J., Gallimore, R., Garnier, H., Givvin, K. B., Hollingsworth, H., & Jacobs, J. (2003). *Teaching mathematics in seven countries: Results from the TIMSS 1999 video study (NCES 2003-013).* U.S. Department of Education. Washington, DC: National Center for Education Statistics.

Hino, K. (2003). Nichigo no sugakuka jyugyo ni okeru Jiriki-kaiketsu bamen no imi [Meaning of "Jiriki-kaiketsu" (solving problems by oneself) situation in the mathematics lessons in Japan and Australia: Analysis of two classrooms]. In Y. Shimizu (Ed.), *Research on cross cultural comparison of teaching and learning process in mathematics lessons* (Report on collaborative research between Japan and Australia by Japan Society for the Promotion of Science) (pp. 51-70). Tokyo Gakugei University.

Jacobs, J. & Morita, E. (2002). Japanese and American teachers' evaluations of videotaped mathematics lessons. *Journal for Research in Mathematics Education, 33*(3), 154-175.

Kato, H., & Senuma, H. (2005). TIMSS 1999 jyugyo video ni miru zukei no seishitsu ni kansuru nihon no jyugyo no tokucho [Characteristics of Japanese lessons on the property of geometrical figures in the TIMSS 1999 video study]. *Proceedings of the 38th Annual Meeting of Japan Mathematical Education Society,* 529-534.

Nakamura, K. (2002). Sugaku jyugyo ni okeru koteki na discourse to kojinteki na discourse no kakawari [Relationship between public discourse and private discourse in the mathematics lesson]. *Proceedings of the 35th Annual Meeting of Japan Mathematical Education Society,* 563-568.

Shimizu, Y. (1999). Aspects of mathematics teacher education in Japan: Focusing on teachers' roles. *Journal of Mathematics Teacher Education 2,* 107-116.

Soma, K. (2003). TIMSS 1999 sugaku jyugyo video kenkyu ni okeru nihon no chugakko sugaku jyugyo no kosatsu [Analysis of Japanese eighth-grade mathematics lessons in the TIMSS 1999 video study]. *Proceedings of the 27th Annual Meeting Japan Society for Science Education,* 233-234.

Stigler, J. W., & Hiebert, J. (1999). *The teaching gap.* New York: Free Press.

Stigler, J.W., Gonzales, P., Kawanaka, T., Knoll, S., & Serrano, A. (1999). *The TIMSS videotape classroom study: Methods and findings from an exploratory research project on eighth-grade mathematics instruction in Germany, Japan, and the United States.* (NCES 1999-074) U.S. Department of Education. Washington, DC: National Center for Education Statistics.

Yoshida, M. (1992). Nihon no sansu/sugaku no jyugyo ni tsuite no oboegaki [A note on mathematics lessons in Japan: From a comparison of lessons with a common topic between Japan and U.S.]. In T. Miwa (Ed.), *Teaching of mathematical problem solving in Japan and the United States* (pp. 188-221). Tokyo: Toyokan.

Keiko Hino
Faculty of Education,
Utsunomiya University
Japan

TERRY WOOD, SOO YEON SHIN AND PHU DOAN

CHAPTER FIVE

Mathematics Education Reform in Three US Classes

INTRODUCTION

The primary focus of the reform in mathematics education in the schools in the United States over the past 17 years has been to improve the quality of students' mathematical knowledge through substantial changes in curriculum, teaching and learning practices. These changes are delineated in several 'reform documents' (e.g., National Council of Teachers of Mathematics 1989, 2000). The controversy about school mathematics instruction in the US resides in concerns about the level of student achievement at all grade levels, which are measured by national and international exams. The proposed changes are premised on the assumption that focusing on the conceptual nature of school mathematics results in better student learning and ultimately a better prepared work force. Yet,

> . . . for the most part, with the reform movement well into its second decade, its goals have yet to be realized in the large majority of school districts in North America (Van de Walle, 2004, p. 9).

This statement is supported by recent research (Weiss, Palsey, Smith, Banilower & Heck, 2003; Jacobs, Hiebert, Givvin, Hollingsworth, Garnier & Wearne, 2006), which examined teaching in US eighth-grade mathematics classes. The findings indicate that the teaching observed did not reflect the innovative pedagogical practices believed to be essential to quality mathematics learning advocated in the reform documents and by current research in the field. More to the point the teaching practices observed are still illustrative of traditional instruction that emphasises knowledge of computational procedures in lieu of conceptual understanding. As Weiss et al. (2003) comment, "Observations conducted for the Inside the Classroom study suggest that the nation is very far from the idea of providing high quality mathematics. . .education for all students" (p. xiii).

The analysis of the Third International Mathematics and Science Study (TIMSS) 1995 and 1999 Video Study data by Jacobs et al. (2006) and Leung (2005) reveal characteristics of US mathematics classes that distinguish them from their counterparts. US lessons remain – traditional – in which 55% of the content is at a moderate or elementary level and the presentation of mathematical ideas are

D. J. Clarke, C. Keitel & Y. Shimizu (Eds.), Mathematics Classrooms In Twelve Countries: The Insider's Perspective. 75–85. © 2006 Sense Publishers. All rights reserved.

underdeveloped. Of concern is the significant shift from 1995 to 1999 from solving problems involving stating concepts to giving results only. Added to this is the fact, that in these TIMSS lessons, eighth-grade students did not engage in developing a rationale, generalisation, or counterexample nor did they engage in deduction – a hallmark of mathematics.

Even though the TIMSS Video Study findings can be taken to be representative of teaching in a country, the finding that traditional pedagogy still dominates makes analysis of the classroom data from the Learner's Perspective Study (LPS) essential for two reasons. Firstly, the three classes in the US data contain several classroom features thought to be important in reform instruction, such as group learning, the use of curricula grounded in problem solving and containing quality mathematical content, and teachers who were thought to reflect attempts to teach in ways compatible with reform ideals (National Council of Teachers of Mathematics, 1989, 1991). Secondly, the classroom data consists of a series of consecutive lessons that provides greater detail about the nature of teaching, albeit from a smaller number of classes than the TIMSS Video Study.

In this chapter, we analyse the data collected for the LPS from the three US schools, focusing specifically on US3, for the purpose of examining how these classrooms are realising the goals of the reform in school mathematics instruction. A central premise of the LPS is that just as there are 'teacher practices' in mathematics classes there must also be specific 'learner practices' in these mathematics classes. These 'learner practices' form a coherent body that is as important to identify as are 'teacher practices. The analysis we report in this chapter attempts to provide not only descriptions of specific teacher practices but also insight into issues for the learners.

CONCEPTUAL FRAMEWORK

In earlier research Wood and Turner-Vorbeck (1999, 2001) analysed reform-oriented mathematics classes of 7 and 8 year olds and empirically identified differences among reform-oriented classes in terms of the social features and quality of students' thinking. The analysis of the data revealed that reform-oriented classes fell into two major types categorised as *strategy reporting* and *inquiry/argument*. These types were assumed to represent the ways in which the culture of the classrooms was established and differed (Wood, 1994, 1996, 1998; Wood and Turner-Vorbeck, 1999, 2001). Wood and Turner-Vorbeck (1999, 2001) then generalised the differences they found in classroom interaction and discourse consisting of two dimensions: student participation and student thinking.

The participation dimension consists of the extent to which teachers, through the kinds of social norms established for interaction with students, made it possible for all pupils to participate actively in the interaction and discourse. There was a difference in student participation between the two reform cultures in terms of the teacher's expectations for students, both in giving their explanations and in asking questions of the explainer. From these empirical findings, a theoretical connection was made between the social norms constituted in a class and the social interaction patterns that evolve. The student thinking dimension consisted of the extent to which the quality of student thinking expressed during interaction increased to include reasoning and justification of mathematical ideas. Theoretically, these levels of thinking were hypothesised to differentiate among reform cultures in terms of a deepening in thought processes and as a means to particular types of knowledge outcome. Mutually, the two dimensions provide a conceptual framework for describing the differences among reform classroom in terms of the relationship between social interaction (participation) and student thinking (cognition). This conceptual framework was used in the analysis of the LPS data.

METHODOLOGY

Characteristics of Classroom Data

Classroom data for the three US schools are shown in Table 1. As can be seen, the schools are diverse with respect to student achievement as measured on the International Benchmark Test (IBT)[i] and the grouping of students (by ability or not).

Table 1 Student Achievement on International Benchmark Test

Characteristics	US1	US2	US3
Teacher Age/Gender	33/M	45/F	40/F
Number of students	28	36	33
IBT %;	99-90(2)	99-90(1)	96-90(7)
(number of students)	89-80(0)	89-80(6)	89-80(10)
	79-70(0)	79-70(14)	79-70(9)
	69-60(1)	69-60(6)	69-60(7)
	59-50(7)	59-50(9)	
	49-40(10)		
	39-30(7)		
	29-20(1)		
	<50(N=18)	<50(N=0)	<50(N=0)
Ability Group	Mixed ability (not grouped)	Average to high ability (grouped)	High ability (grouped)
Class Length	48 minutes	100 minutes every other day	60 minutes
Name of Math Course	Pre-Algebra	Mathematics Course I	Algebra
Curriculum	*Pre-Algebra* (McGraw Hill/Glencoe)	*Integrated Mathematics-Algebra* (McDougal Littel/ Houghton Mifflin)	*Algebra* (College Preparatory Mathematics-CPM)
Data Collection Period	7th to 21st Nov (midway first semester)	17th to 29th Jan (beginning second semester)	12th to 26th Mar (midway second semester)

Students from US1 are the lowest achieving with 64% scoring below the 50th percentile on the IBT. They also have the least amount of instructional time and a novice teacher. US2 and US3 classes have no students scoring below the 50th percentile and students from US3 are the highest achieving. US2 is ability grouped but also ungraded meaning that the age range of students in the class is broader than pupils from US1 and US3. Additionally, the classroom organisation (e.g., time allocated for instruction) and curriculum are different across the three classes as shown in Table 1.

Lesson Selected

Intially, the first author viewed the 10 video records and transcripts of the lessons for each US school[2]. Following this, specific lessons for each US school were selected and the transcripts analysed using the previously established coding categories by the three authors. For US1, 5 lessons (L01, L02, L05, L10), for US2 5 lessons (L01, L02, L05, L10), and for US3 all 10 lessons were selected for analysis. The data analysed from these lessons consisted of video records of mixed images or split screen of teacher and Focus students' actions during the lesson. The lessons were then transcribed to form an integrated text that combined Teacher Camera dialogue and Student Camera dialogue into one transcript (see Chapter 2 for a more complete discussion of the project methodology).

Analysis

The method of analysis was based on a qualitative research paradigm in which a previously developed coding scheme for the analysis of interaction patterns was used to interpret each of the transcribed videotaped lessons for the three US schools. The procedure for analysis of the nature of the interaction consisted of 1) initially segmenting the transcript by the structure or lesson organisation (checking homework, discussing homework, introduction to lesson topic, group work, whole class discussion); 2) conducting a line-by-line coding of the dialogue for those lesson events that involved whole class instruction using a coding scheme described in Wood, Turner-Vorbeck, Walker, Larsen, Fox and de Souza (1999), and then, 3) identifying and sectioning the distinct patterns of the interaction that existed within each segment; 4) identifying the consistent and repeatable interaction patterns across the lessons and assigning a label or name to these patterns. The interaction patterns were given labels that corresponded to the perceived function, intention, or goal of the interaction.

RESULTS

Overview of Mathematics Practices in Each School

US1. The pedagogy that dominated US1 primarily consisted of teacher-led whole class instruction. As an overview, the lessons consisted of the following segments:

a warm up/review of homework, review of previously taught topics, teacher explanation as the central mode of instruction for the main topic of the lesson and assignment of homework. Student assessment occurred during two lessons, in L04 the assessment consisted of reviewing items from a previous quiz and taking a new quiz (7 minutes), in L07 a formal assessment designed as a diagnostic test was given (whole class period). Following the diagnostic test, L08, L09, L10 consisted of discussion of the test items and reteaching that consumed the whole class period during each lesson.

As can be seen in Table 1, the teacher in US1 was confronted not only with the widest range in student achievement but also the highest number of low achieving students. Examination of L08, L09 and L10 using the coding scheme provides insight into the situation in which the teacher found himself. In attempting to discuss with students the results of a diagnostic test, the teacher found that the students understanding of fractions was far less than he realised. In order, to attend to the students' lack of knowledge, he moved from his initial discursive approach of involving students in participating in whole class discussion to a traditional form of interaction that involved long instances of interaction described as *Teacher Explain* and student questioning with *Test Questions* as exemplified in the short episode below. In several instances, however, the teacher tried to ask questions to prompt student reasoning but with little or no success.

```
Teacher:    Seven divided by seven- so, the numerator and the
            denominator, like we were saying yesterday, right? We said
            numerator- if that was the same number and the
            denominator's the same, the result of which would be one,
            right? Okay. Now, remember we were talking about this form
            of expressions also, the old- I guess you wanna say the
            old style of representing division with this sign right
            here. Okay? [points at division sign]. Lemme ... help you
            out a little bit [writes equation as fraction]. Is that
            true?
Alaine:     Yes.
Teacher:    That would be the way that we would generally represent
            things now, right? We no longer use the division sign with
            the two dots- that we'd write it like a fraction, correct?
            Tell me about that situation now, Gerrison.
Gerrison:   Equals one.
Teacher:    Why? Why is it the same?
Gerrison:   (Because the denominator and the numerator) ( ).
Teacher:    Well, denominator and the numerator- now, in this case,
            what might be a little more confusing (right now)- I'm
            just trying to read your mind a little bit here, but ...
            that's a number, isn't it? A single number. Okay? Is this
            a single number? What might be a little more confusing to
            you, I think, is probably that it's a fraction, right?
            You've got a fraction and then you divide it by another
            fraction? And so you've got all these fractions, right?
            But ultimately, one-third represents what? A number. Even
            though it does have its own numerator a- and denominator
            within the fraction, right? But ultimately, when I put it
            this way, it seemed to make a little more sense, didn't
            it? Now you know that that's- we could even do it like
```

```
                    this [writes on board]. Remember I was saying sometimes
                    parentheses kinda help clarify certain ideas? Just to help
                    you see things clearer sometimes. You notice here, seven
                    over seven. Same number over the same number. One third is
                    the same number over the same number, right? Does this
                    help everybody, in general, to see it that way?
Class:              Yes.
Amelia:             I don't know what we're talking about.
Teacher:            Okay. Uh, we're just talking about division here on this
                    first one; why this is one. We did remember from
                    yesterday. Again, we had seven over seven. Here, it's one-
                    third over one-third. Ultimately guys, what you're going
                    to have to come down to and- and really think about ...
                    okay, and Gerrison kinda pointed out ... "hey, this is the
                    same number and this is the same number". Ultimately, the
                    same number divided by the same number is ...? gonna be
                    one. Let me ask you a quick question, then we'll move on
                    ( ). If I did one-third and divided it by two-thirds,
                    would that be equal to one?
Class:              No.
Teacher:            Why not? Why not? Why wouldn't that be equal to one?
                    Esperanza? You have an idea? No? Okay. Clerk? You have an
                    idea why that wouldn't be equal to one? Belinda? Are you
                    sure? Eaton?
Eaton:              Two-thirds is ... (bigger than one-third) ( ).
Teacher:            Two-thirds is bigger than one-third? //Is that what you're
                    saying? Okay. So are they- ultimately, are they the same
                    number?
Class:              No.
Teacher:            And the result of which is gonna give me another answer
                    besides one, right?
```

Key to symbols used in transcripts in this chapter
...   A short pause of one second or less.
//   Marks the beginning of simultaneous/overlapping speech.
( )   Indecipherable speech.
(text)   A plausible interpretation of speech that was difficult to decipher.
[text]   Comments and annotations, often descriptions of non-verbal action.
....   Indicates that a portion of the transcript has been omitted.
–   Indicates self-interruption.

As Balfanz, MacIver, and Byrnes (2006) found in their attempts to improve the level of middle-school student achievement in US schools with populations similar to US1, the teaching and learning process is inhibited.

When many students in a school are behind grade level, without effective extra help available, classroom teachers are left with two undesirable choices. Either they can teach grade-level, standards-based lessons to those students who have the prior preparation to succeed and hope the others can catch on, or they can slow down the pace of their instruction and attempt to provide additional support to the students who are struggling but fail to cover all the grade level material that is assumed by the next year's curriculum (p. 59).

US2. US2 consisted of data from only five days in which the lessons are 100 minutes long or essentially 'double' with a Part A and B. Part A of the lesson always began with a warm up activity involving individual seatwork followed by student sharing and teacher explanation. Following this, Part A and Part B of a lesson used a wide variety of organisational formats; ranging from teacher explanation, teacher-led whole class with student participation, small group work filling in worksheets followed by whole class sharing of answers, or working together. These activities typically consisted of matching or recognition tasks as a means of practice and students reported in the interviews that in prior lessons group work typically involved traditional textbook problems. However, in the final lesson the teacher gave the students an open-ended task to work on in their groups; this group activity was not completed and it was not clear if students returned to it the next lesson. No summative or formative assessment was observed during these lessons.

US3. The classroom organisation found in US3 is similar to that described as 'active teaching' by Good, Grouws, and Ebmeier (1983); the events consisted of a review of homework, teacher introduction and presentation of the lesson topic with opportunities for students' to practise the new ideas, followed by group tasks that were occasionally open-ended (e.g., L02). However, assessment (preparing and taking a group test and individual test) was more prevalent in US3 than in either US1 or US2. Learners also more frequently discussed their grades. Students took two tests, a group test and an individual test. In L03 the teacher spent 45 minutes reviewing for the group test. One class period was spent taking the group test (L04), and, in L05, 15 minutes going over the test answers. In L08 students spent 60 minutes taking an individual test for which they could use their Tool Kits (notes on procedures) and calculators.

Close examination of teaching and learning in US3 revealed that many of the characteristics of reform were observed in the class. Students did work in groups, and were encouraged to be autonomous and responsible for their learning. This was evidenced in the ways in which the students were given autonomy for checking their homework and clarifying solutions they did not understand. These explanations for homework solutions were left to other students who had gotten the problem correct. Students often worked together on class assignments. However, the quality of the mathematical experiences of the students was generally focused on the procedures needed to solve the problems. This was evident in the students' explanations for the homework problems and in the talk during the last part of the lesson, which was devoted to completing practice problems. In general, the teacher spent the majority of the lesson in routine activities (e.g., taking attendance or checking whether students' homework was completed), while students were engaged in group work. The teacher did not typically observe or listen as students talked during the group work. It was evident in the examination of the form of teaching that was used that listening to students was not an essential activity for teaching because a majority of the teacher's actions did not require knowledge of students' thinking. Teaching was of two types, explicit explanation of the procedures that students were to put in their 'tool kits' or a sequence of cyclic events that involved: the teacher's explanation of techniques, then students practising the technique in groups, followed by students checking their answers as the teacher does each practice problem. The following discourse illustrates the way in which the teacher explained procedures before assigning practice problems for group work.

```
Teacher:   How do you find out the slope given two points? Judy?
Judy:      You find the difference between X- the both X's and the
           difference between both Y's.
Teacher:   Okay, so you find the difference between what should go up
           on top. Remember slope is the change in Y over the change
           in X. So you would do four minus six, and three minus
           four. What is four minus six?
Class:     Negative two.
Teacher:   And what is three minus four?
Class:     Negative one.
Teacher:   And what's a negative divided by a negative?
Class:     Two.
Teacher:   So then you could just do Y equals two X plus B,
           substitute in either three and four or four and six and
           then solve for B. So this time let's put in four and six.
           So six equals two times four plus B. So six equals eight
           plus B, subtract- [laughs] subtract eight, subtract eight,
           and you would get negative two equals B. So now you can
           re-rewrite this as Y equals two X minus two. And that
           would be the equation. Yes? Okay, . . .
```

In this class, the teacher rarely provided mathematical reasons for the steps in the procedures she presented. Instead, when students encountered difficulties executing the procedures, she relied on the authority of the text (often explicitly as "they said") for definitions and created rules, such as "you can't do that", to justify the steps in a procedure. Not surprising, learners, following the teacher, presented procedures to the class or in their group work in the same manner.

DISCUSSION

Although all the US teachers were explicitly trying to incorporate some reform ideas, such as group work, and attempting to "balance concepts, problem solving and skills" (Lobato 2003, private communiqué), as the lesson descriptions above depict, the teachers' efforts to incorporate the ideas of reform resulted in a pedagogy that was simply 'mapped on' to the events and ways of teaching typically found in traditional instruction. Moreover, the previous example from US3 reveals that the difficulties faced in an attempt to reform mathematics education in the US consists of more than just a lack of instruction that focuses on conceptual learning. As Nathan and Koedinger (2000) state, most secondary teachers continue to believe that mathematics instruction should emphasise student mastery of symbols and procedures. But more seriously, the instructional practices "tend to focus on the mechanics of symbol manipulation, rarely addressing the conceptual underpinnings of those symbols and procedures" (Nathan & Knuth, 2003, p.180). One frequently encounters statements like this about secondary mathematics education, but until explanations of procedures are compared across countries, as is possible with the data from the LPS, one does not truly fathom how procedural the US instructional practice is. What strikes one the most in the analysis of these lessons is that the presentations are divested not only of reasons, but are also completely devoid of any richness of thought that allows the learner to reason and gain insight into what one is doing mathematically when using the procedure. For example, there is little in the example from US3 cited previously that draws the learner into thinking about how and why the procedure works. Consequently, the computational or procedural fluency advocated by NCTM (2000) or 'deep procedural knowledge' proposed by Star (2005) is clearly not seen in these lessons.

The implications of these insights raise issues for the learner and the development of their mathematical learning practices. It is now widely accepted that without teachers' serious committment to change the disciplinary nature of school mathematics, the more autonomous group work setting are perceived by students as an opportunity for socialising rather than situations for expressing mathematical thinking and reasoning important to conceptual learning (Russell, 1999). Perhaps less well known is the extent to which the practices of procedural instruction are also seriously limited, which in turn creates learning practices of superficial memorisation rather than fluency and flexibility. The gap between research on mathematics teaching and learning practices and current school instruction still exists after nearly two decades; it is clear from the analysis of this data that one reason is that the degree of change is more far-reaching than initially perceived.

NOTES

i   The International Benchmark Test for Mathematics (IBT) is norm-referenced and evaluates student achievement on mathematical content for eighth grade. Items are taken from the TIMSS Student Achievement Study (Population 2).

ii During 2003, the first author was a Research Professor at International Centre for Classroom Research, University of Melbourne, Victoria, Australia.

REFERENCES

Balfanm, R., MacIver, D., & Byrnes, V. (2006). The implementation and impact of evidence-based mathematics reforms in high-poverty middle schools: A multi-site, multi-year study. *Journal for Research in Mathematics Education, 36*, 33-64.

Good, T., Grouws, D., & Ebmeier, H. (1983). *Active mathematics teaching.* New York: Longman.

Jacobs, J., Hiebert, J., Givvin, K., Hollingsworth, H., Garnier, H. & Wearne, D. (2006). Does eighth-grade mathematics teaching in the United States align with the NCTM *Standards*? Results from the TIMSS 1995 and 1999 video studies. *Journal for Research in Mathematics Education, 36*, 5-32.

Leung, F. (2005). Some characteristics of East Asian mathematics classrooms based on data from the TIMSS 1999 video study. *Educational Studies in Mathematics, 60*, 199-215.

Nathan, M., & Knuth, E. (2003). A study of whole classroom mathematical discourse and teacher change. *Cognition and Instruction, 2,* 175-207.

Nathan, M., & Koedinger, K. (2000). Teachers' and researchers' beliefs about the development of algebraic reasoning. *Journal for Research in Mathematics Education, 31,* 168-190.

National Council of Teachers of Mathematics (1989). *Curriculum and evaluation standards for school mathematics.* Reston, VA: Author.

National Council of Teachers of Mathematics (1991). *Professional standards for teaching mathematics.* Reston, VA: Author.

National Council of Teachers of Mathematics. (2000). *Principles and standards for school mathematics.* Reston, VA: Author.

Russell, S. J. (1999). Mathematical reasoning in the elementary grades. In L. Stiff (Ed.), *Developing mathematical reasoning in grades K-12. 1999 NCTM Yearbook* (pp. 1-12). Reston, VA: National Coucil of Teachers of Mathematics.

Van de Walle, J. (2004). *Elementary and middle school mathematics: Teaching developmentally.* Boston: Allyn and Bacon.

Weiss, I. R., Pasley, J., Smith, P., Banilower, E., & Heck, D. (2003). *Looking inside the classroom: A study of K-12 mathematics and science education in the United States.* Chapel Hill, NC: Horizon Research, Inc. (Retrieved February 2, 2006 from http://www.horizon-research.com/insidetheclassroom/reports.

Wood, T. (1994). Patterns of interaction and the culture of mathematics classrooms. In S. Lerman (Ed.), *The culture of the mathematics classroom* (pp. 149–168). Dordrecht: Kluwer.

Wood, T. (1996). Teaching to create discussion as mathematical argumentation. In L. Puig & A. Gutierrez (Eds.), *Proceedings of the 20th Conference of the International Group for the Psychology of Mathematics Education* (Vol. 4, pp. 427–433). University of Valencia, Spain: Psychology of Mathematics Education.

Wood, T. (1998). Teaching for conceptual understanding. In A. Olivier & K. Newstead (Eds.), *Proceedings of the 22nd Conference of the International Group for the Psychology of Mathematics Education,* Vol. 4 (pp. 193–200). University of Stellenbosch, South Africa: Psychology of Mathematics Education.

Wood, T., & Turner-Vorbeck, T. (1999). Developing teaching of mathematics: Making connections in practice. In L. Burton (Ed.), *Learning mathematics: From hierarchies to networks* (pp. 173–186). London: Falmer Press.

Wood, T., & Turner-Vorbeck, T. (2001). Extending the conception of mathematics teaching. In T. Wood, B. Nelson, & J. Warfield (Eds.), *Beyond classical pedagogy: Teaching elementary school mathematics* (pp. 185–208). Mahwah, NJ: Lawrence Erlbaum Associates.

Wood, T., Turner-Vorbeck, T., Walker, W., Larsen, S., Fox, M., & de Souza, L. (1999). *Coding scheme for video-tape analysis of classroom episodes.* West Lafayette, IN: School Mathematics and Science Center, Purdue University.

Terry Wood
College of Education
Purdue University
USA

Soo Yeon Shin
College of Education
Purdue University
USA

Phu Doan
College of Education
Purdue University
USA

IDA AH CHEE MOK

CHAPTER SIX

Teacher-Dominating Lessons in Shanghai: The Insiders' Story

INTRODUCTION

Classrooms in Asian regions which are under strong influence of Confucian-heritage cultures were often described as teacher-dominating with passive learners. Despite the fact that the teacher is an important person in a lesson, many pedagogical theories advocate opportunities for students' participation and free expression of ideas and denounce the idea that the teacher should take too much control in a lesson. However, the students in some of these cultures give very good performances in comparative studies such as such as the Third International Mathematics and Science Study (TIMSS) (Hiebert et al., 2003) and the Programme for International Student Assessment (PISA) (OECD, 2004). These results create an unexpected mismatch between the students' performance and the apparent image of teacher-dominating classrooms. Some authors call this phenomenon a paradox (e.g. Biggs & Watkins, 1996; Watkins & Biggs, 2001; Fan, Wong, Cai & Li, 2004). Attempts in solving this paradox lead to more appreciation about the nature of Chinese teaching and learning. For example, Ma (1999) found that the Chinese teachers in her study demonstrated profound understanding of the subject matter; Leung (2001) explained the East Asian identities in terms of dichotomies between the East and the West; Gu, Huang and Marton (2004) described teaching with variation as a Chinese way of promoting effective teaching; Lopez-Real, Mok, Leung and Marton (2004) studied a Shanghai teacher's teaching style and found varying patterns. In summary, these studies suggest very clearly that the simple phrase "teacher-dominating" tells too little to inform how the nature of the teacher's intervention may contribute to learning. Furthermore, the lack of investigation into how students react to the teacher's intervention also suggests an important gap in the understanding of the real picture.

The teacher and the students are the key people in lessons and they are called the 'insiders' in this chapter. The Learner's Perspective Study (LPS) design built in ways of documenting data to capture the views of both parties. Attempting to produce the insiders' picture of the lesson by combining accounts from both parties, I will present in this chapter a detailed analysis of the LPS data of a

D. J. Clarke, C. Keitel & Y. Shimizu (Eds.), Mathematics Classrooms In Twelve Countries: The Insider's Perspective. 87–97. © 2006 Sense Publishers. All rights reserved.

Shanghai school (SH2), which includes the lesson videos, and the teacher and student interviews.

In the remainder of the chapter, I will first present some preliminary analysis of the lesson videos to give a general picture of the sequence of the lessons. Next, the teacher's and the students' perceptions of their mathematics lessons based on the analysis of the interviews will be presented. Then, their reflection and comments upon the same lessons will be compared. Finally, integrating the views from both parties with what happened in the lesson, the nature of the teacher-dominating lessons will be discussed.

CONTEXTUAL DATA

The teacher Mr. X is a very experienced teacher with 17 years of teaching experience. He is a well-recognised competent teacher and was awarded the title "Senior Lecturer in Secondary School" by the Shanghai Senior Academic Title Appraisal Group in 1997. The research team recorded a sequence of 14 grade-7 lessons on algebra topics which included linear inequalities, linear equations in two and three unknowns, the solution of the equations, the methods of substitution and elimination, and application to word problems. The class size was 46 students.

In the analysis, we categorised events in the lesson into a) exploratory, b) directive, c) summarising, d) exercises and practice and e) assigning homework. The focus of 'exploratory events' is usually on a relatively open or difficult problem which has more than one possible answer and the problem may be explored in either pair discussion or whole class discussion facilitated by the teacher. The term 'directive events' was used to refer to moments in which the teacher played an explicit directive role and they may serve a range of purposes such as: stating what should be done; emphasis on convention and precise language; recapitulation of what has been learned for a foundation of what follows; clear and directive definition of a concept after discussion; direct explanation by the teacher; or probing for a teacher-expected answer. In the entire sequence of 14 lessons, the main elements in terms of time were exploratory (17.7%), directive (10.1%) and exercise and practice (70.1%). These events form blocks with a variety of sequences and the distribution does not suggest a routine pattern at the level of the lesson (Mok & Lopez-Real, Chapter 16, this volume).

At this stage, the analysis of the lessons from SH2 shows some features which are also shared by another Shanghai school, SH3 (Mok, in press). The lessons were strongly guided by the carefully designed examples and exercises. The students were consistently attentive and followed the teacher's instruction, and there were no instances of either inattentiveness or off-task behaviour. The explication of the contents was always made clear in the teacher's directive discourse. Therefore, it is quite fair to describe the lessons as 'teacher-dominating' if we subscribe to the traditional interpretation of teacher-dominance. That is, the teacher takes very much control of the class activities.

THE TEACHER'S PERCEPTIONS OF HIS OWN LESSONS

The teacher Mr. X was interviewed three times. In the interviews, he was well aware that his teaching style was not the same as the traditional model of teaching which placed emphasis on practice and students imitating the teacher's work. In his own model, he let his students experience as much as possible, aiming to let them understand by their own comprehension instead of memorising ideas. He obviously had a clear picture of a traditional model in mind and he often described his own teaching style in contrast with this traditional model.

Transcript 1: SH2-Teacher Interview 1

```
T        My way of teaching is unlike the conceptual way of teaching
         traditionally. I let the students to experience as far as
         possible. Otherwise if the teacher teaches mechanically,
         right, the students cannot recite that even if they have tried
         to recite for ten times. Like these concepts, right? Students
         just need to understand them. So you have to let them figure
         it out, there is no need to require them to recite every word,
         they just need to understand. This is conceptual teaching,
         which allows the students to sort out and apprehend according
         to what they have experienced.
```

Key to symbols used in transcripts in this chapter
...   A short pause of one second or less.
[text]   Comments and annotations, often descriptions of non-verbal action.
. . . .   Indicates that a portion of the transcript has been omitted.

Objectives emphasis on the content

The teacher's objectives were much emphasised in his delivery of the content. In his elaboration of his objectives, he included a clear analysis of the topic taught in the lesson. He explained whenever a part of the lesson was about the foundation or conceptual development. There were some occasional references to developing a general capacity: "the way of thinking" and "let student understand," This was the rationale of his design of the lesson but may as well have been seen as his general approach to teaching. He describes below the objective of one of his lessons.

Transcript 2: SH2-Teacher Interview 1

```
T        There are two main topics in this lesson. The first one is
         about systems of linear equations in two unknowns, their
         solutions, and the solving of systems of linear equations in
         two unknowns. The second one is about how to solve the systems
         of linear equations in two unknowns. The lesson is divided
         into two parts. The first part is about the concepts, and the
         second part is more concrete, it is about the solving of
         linear equations in two unknowns....
T        The lesson lets the students understand that, why it can be
         transformed, how to transform, this is the key, therefore
         during the whole teaching process, much time has been spent on
         the first part, much time has been spent on the first part,
         there are two functions. First, it can be the foundation, and,
         second, it can be the exploring tool for solving systems of
         equations in the second part. Isn't that so?" I asked students
```

89

```
        that why y equals two x in the first equation can be
        substituted into the second equation and in the second
        equation, y also equals two x. The student said, didn't you
        mention that just now? The solutions of systems of equations
        are the common solutions of these two equations. So if the
        first equation equals two x, then the second should also be
        equal to two x. He knew this principle, and then he can solve
        this problem. Otherwise, even if I spend all of my efforts in
        teaching them, they will not be able to know that, right?
        Therefore I think, here, it is not only the teaching of
        knowledge, the key is to teach such a way of thinking to the
        students.
T       The lesson today involved, that was, which unknown should be
        selected of which the algebraic expression represents another
        unknown. This is the crux.
```

What the teacher saw as important

In the interview, the teacher selected instances in the lesson which he saw as important and explained his interpretation of these instances. Throughout the interviews, the teacher showed a very clear analytical picture of how his planning was carried out in the lesson and how he reflected upon these instances. In his evaluation of his own lesson, the teacher had an explicit philosophy of helping students build a foundation for learning mathematics, mastering the formal mathematical language, teaching the concepts and encouraging the students' thinking and trials.

Foundation and linkage. Basic knowledge was often brought into focus in the early part of the lessons and the instances were pointed out by the teacher in the video. The rationale for its importance did not lie only on the piece of knowledge in itself but for the foundation of further development in the later part of the lesson or in later lessons. This rationale was clearly articulated in his own words, which are translated below.

Transcript 3: SH2-Teacher Interview 1

```
T       That is, the using of the algebraic expression of an unknown
        to represent another unknown. During the whole process of
        teaching the method of substitution, this is the basic
        knowledge [知识点], therefore, we have laid the foundation
        before this lesson. This is not only for the sake of this
        lesson, but also for the next lesson and the lessons that
        follow. It is even for the sake of the solving of systems of
        linear equations in three unknowns. It is the basic knowledge
        which we expect the students should manage very well.
```

Let the student think. Providing opportunities for student to think and try was one of the most emphasised aspects of the teacher's self appraisal of the lesson. It was mentioned more than once in the interviews that the teaching process should let students "think" and "brainstorm" and criticised the teaching of imitation or "simulation" in his words. However, by encouraging students to think more, the teacher did not mean free exploration of the work. What he referred to is a kind of framed exploratory work based on a specific task or question with limited options. The teacher made it clear that the feasibility of his ideal lay in the design of his lesson, in other words, his choices of tasks. Based on what he observed, he was quite happy with what his students did.

Transcript 4: SH2-Teacher Interview 1

T   [Referring to a specific question in the lesson] I let the students consider that: how many chickens and rabbits should there be respectively? So, there are many solutions for the first equation, there are also many solutions for the second. Then, I wrote three solutions purposely. Actually there are more than ten solutions. After students analyse these solutions, the students found the results.

T   [Referring to a part near the end of the lesson] It is now forty-minutes, uh, I prepared...the exercises of the last two equations. Because of limited time, we did just one of them. I thought that during the whole teaching process, the students attempted to do everything themselves, right? So to do one or two questions in the later part of the class is no big deal. …. Now we were adopting this kind of method, I let the students…brainstorm, and solve the questions themselves. Then it is not the problem of doing more or fewer exercises of a certain kind. Uh, so, I selected one question here. After doing this question [watching video], from the exercise, the students were doing…doing quite well. I inspected around then, what they did was okay.

Standardised language. In the interview, the teacher showed that he was well aware of the form of language which his students used. On the one hand, he appreciated students expressing their ideas in their own language in the stage of exploratory work. On the other hand, he had expectations of the use of standardised language linked to the meaning and understanding of a mathematical concept. Below is how he commented on the students' language in an episode in a lesson:

T   The language of the students was not standardised, but from here, the students' language was standardised. This is compatible with the first equation, but not the second, so basically, the student used the standardised language. Finally, he said, 'it can be compatible with the first and also the second, therefore they are the common solutions for these two equations', so I had to link them as a system of equations. Uh, from this example, we are not solely guided by rules, why we have to link the equations, because they are of the same question, there is the need to link as a system of equations, so the reason of calling them systems of equations stands apart. The question why it is called a system of linear equations in two unknowns was the focus of the following discussion. After the systems of equations have appeared, I asked students why it was called a system of linear equations in two unknowns. It permitted the students to answer on their own. The student answered why it was called a system of equations according to their characteristics.

THE STUDENTS' PERCEPTIONS

Two students were interviewed after each lesson and a total of 28 students were interviewed from this class. There were 27 records of complete interviews because of a technical failure in one case. Overall, the students were extremely positive towards their lessons.

Good lessons

All 27 students said that the lessons were good lessons. Their reasons for good lessons made references to the teacher (15), the student(s) (7) and the content (8). The reasons referring to what the teacher did are very positive, for example, "the teacher taught very clearly and in great detail", "the teacher let us think on our own" and "when we were working on the exercises, he came and looked at what we wrote, that can tell us what's wrong, and then we have to avoid making the mistakes again". Their reasons suggest that they highly valued the teacher's input in the lesson. The reasons referring to the student factor were less explicit. Some simply mentioned that the student was a factor without further elaboration and some mentioned that classmates paying attention collectively in class was a factor. The reasons referring to content varies from fairly general comments such as "in this lesson we can find that mathematics is fun" to very specific one such as "this lesson showed us the ways to solve the linear equations in three unknowns".

The students liked the lessons

All 27 students said that they liked the lessons. Twelve said that they liked the lessons because of the teacher and made explicit compliments such as "the teacher taught well" and "the teacher taught very clearly and I could understand right away". Seven comments were content-related such as "learnt much new knowledge" and "the topics are related to real life", six mentioned student for a reason without explanation. Three mentioned atmosphere for a reason and they used descriptions such as "interesting", "energetic", "the teacher taught vividly" and "not much pressure".

What the students saw as important in the lessons

With respect to the general question about what things were important in mathematics lessons, seven students mentioned the word "knowledge", and four mentioned the word "concept" without much explanation. Eighteen referred to skills such as "how to do the elimination" and "how to solve the questions".
During the interviews, we invited the students to pause the video to comment on the instances which they thought were important. The students made an average of four pauses per interview but there is a variety in the number of pauses. Some students only stopped the video at one or two instances to make brief reflections and some made very detailed reflections on the lesson (see Table 1).

Table 1. The number of pauses made by the students

No. of pauses	1	2	3	4	5	8	18
No. of students	5	7	6	1	3	5	1

All students paused the video with comments referring to what the teacher said and the content of the lesson. Their comments about content were often restating what was taught, for example, "the steps of solving the system of linear equations in two unknowns". It seems that they saw the content as important by itself. When they referred to the instances of teacher talk, these instances included all sorts of teacher activities such as a review of what was taught before, an explanation for a concept or how to do a question, checking students' answers, giving feedback on students' work including correction of mistakes, and making summaries for the lesson.

Besides referring to the teacher and content as important, some also pointed specifically to the exercises, the discussion with classmates and mistakes. Ten students referred to the exercises for example,

He let us work on the exercises, then we would remember what we had learned yesterday. I was thinking about the questions at that time.

Four students referred to the discussion with classmates, for example,

The teacher let us have the discussion, and we actively participated, so the lesson became more lively, more lively, and that can help me learn more about this method.

Three students referred to the teacher's feedback to mistakes and they made lengthy comments showing that they appreciated the teacher's correction and effort in correcting their mistakes as well as their writing, for example,

We should not only be acquainted with the solution of the questions, but have to know the format of the question, or the choice of words, language and other things.

A JUXTAPOSITION OF THE TEACHER'S AND THE STUDENTS' REFLECTION OF A LESSON

In this section, the teacher's comments on one of his lessons are juxtaposed with the two focus students' comments from the interviews. The time-codes for the pauses of the video are listed in Table 2. This was a lesson on simultaneous equations in two unknowns. According to the teacher, students had to learn the concept of the system of equations and its solution, and how to solve a system of two equations.

Table 2. Juxtaposition of the teacher's and students' pauses of the video for SH2-L03

Content / Activities	Teacher	S1	S2
Review			00:00:49:25 review 00:01:36:09 seatwork, exercise, review 00:03:03:29 teacher explanation
Given the linear equation in two unknowns 7x-2y=-5. Use the equation in y to represent x=. Use the equation in y to represent y=.	00:04:01:00 basic knowledge and foundation		00:06:39:05 teacher explanation
Guess it: How many chickens and rabbits are there? There are x rabbits and y chickens in a cage. There are altogether twelve heads, and forty legs. How many rabbits and chickens are there in the cage?	00:07:56:00 Trial teaching.		00:09:48:15 discussion 00:12:22:62 teacher explanation
The meaning of a system of equations and its solution	00:13:01:13 Conceptual teaching		00:13:14:23 teacher questions 00:15:28:09 concept 00:16:49:09 teacher explanation 00:18:04:09 teacher conclusion
Try it: Solve the system of equations $\begin{cases} y = 2x \\ x + 2y = 15 \end{cases}$	00:18:21:00 comment on students' work	00:19:52:09 content	00:19:16:23 seatwork, exercises 00:19:52:09 teacher explanation 00:22:04:15 teacher feedback 00:24:45:11 teacher explanation 00:26:30:17 seatwork, exercises
More examples and exercises	00:32:00:00 explain how he chose the examples		00:33:17:21 Teacher questions 00:34:25:01 teacher explanation
Summary	00:37:35:00 explain why students were doing fewer exercise		00:43:46:03 teacher questions

In the interviews, the teacher usually stopped at the beginning of a task and explained very clearly his rationale for his teaching strategies and examples. The comments were long and very often referred to his observation of the students in the lesson. His comments were more detailed for the first 30 minutes of the lessons.

The two students happened to be two extreme cases in terms of the level of detail in their comments on the lesson. S1 only commented on one instance which involved a board representation of procedures, "Because the steps to solve the system of linear equations in two unknowns are written on the board". S2 in contrast was the student in the class who gave the most detailed reflections on the lessons. S2 traced nearly all actions of the teacher and her comments very often showed an appreciation of the teacher's strategy and the content. For example, in the case of a review problem, the teacher's comment was that that part was important basic knowledge and foundation, whereas S2 said,

> It was a revision of what we learnt in our previous lesson. It was a transition for what we were about to learn today.

In the case of the rabbit and chicken problem for which the teacher purposefully let the students discuss and explore, S2 said,

> We were having a discussion; that was what I said that just now, it allowed us to know how other people would interpret the question, um…I was thinking.

The juxtaposition tells more about the possible variety of the students' attentiveness to the lesson. S1 made only one pause but the students' comment suggested that the most important in the student's mind was the content of the lesson. S2's case showed the behaviour of a very attentive student who saw nearly all the teacher's input important. Both cases showed a demand for a detailed elaboration of the content which made a nice match with what the teacher prepared to give.

DISCUSSION

Are the lessons teacher-dominating?

Simply put, the answer is yes. From the dictionary, "to dominate" means "to control, govern, or rule by superior authority or power" or "to exert a supreme, guiding influence on or over". Also reported in studies of classrooms in some Asian regions, there are prevailing features of whole class instruction in which the teacher takes up an influential role (e.g., Paine, 1990; Mok and Morris, 2001). Analysis showed that this Shanghai teacher shows many such features. Nevertheless, the teacher-dominance has a clear rationale. The teacher has a clear understanding of the subject matter at a level of subtle detail and tries hard to make his students understand the same level of detail. He gives the student opportunities for discussion but he controls their activity by choosing tasks with limited options so that students will see what he expects. He welcomes students to express ideas in his own words but he corrects their language to the standardised language.

How do the students perceive the lessons?

The students obviously like their lessons and the teacher. Upfront in their comments in the interviews, they want to learn knowledge about the topic and methods to solve problems. They appreciate the teacher's direct or indirect ways to help them understand the content of the lessons and reach the required standards. Consequently, their focus is mostly on the content and teacher's actions. The teacher's explanation, questions, and feedback regarding mistakes are all means to help them learn. To a certain extent, this is a kind of matching between what the teacher wants to give and what the students want to receive.

Is this a match or mismatch?

Comparing the teacher's and the students' comments, there is a consistent match between the expectations of both parties. The emphasis on the content, the teacher's directive nature and the appreciation of the teacher's intervention are the predominating attitudes and behaviours that characterise this group. The teacher showed in the interview that he has a strong belief that teaching should be providing students with opportunities to try, comprehend and think, rather than imitating the teacher's ideas and actions. His model in this view is to some extent under the influence of pedagogical philosophy from the West. It is obvious that he denounces the traditional transmission model. Nevertheless, the teacher's suggestions of "to increase attempts and exploration as much as possible" or "to let students brainstorm" probably prompt readers of this chapter to imagine a picture of students enjoying free exploration, which is a mismatch with what actually happened in the reality of the classroom. To avoid any misunderstanding caused by the naming of models, the lessons may be better described as the teacher's own synthesised pedagogy based on his own understanding of the traditional models, the western models and his own students.

As mentioned earlier, there is a match between the teacher's and the students' expectations. However, there is also a mismatch between the rhetoric conceptions of western models and the teacher's actual implementation in the lessons. The phenomena of such match and mismatch tell us an important research agenda for comparative studies. The objective of comparison has never been the search for prescriptive models for importing purposes but rather seeking an understanding of the holistic nature of mathematics teaching and our own practice from different perspectives and to different depths. As the teacher in this chapter has demonstrated, a pragmatic practice will not be a stagnant copy of a prescriptive model but a synthesis between different models to meet the expectation of the insiders.

ACKNOWLEDGEMENTS

The work described in this chapter was supported by a grant from the Research Grants Council of the Hong Kong Special Administrative Region, China (Project No. HKU 7192/02H) and the Small Project Funding from the Committee for Research and Conference Grants of the University of Hong Kong (Funded Project: A comparison between the mathematics teachers' approach between Australia and Shanghai).

REFERENCES

Biggs, J. B., & Watkins, D. A. (Eds.). (1996). *The Chinese learner: Cultural, psychological, and contextual influences*. Hong Kong: CERC & ACER.

Fan, L., Wong, N. Y., Cai, J., & Li, S. (Eds.). (2004). *How Chinese learn mathematics: Perspectives from insiders*. Singapore: World Scientific.

Gu, L., Huang, R., & Marton, F. (2004). Teaching with variation: A Chinese way of promoting effective mathematics learning. In L. Fan, N. Y. Wong, J. Cai & S. Li (Eds.), *How Chinese learn mathematics: Perspectives from insiders* (pp. 309-347). World Scientific Publishing Co.

Hiebert, J., Gallimore, R., Garnier, H., Givvin, K. B., Hollingsworth, H., Jacobs, J., Chiu, A. M.-Y., Wearne, D., Smith, M., Kersting, N., Manaster, A., Tseng, E., Etterbeek, W., Manaster, C., Gonzales, P., & Stigler, J. (2003). *Teaching mathematics in seven countries: Results from the TIMSS 1999 video study (NCES 2003-013)*. U.S. Department of Education. Washington, DC:

Leung, F. K. S. (2001). In search of an East Asian identity in mathematics education. *Educational Studies in Mathematics 47*, 35-51.

Lopez-Real, F. J., Mok, I. A. C., Leung, F. K. S., & Marton, F. (2004). Identifying a pattern of teaching: An analysis of a Shanghai teacher's lessons. In L. Fan, N. Y. Wong, J. Cai & S. Li (Eds.), *How Chinese learn mathematics: Perspectives from insiders* (pp. 382-412). World Scientific Publishing Co.

Ma, L. (1999). *Knowing and teaching elementary mathematics: Teachers' understanding of fundamental mathematics in China and the United States*. Mahwah, New Jersey: Lawrence Erlbaum Associates.

Mok, I. A. C. (in press). Shedding light on the East Asian learner paradox: Reconstructing student-centredness in a Shanghai classroom. *Asia Pacific Journal of Education*.

Mok, I. A. C., & Morris, P. (2001). The metamorphosis of the 'Virtuoso': Pedagogic patterns in Hong Kong primary mathematics classrooms. *Teaching and Teacher Education: An International Journal of Research and Studies, 17*(4), 455-468.

OECD (2004). *Learning for tomorrow's world: First results from PISA 2003*. OECD, Organisation for economic co-operation and development.

Paine, L. W. (1990). The teacher as Virtuoso: A Chinese model for teacher. *Teachers College Record, 92(1), fall issue*, 49-81.

Watkins D. A., & Biggs J. B. (Eds.). (2001). *Teaching the Chinese learner*. Hong Kong: Comparative Education Research Centre, The University of Hong Kong.

Ida Ah Chee Mok
Faculty of Education
University of Hong Kong
Hong Kong SAR
China

BERINDERJEET KAUR, LOW HOOI KIAM AND SEAH LAY HOON

CHAPTER SEVEN

Mathematics Teaching in Two Singapore Classrooms:
The Role of the Textbook and Homework

INTRODUCTION

There appears to be a lack of focused research studies on the teaching of mathematics in Singapore schools. The few studies (Kaur & Yap, 1997; Chang, Kaur, Koay & Lee, 2001; Kaur, Koay & Yap, 2001), that have been done to date have focused on the teacher and looked at classroom teaching at the macro level. These studies have listed among their findings that the textbook is an essential teaching aid of the teacher in most if not all classrooms and that homework is assigned at the end of almost every lesson. However, neither the link between the textbook and homework nor the role of the textbook and homework both from the perspectives of the teachers and students were ever explored. This paper aims to explore these questions at the micro level using the data collected as part of the Learner's Perspective Study (LPS) (Clarke, 2002a; 2002b) in Singapore.

Singapore's participation in the LPS has given us the opportunity to examine in considerable depth the practices in three 'well-taught' Grade 8 mathematics classrooms from three secondary schools in an integrated and comprehensive manner. A preliminary analysis of the data from School 1 (SG 1) has shown that the learning of mathematics in this classroom appears to extend beyond the classroom into the home and that the textbook drove both the classroom instruction as well as the home assignments (Kaur, Seah, & Low, 2005). In this chapter the data from two 'well-taught' Grade 8 classrooms in two schools will be used to explore the roles of the textbook and homework from the perspectives of both teachers and students.

METHODOLOGY

This chapter reports the analysis of a part of data from sequences of lessons that are consistent with the protocol of data collection of the LPS (Clarke, 1998; 2001). In the LPS, three video cameras documented teacher and learner actions for sequences of at least ten consecutive lessons and these video records were supplemented by post-lesson reconstructive video-stimulated interviews with

D. J. Clarke, C. Keitel & Y. Shimizu (Eds.), Mathematics Classrooms In Twelve Countries: The Insider's Perspective. 99–115. © *2006 Sense Publishers. All rights reserved.*

teacher and students, together with test and questionnaire data and copies of written material produced in class and during the interviews. The teachers, one from each school, were selected for their locally-defined teaching competence in mathematics and ten consecutive lessons of each were recorded. From the corpus of data from the two schools, relevant data relating to textbook and homework were extracted and are presented in this chapter.

TEXTBOOK

Background

Textbooks are a kind of document that embody curriculum visions, aims and goals (Schmidt et al., 1996). How consistent textbooks are with the intended curriculum of any system depends to a large extent on how the books are conceptualised, developed, designed and perhaps even guided by research. In some systems textbooks are official publications of the government or educational system. In other systems they may be produced commercially and compete for official adoption or have absolutely no official status and develop totally as commercial projects. In Singapore, textbooks are produced commercially and compete for official adoption by the Ministry of Education (MOE). Hence they adhere very closely to the intended curriculum – the Mathematics syllabus issued by the MOE (Ministry of Education, 2000). The Mathematics syllabus comprises two sections. The first outlines the aims of mathematics education in schools, the framework of the mathematics curriculum, the objectives of the secondary mathematics curriculum and the instructional programme, detailing curriculum time, schemes of work, teaching approaches, an effective mathematics programme and assessment. The second is, as shown in Table 1, the mathematical content presented as a detailed list of topics, outcomes and teaching notes. The textbooks used in Singapore schools amplify the syllabuses as shown in Table 2 and provide an explicit link between what is intended and implemented. The American Institutes for Research (AIR) study (Ginsburg, Leinwand, Anstrom, & Pollock, 2005) found that Singapore's textbooks build deep understanding of mathematical concepts through multi-step problems and concrete illustrations that demonstrate how abstract mathematical concepts are used to solve problems from different perspectives while traditional US textbooks rarely get beyond definitions and formulas, developing only students' mechanical ability to apply mathematical concepts.

Tables 1 and 2 respectively show an extract of curriculum content from the syllabus and the corresponding content in a textbook. This textbook, *New Syllabus Mathematics 2* (Teh & Looi, 2002), is used by Teacher 1 (T1) from School 1 (SG1) and Teacher 2 (T2) from School 2 (SG2) in the study. Textbook content helps to guide what individual teachers do in their classrooms. A textbook that has been approved by the MOE for use in schools for a stipulated period of time must display a seal like the one shown in Figure 1 on the same page of the book as the copyright and publication details.

Table 1. Extract of syllabus content

Source: Lower Secondary Syllabuses – Secondary Two (MOE, 2000)
Topic: Algebra – Algebraic manipulation and formulae

Topics / Outcomes Pupils should be able to:	Teaching Notes
o Expand products of simple algebraic expressions	o Examples: - $(ax + b)(cx + d)$ - $(ax + by)(cx + dy)$ (a, b, c, and d are integers)
o Factorise algebraic expressions of the form - $ax + ay$ - $ax + by + kay + kby$ - $a^2x^2 - b^2y^2$ - $a^2 \pm 2ab + b^2$ - $ax^2 + bx + c$	o Emphasise the process of deducing the factorisation from algebraic rules

---

Table 2. Extract of textbook's content

Source: New Syllabus Mathematics 2 (Teh & Looi, 2002)

Chapter 5 Expansion and Factorisation of Algebraic Expressions
   o Expansion of Algebraic Expressions
   o Further expansions
   o Factorisation
   o Perfect Squares and Difference of two squares
   o Factorisation by grouping
   o Summary
   o Review Questions 5
   o Problem Solving

The content for the sub-topic Expansion of Algebraic Expressions

Example 1 (worked examples)
Expand (i) $(a + b)(a + b)$; (ii) $(a - b)(a - b)$; (iii) $(a + b)(a - b)$
Results of Example 1 are generalized as follows:
1. $(a + b)^2 = a^2 + 2ab + b^2$
2. $(a - b)^2 = a^2 - 2ab + b^2$
3. $(a + b)(a - b) = a^2 - b^2$

Example 2 (worked examples)
Expand (i) $(2a + 3b)^2$; (ii) $(5x - 3y)^2$; $(3x + 2y)(3x - 2y)$

Example 3 (worked examples)
Use the algebraic results shown to evaluate the following:
(i) 201×199; (ii) 797^2 ; (iii) $39^2 + 78 + 1$

Exercise 5a (Pupils to practice)
Expand the following expressions Q 1(a) – (x)
Use the algebraic results obtained to evaluate each of the following Q 2 (a) – (l).

Figure 1. Approval Seal.

Zhu and Fan (2002) in their study of the use of textbook by mathematics teachers at grades seven and eight in Singapore found that textbooks were the major print resource of worked examples and exercises for mathematics teachers. Teachers often supplemented their collection of worked examples from other sources such as reference books, the internet or CD-ROMs for use in the classroom, but they almost always used only the exercises in the textbook for homework assignments. The study also revealed that the academic status of the school, the teaching experience of the teacher and the gender of the teacher did not affect the ways in which the teachers used the textbook. The Third International Mathematics and Science Study (TIMSS) conducted in 2003 reported that all the teachers who taught the participants of the study from Singapore at the Grade 8 level indicated that they used a textbook to teach mathematics, either as a primary basis (74%) or a supplementary resource (26%) (Mullis, Martin, Gonzalez, & Chrostowski, 2004).

The Role of Textbooks in the Two Classrooms

From the corpus of data of schools SG1 and SG2, three sources of data, namely the lesson tables, interviews with teachers and interviews with students, were used to ascertain the role of the textbook in the classrooms. A lesson table is a chronological narrative account of activities that take place during the lesson. This table also details all the tasks (learning, practice and homework) that the teacher uses during the lesson, and their source. A learning task (Mok, 2004) is a task that the teacher uses to teach the students a new concept or skill. Practice tasks are tasks used during the lesson either to illuminate the concept or demonstrate the skill further, and tasks the teacher asks students to work through during the lesson either in groups or individually. Homework tasks are tasks assigned to be done at home or during out of class time.

From the lesson tables of T1 and T2, the number and sources of learning and practice tasks were traced. Similarly the sources of homework assignments were also traced. Tables 3 and 4 show the data from the two schools. It is evident from Tables 3 and 4 that the textbook is a significant source of the mathematical tasks that the teachers used in their sequences of lessons. It is also apparent from the tables that the textbook is not the only source that both the 'competent' teachers rely on for their lessons. These findings concur with that of Zhu and Fan (2002) and Kaur, Seah and Low (2005).

Table 3. Source of Learning and Practice Tasks

School 1 Lessons	Learning tasks		Practice tasks		Total
	Source		Source		
	Textbook	Other	Textbook	Other	
SG1-L01	5	3	11	-	19
SG1-L02	2	10	3	-	15
SG1-L03	1	-	11	1	13
SG1-L04	1	1	24	-	26
SG1-L05	4	1	14	-	19
SG1-L06*	-	-	-	-	-
SG1-L07	1	1	2	-	4
SG1-L08	1	-	16	-	17
SG1-L09**	-	-	-	-	-
SG1-L10	2	1	5	-	8
Total	17	17	86	1	121

School 2 Lessons	Learning tasks		Practice tasks		Total
	Source		Source		
	Textbook	Other	Textbook	Other	
SG2-L01	3	4	6	9	22
SG2-L02	3	-	7	3	13
SG2-L03	5	1	14	2	22
SG2-L04	4	-	5	-	9
SG2-L05	6	-	3	-	9
SG2-L06	4	-	1	1	6
SG2-L07**	-	-	-	-	-
SG2-L08**	-	-	7	5	12
SG2-L09	2	-	4	-	6
SG2-L10	3	-	4	-	7
Total	30	5	51	20	106

*   class was having a test
** teacher went through the test corrections

In Table 3, for SG1-L02, T1 used 10 learning tasks that were not from the textbook. These tasks were crafted by the teacher on the spot and they involved simple rearrangements of numerals and symbols such as 3^6, 2.1^5, 4.5×10^{-6}, and so on, to demonstrate the use of a calculator to compute. These 10 very simple tasks resulted in making the percentage of learning tasks taken by T1 from the textbook and other sources over the period of the ten lessons equal. As for T2, who was mainly teaching algebraic expressions, equations and simultaneous equations over the period of the ten lessons, it was not easy to craft coherent tasks on the spot. Therefore the high number of non-textbook learning tasks used by T1 may be considered an anomaly. Having accounted for the anomaly, T2 used more tasks from other sources in total for learning and practice tasks. From Table 4 it is evident that, over the period of the ten lessons, T1 gave fewer homework assignments than T2. Furthermore T1 based her homework assignments either from the textbook or other sources, with most of them from the textbook, while T2 not only gave more homework assignments during the same period but also homework assignments from either the textbook or other source or both. Also, T2 gave a higher proportion of assignments from other sources than T1.

Table 4. Source of Homework Assignments

School and Lesson	Homework Assignment Source		School and Lesson	Homework Assignment Source	
	Textbook	Other		Textbook	Other
SG1-L01	no homework assigned		SG2-L01	√	-
SG1-L02	√	-	SG2-L02	√	√
SG1-L03	-	√	SG2-L03	√	-
SG1-L04	√	-	SG2-L04	√	-
SG1-L05	√	-	SG2-L05	√	√
SG1-L06	no homework assigned		SG2-L06	√	√
SG1-L07	no homework assigned		SG2-L07	-	√
SG1-L08	no homework assigned		SG2-L08	no homework assigned	
SG1-L09	-	√	SG2-L09	√	-
SG1-L10	√	-	SG2-L10	√	√

The second source of data was the teacher interviews. It was during the fourth teacher interview that the teachers, T1 and T2, were asked specific questions related to their planning of lessons, vis-à-vis selection of tasks for learning, practice and homework. The interview data was consistent with the data from the lesson tables (see the following interview transcripts). In Episodes 1–3, T1 stated that she used the textbook in planning her lessons and also as a source for learning, practice and homework tasks. In Episode 4, T2 stated that her first source of learning and practice tasks was also the textbook. Episode 5 shows that T2 appears to differ from T1 in her homework tasks. While T1 takes all her homework tasks from the textbook, T2 takes only about 50 % of her homework tasks from the textbook. T2 allows her students to self-evaluate their homework tasks from the textbook, as the textbook has an answer key. However, she does solve on the white board tasks they are unable to do.

Episode 1: T1

Int: Can you ah share with ah me how you //planned your lesson?
T1:   //Yes.

```
T1:    Er first we have to look at the syllabus that is laid out by
       the ministry.
Int:   Mm.
T1:    And then we look at our scheme of work.
Int:   Mm.
T1:    And after that er we will refer to er textbooks and so we look
       at the students ability,
Int:   Mm.
T1:    and um whatever er what are their previous knowledge (and/in)
       their background.
```

Key to symbols used in transcripts in this chapter

T1	Teacher 1.
Int	Interviewer.
//	Indicates the beginning of simultaneous speech.
/	Indicates an interruption, where one speaker 'cuts in' before another has finished.
-	Hyphen at the end of a word indicates self-interruption.
...	A short pause.
( )	Indecipherable utterance.
(text)	'Best guess' for speech that was difficult to hear.
(text 1/text 2)	Alternative possible readings of speech that was difficult to hear.
[text]	Annotations, comments by transcriber or author, or descriptions of non-verbal actions.
....	Indicates that some words have been omitted from the transcript.

Episode 2: T1

```
Int:   Okay. So how do you actually select these- um I'm going to use
       the word learning task.
T1:    Mm mm mm.
Int:   How do you select them?
T1:    Okay the learning task must be related to the objective of the
       lessons
Int:   Okay.
T1:    (Obviously) related to the objective of the lesson and er as I
       said we try to select task you know that they are familiar
       with
Int:   Mm. //Okay.
T1:    //that they can relate to and mm … some of it will come from
       textbook
Int:   Mm-hm.
T1:    which every child has.
Int:   Mm.
T1:    Um some of it of course will come from other sources like the
       internet you know,
Int:   Mm.
T1:    or from a … yeah from the internet most of it.
T1:    Yeah either from the internet or the textbook.
```

Episode 3: T1

```
Int:   Where do you usually take your homework questions from?
T1:    For homework is normally the textbook. We use the textbook to
       (   )
Int:   Mm.
T1:    Yeah. But for holiday assignment, er we took our questions
       from- we either make use of other schools exam papers,
Int:   Mm.
```

```
T1:      or sometimes um we could use er questions from foreign
         textbooks.
```

Episode 4: T2

```
Int:     So um how do you select the learning task for your lessons?
         Here learning task means
T2:      Mm
Int:     Like the questions you will use in class to introduce the
         concepts. How do you select these questions?
T2:      Er … usually I will look at the the textbook because every
         student they have compulsory for them to buy a textbook so … I
         will view through the examples in the textbook. If it's er not
         sufficient or not suitable then I will find other sources
         through past year questions, ten years series or test papers.
```

Episode 5: T2

```
Int:     So do you see any need of er giving assignment a homework
         assignment after every lesson?
T2:      Oh we still need, we still need.
Int:     Mm.
T2:      But I think er … er I do it this way like a … for example the
         textbook, if er about 50 percent,
  . . . .
Int:     Mm.
T2:      But it self evaluate.
Int:     okay
T2:      Because they have answers what.
Int:     Mm.
T2:      So if they can't get the answers or whatever,
Int:     Mm.
T2:      They can discuss with their their group group er … er members.
Int:     Mm.
T2:      If er they can't solve it ah then they will bring it up during
         the lesson and then I will solve it on the white board.
Int:     Mm.
  . . . .
T2:      But assignment is I prepare myself and they don't have the
         answers.
Int:     Mm.
T2:      Because usually the students do the er assignment they will ve
         have er the tendency to check the answer.
Int:     Mm.
T2:      So er … er this one I I don't give them the answers but
         selectively from other sources
```

The third source of data was the student interviews. Episodes 6–9 are extracts from student interview transcripts. In addition to the use of the textbook by the students to complete homework assignments, Episodes 6–9 highlight other roles of the textbook from the student's perspective. In Episode 6 it appears that the student uses the textbook as a first-aid to clarify his thinking and self assess his knowledge and application of concepts and skills the teacher has taught in class. Similarly, in Episode 7 too, the student turns to her textbook when she does not understand a chapter and reads it to clarify her thinking before working through the examples in the book. Both Episodes 8 and 9 show that the students use the textbook to self-assess themselves. In Episode 8, the student does it in class while the teacher is teaching and in Episode 9 the student uses the textbook to self-assess her achievement in mathematics. Hence, it may be said that students not only use the textbook to complete work assigned by their teachers but also as a first-aid to clarify their understanding (self-study) and self-assess their knowledge and application of concepts and skills that the teacher has taught in class.

Episode 6: SG2-L05, Musli (M)

```
Int:    Mm-hm. So do you do very much mathematical work at home?
M:      Um like … if like some if like the teacher teach teach us
        something new right then if like I don't know then I go home
        then I then I try the … try the sums that she has give give us
        again ah.
Int:    Mm … I see.
M:      So that get get the hang of it yeah.
Int:    Mm-hm. But do you do any additional work like finding extra
        questions in the textbook or assessment //book to try?
M:      //Yeah. Yeah the er in the textbook there're there're some
        there's a lot of useful stuff there inside
Int:    Mm …
M:      it's like if like you don't know then you flip open the
        textbook then you try out the questions inside ah.
```

Episode 7: SG1-L08, Sharifa (S)

```
Int:    You will ask your friends. Okay. So do you do a lot of maths
        homework at home?
S:      Yeah.
Int:    Besides the school er homework.
S:      Er no.
Int:    Oh you don't do anything besides the school homework?
S:      Just read the textbook only.
Int:    Oh you you do read the textbook on your own?
S:      Yeah. [nods]
Int:    I see.
S:      On the chapters I don't understand.
Int:    Oh okay. How do you read?
S:      I read like- then after I read then I read the eh I try the
        examples.
Int:    Oh you read and then you try practice on the examples?
S:      Yeah.
```

Episode 8*: SG1-L01, Guangjun (G)*

```
Int:    Okay, what were you doing at that time? You're writing
        something right?
G:      I was like er calculate the question on the exercise book.
Int:    Oh. Is it the same question that teacher asked you to do?
G:      Oh no. Different.
Int:    Oh you were just working out a question that you pick
        yourself?
G:      I want to see whether I really understand … um what teacher
        have teach me ah.
Int:    Oh so while the teacher is teaching standard form, you were
        actually trying out a question on the textbook.
G:      Yeah
```

Episode 9: SG2-L07, Jarinta (J)

```
Int:    Okay. Ah-huh. So what are the … what are the sorts of things
        that help you to tell how well you are achieving in maths?
J:      Oh first of all I think test is important,
Int:    Okay.
J:      then at home I try to do all those exercises in the textbook
        also and I check the answers I when I get them correct I think
        I can do well in maths.
Int:    Mm.
J:      Then when I get them wrong it's like I'm in danger.
```

HOMEWORK

Background

In most mathematics lessons in Singapore schools teachers assign their students homework on a regular basis, that is, after every lesson or after every two or more lessons. Homework is usually meant to be done alone but at times teachers do assign tasks for a group of students to do as a homework assignment. Homework is most frequently done at home but it may be done in school during study periods or after school hours in the library. According to Hong and Milgram (2000) there are three kinds of homework. The first is to assure that students review, practice, and drill material that has been learned at school. The second is to provide students with the opportunity to amplify, elaborate, and enrich previously learned information and the third is to prepare, in advance, material to be learned in subsequent lessons.

Cooper (2001) describes homework assignments as either same-day-content or distributed. Same-day-content assignments require students to do tasks that pertain only to material presented in class on the day the tasks are assigned and the purpose is similar to the first kind of homework described by Hong and Milgram (2000). Distributed homework assignments include the introduction of material that has not yet been covered in class or that was covered in lessons prior to the current day and is similar to the second and third kinds of homework described by Hong and Milgram (2000). Both kinds of homework, same-day-content and distributed, are often assigned by mathematics teachers in Singapore schools.

When homework assignments are brought back to school, teachers review them to check for correctness and indication of whether their students have or have not grasped the concepts and skills taught during the lesson corresponding to the homework assignment. At times, if the homework is of the third kind according to Hong and Migram's (2000) classification, they may use the homework as an impetus for the development of a new concept or skill for the day's lesson. TIMSS conducted in 2003 reported that 89% of the eighth grade participants said that their teachers always or almost always monitored whether or not they had completed their homework and 87% said that their teachers always or almost always corrected their homework assignments and gave them feedback. Participants also reported that in a typical week 11% of time in mathematics lessons was spent reviewing homework (Mullis, Martin, Gonzalez, & Chrostowski, 2004).

The Role of Homework

From the corpus of data of schools SG1 and SG2, three sources of data, namely the lesson tables, interviews with teachers and student questionnaires, were used to ascertain the role of homework from the perspectives of the teachers and students in the two classrooms. From the lesson tables, the nature of homework was established. Table 5 shows the types of homework assigned to the students during the sequence of ten lessons each by the two teachers T1 and T2.

Table 5. Types of Homework Assignments

School and Lesson	Homework Assignment Type			School and Lesson	Homework Assignment Type		
	I	II	III		I	II	III
SG1-L01	No homework assigned			SG2-L01	√	-	-
SG1-L02	√	-	-	SG2-L02	√	√	-
SG1-L03	√		-	SG2-L03	√	√	-
SG1-L04	√	√	-	SG2-L04	√	-	-
SG1-L05	√	-	-	SG2-L05	√	√	-
SG1-L06	No homework assigned			SG2-L06	√	-	-
SG1-L07	No homework assigned			SG2-L07	√	√	-
SG1-L08	No homework assigned			SG2-L08	No homework assigned		
SG1-L09	-	√	-	SG2-L09	√	-	-
SG1-L10	√	-	-	SG2-L10	√	√	-

Type I - Same-day-content
Type II - Amplify, elaborate & enrich previously learned information
Type III - To prepare, in advance, material to be learned in subsequent lessons.

From Table 5, it is evident that the homework assigned by both teachers T1 and T2 was mainly of Types I and II. On the two occasions when T1 gave her students homework assignments of Type II, they were intended in SG1-L04 to prepare the students for the upcoming test and in SG1-L09 to remediate their poor performance in the class test held during SG1-L06. Other than these two occasions T1 mainly gave her students homework of Type I – for practice of same-day-content. It appears that there are two main reasons for this. First, homework appears to be an extension of the lesson during which the students engage in seatwork, and second, T1 believes that "reinforcement of memory" (Episode 10: T1) is important, hence the homework is meant to hone the concepts and skills taught during the lesson. Homework assignments given by T2 on many occasions comprised Types I and II. T2 gave Type I homework for very similar reasons as T1. She believed that "practice makes perfect" (Episode 11: T2). However, she assigned students Type II homework to enable them to consolidate their learning and engage in challenging tasks, taken mainly from non-textbook sources. She only graded assignments of Type II, to assess her students' progress and give feedback.

The second source of data was the teacher interviews. During the fourth teacher interview the teachers, T1 and T2, were asked a specific question: "Do you think homework is important to students? Why?" Episodes 10 and 11 are extracts taken from their interview transcripts. These transcripts further establish the role of homework from their perspective that has been discussed in the above paragraph.

Episode 10: T1

```
Int:   Okay. Now do you think homework is important to students?
T1:    Er... personally I think yes. It is it is important because is a
       reinforcement of memory and is also er an assessment whereby
       the teacher can see whether the child has er understand the
       concept or not.
```

Episode 11: T2

```
Int:   Mm. Okay. Er do you think homework is important to students?
T2:    Er... actually it depends on the student again.
Int:   Mm.
T2:    Because I I still prefer small little tests rather than giving
       them homework because usually they will copy their homework.
       .  .  .  .
Int:   Okay. And so what is your main purpose of assigning homework?
T2:    Main purpose oh I just want to let them have more practice.
Int:   Keep them/
T2:    /er keep them
Int:   practicing.
T2:    Yes yes because practice makes perfect.
```

The third source of data was student responses to the question "Do homework assignments given by [name of teacher] help you in the learning of mathematics?" which was a part of the student questionnaire. All 37 of the students from SG1 and 38 of the 40 students from SG2 completed the questionnaire. The qualitative responses to the question were analysed. Thirty-six of the students from SG1 and all 38 students from SG2 indicated in their responses that homework assignments given by their teachers assisted them in their learning of mathematics. Shown below are five sample responses to the question. The phrases in italics were used to infer the functions of homework.

SG1-S9 (Doreen)

The homework assignments given by [name of teacher] allows us to *practice more questions*, therefore having *better understanding of maths*, and is useful. However, I feel that homework should not only be given from the textbook, and the quantity of homework should be reasonable.

SG1-S18 (Nurul)

It helps me practice for the tests.

SG2-S15 (Leshan)

Yes. The homework assignments are usually more challenging than what [name of teacher] teach. Therefore, I have to fully understand the basic before proceeding with the assignments. If I'm stuck at a question, I will revise my notes and again, go through what [name of teacher] taught earlier on.

SG2-S26 (Raihanah)

It helps me in the learning of mathematics because I get to practice maths and *learn new methods on how to do the questions*.

SG2-S33 (Linrong)

Yes. I get to encounter more challenging problems and I get to *learn how to solve a problem using different approaches*.

A qualitative analysis of the all written responses was carried out. Table 6 shows the six functions of homework that were inferred from the students' responses.

Table 6. Functions of Homework

Function	Descriptors
Improving/ Enhancing understanding of mathematics concepts	Help to improve understanding on the subject / Better understanding of maths / Understand topic taught / Understand well the subject / Help in understanding the concepts better / Improve maths / Better understanding / Understand more about the topic taught / Further understand formulas and concepts taught.
Revising/ practicing the topic taught	Help to revise daily / Practice topic taught / Recap the topic taught / Practice in areas that are unfamiliar/not good at / Practice makes perfect / Revise works / Revise and practice topic taught / Practice methods taught / Practice on the type of questions for that topic / As a revision / Give ample practice / Practice is important in mathematics / Refreshes memories so that can remember better / Remember the method of solving problems.
Improving problem-solving skills	Become more fluent in doing sums through practice / Able to solve problems / Help to master the skills of mathematics / Learn how to apply formulas in different questions / Able to do higher order questions and assignments / Reinforce the ways of solving the questions / Familiarize formulas taught / Help to understand the formulas and put in good use / Learn how to solve a problem using different approaches.
Preparation for test/exam	Practice for the tests / Know what kind of questions are coming out for exams / Will not panic if some challenging questions come out for exam.
Assessing own understanding/ knowing own mistakes	Able to learn from mistakes made in the homework / Assesses how much have learnt about the topic taught / Acts as a gauge to see whether can understand the concepts taught / Help to see whether understand the lessons / Assesses level of understanding through practice / Assesses level of understanding on how to apply certain formulas to some questions / It would determine whether we understand the topic or not. If not, we are "forced" to learn it in order to do the questions / Challenges our mind so when confronted with easier questions, able to do with ease.
Extension of mathematics knowledge	Exposes to different types of questions / Overview of the setting of questions / Exposes to how different types of questions are being phrased / Broadens knowledge / Learn new methods on how to do the questions / Exposes to more challenging questions.

It appears that all of the six inferred functions of homework are direct consequences of the type of homework assigned by the teachers. The function 'extension of mathematical knowledge' was solely inferred from the responses of students from SG2. Unlike students in SG1, students in SG2 were exposed to 'challenging' tasks taken from non-textbook sources. This may have provided them with opportunities to extend their mathematics knowledge.

DISCUSSION

Unlike past studies (Kaur & Yap, 1997; Chang, Kaur, Koay, & Lee, 2001; Kaur, Koay, & Yap, 2001) conducted on classroom teaching in Singapore schools that have listed their findings about the textbook and homework in an unrelated manner, this chapter has attempted to establish the link between them. The findings of this chapter affirm that homework is assigned on a regular basis in both the classrooms, which are the focus of this chapter, and the textbook is a source of homework tasks. Most importantly, this chapter has also examined the roles of the textbook and homework from both the perspectives of teachers and students in the two classrooms. The rich data available has allowed us to triangulate our findings.

From the perspective of the teachers, the textbook is a resource for their teaching but is not the only one. They draw learning, practice and homework tasks from the textbook for their lessons. It appears that both teachers use the textbook as a tool to implement the curriculum. In planning her lessons, T1 mentioned that she refers to the 'intended curriculum' (which is the syllabus) and then the textbook. Bearing in mind the alignment of the textbooks with the intended curriculum, there is often no need to refer to the syllabus when planning lessons. However, in the interview she may have felt that she had to give a 'proper' answer. Bearing in mind the validity of data in the form of self-reports, which form a significant part of the data used in this paper, this sequence of actions may not be common amongst teachers. Furthermore, the topic the teachers were teaching at the time of the data collection may have some bias on the source of tasks used. At the time of data collection, T1 was teaching 'Everyday Arithmetic, Standard Form and Scale Drawings' and T2 'Algebra'. For both teachers, the role of the textbook was consistent and in line with the findings of TIMSS in 2003 (Mullis, Martin, Gonzalez, & Chrostowski; 2004). Though the textbook was not the only resource used by the teachers it was a significant one. This may perhaps be due to the fact that the textbook must guide the students' learning outside of school time. As the textbooks have the MOE's approval for use in schools the teachers are assured that it is the best tool for their use. Furthermore, from the outsider's perspective, as affirmed by the AIR study (Ginsburg, Leinwand, Anstrom, & Pollock, 2005), textbooks used in Singapore schools are perfectly aligned to the intended curriculum and provide for deep understanding of mathematical concepts through multi-step problems and concrete illustrations that demonstrate how abstract mathematical concepts are used to solve problems from different perspectives. In addition, every chapter in the textbook is complete with graded practice tasks that range from simple exercises to challenging problems.

In addition to being a source from which students draw their practice and homework tasks, the textbook is also a first-aid for students to clarify their understanding and self-assess their knowledge and application of concepts and skills they were taught in class. The students who participated in the study were in grade eight and of above average ability. It is interesting to note that, during the interviews, a student reported "reading the textbook". This student apparently had the mathematical language skills to make sense of the worked examples (tasks presented with detailed solutions) found in significant numbers in the textbook. It appears that the perspectives of the role of the textbook for both teachers and students stem from the centrality it has as a means to concretise the curriculum.

"Practice makes perfect" appears to be an underlying belief that drives the initiative of homework from both the perspectives of the teachers and students. From the perspective of the teachers it is obvious that they view homework as a means to hone the concepts and skills taught during the lesson and to provide practice in lieu of tests and examinations. This is evident from the types and frequency of homework assignments. The role of homework from the perspective of the students appear to derive to a large extent from the types of homework assigned by the teachers. Students perceive that the role of homework is to help improve/enhance their understanding of concepts taught, revise/practice what has been taught and prepare for tests/examinations, assess their own understanding, improve their problem-solving skills and extend their mathematical knowledge. It appears that the role of homework from the perspectives of the students match the intended outcomes of homework from the perspectives of the teachers. However, such an ideal student-teacher relationship may be a consequence of students having internalised what their teachers tell them about the usefulness of homework.

CONCLUSION

This chapter has certainly added depth to the findings related to textbook and homework in mathematics classrooms of Singapore schools. In all three previous studies (Kaur & Yap, 1997; Chang, Kaur, Koay, & Lee, 2001; Kaur, Koay, & Yap, 2001) that looked at mathematics classroom teaching at the macro level, the focus was the teacher and general instructional patterns were documented. In all three studies, amongst the findings it is merely listed that the textbook is an essential aid of the teacher and that homework was assigned at the end of almost every lesson. The data from the two 'well-taught' Grade 8 classrooms presented and analysed in this chapter have resulted in findings about textbook and homework from two perspectives, the teacher and students. In addition, the two perspectives have also helped in connecting the findings. For example the role of the textbook from the perspectives of both the teachers and students appears to stem from the centrality it has as a means to concretise the curriculum. Similarly, the findings for homework from both perspectives appear to stem from an underlying belief that "practice makes perfect". The findings have also shed light on the link between the textbook and homework, which is that both are permanent features of mathematics lessons and that the textbook is an important source for homework assignments.

As discussed in the chapter some findings on textbook and homework were similar to those in Zhu and Fan (2002); and Mullis, Martin, Gonzalez and Chrostowski (2004); and Kaur, Seah and Low (2005). As it is neither possible nor the intention of this chapter to make any generalisations, the findings of this chapter have made a significant contribution towards providing a basis for further studies at the micro level relating to textbook and homework issues at the national level. The findings of this chapter may also provide a forum amongst the member countries of LPS leading to comparative studies. The rich source of data of the LPS may be used to explore the role of textbook and homework from the perspectives of both teachers and students in classrooms of the LPS members. There is also scope for in-depth analysis of tasks assigned for homework, from textbooks and other sources respectively, for dimensions such as depth of knowledge, knowledge manipulation and criticism.

ACKNOWLEDGEMENTS

This chapter was based on the funded project, CRP 3/04 BK, at the Centre for Research in Pedagogy and Practice, National Institute of Education, Nanyang Technological University, Singapore. Thanks are due to Keiko Hino and Soledad A. Ulep for their feedback on this chapter.

REFERENCES

Clarke, D. J. (1998). Studying the classroom negotiation of meaning: Complementary accounts methodology. In A. Teppo (Ed.) *Qualitative research methods in mathematics education. Journal for Research in Mathematics Education, Monograph No. 9* (pp. 98-111). Reston, VA: NCTM.

Clarke, D. J. (Ed.) (2001). *Perspectives on meaning in mathematics and science classrooms.* Dordrecht, Netherlands: Kluwer.

Clarke, D. J. (2002a, April). The learner's perspective study: Methodology as the enactment of a theory of practice. In D.J. Clarke (Chair), *International perspectives on mathematics classrooms.* Symposium conducted at the Annual Meeting of the American Educational Research Association, New Orleans.

Clarke, D. J. (2002b). Developments in international comparative research in mathematics education: Problematising cultural explanations. In S. L. Y. Yam & S. Y. S. Lau (Eds.) *ICMI Comparative Study Conference 2002: Pre-Conference Proceedings* (pp. 7-16). Hong Kong: University of Hong Kong.

Chang, A. S-C., Kaur, B., Koay, P. L., & Lee, N. (2001). An exploratory analysis of current pedagogical practices in primary mathematics classrooms. *The NIE Researcher, 1*(2), 7-8.

Cooper, H. (2001). T*he battle over homework: The common ground for administrators, teachers and parents.* Thousand Oaks, CA: Sage.

Ginsburg, A., Leinward, S., Anstrom, T., & Pollock, E. (2005). *What the United States can learn from Singapore's world-class mathematics system and what Singapore can learn from the United States: An exploratory study.* Washington, DC: American Institutes for Research.

Hong, E., & Milgram, R. M. (2000). *Homework: Motivation and learning preference.* Westport, CT: Bergin & Garvey.

Kaur, B., Koay, P. L., & Yap, S. F. (2001). *IPMA report (NIE-Exeter Joint Study). Year two (January-December 2000).* Singapore: National Institute of Education.

Kaur, B., & Yap, S. F. (1997). *Kassel project report (NIE-Exeter Joint Study) Second phase (October 1995-June 1996)*. Singapore: National Institute of Education.

Kaur, B., Seah, L. H., & Low, H. K. (2005, May/June). A window to a mathematics classroom in Singapore. In B. Kaur (Chair), *Multiple approaches to studying mathematics classrooms in Australia, Singapore, Hong Kong and Japan*. Symposium presented at the International Conference on Education, Singapore.

Ministry of Education. (2000). *Mathematics syllabus – Lower secondary*. Singapore: Ministry of Education, Curriculum Planning and Development Division.

Mok, I. A. C. (2004, April). Learning tasks. In D.J. Clarke (Chair), *Lesson events as the basis for international comparisons of classroom practice*. Symposium conducted at the Annual Meeting of the American Educational Research Association, San Diego.

Mullis, I. V. S., Martin, M. O., Gonzalez, E. J., & Chrostowski, S. J. (2004). *TIMSS 2003 International mathematics report*. Boston, MA: TIMSS & PIRLS International Study Center, Boston College.

Schmidt, W. H., Jorde, D., Cogan, L. S., Barrier, E., Gonzalo, I., Moser, U., Shimizu, K., Sawada, T., Valverde, G. A., McKnight, C., Prawat, R. S., Wiley, D. E., Raizen, S. A., Britton, E. D., & Wolfe, R. G. (1996). *Characterizing pedagogical flow: An investigation of mathematics and science teaching in six countries*. Dordrecht: Kluwer.

Teh, K. S., & Looi, C. K. (2002). *New syllabus mathematics 2*. Singapore: Shinglee Publishers.

Zhu, Y., & Fan, L. (2002). Textbook use by mathematics teachers at lower secondary level in Singapore. In D. Edge, & B. H. Yeap (Eds.), *Proceedings of second East Asia Conference of Mathematics Education* (Vol 2, pp. 194 - 201). Singapore: National Institute of Education.

Berinderjeet Kaur
National Institute of Education,
Nanyang Technological University,
Singapore.

Low Hooi Kiam
National Institute of Education,
Nanyang Technological University,
Singapore.

Seah Lay Hoon
National Institute of Education,
Nanyang Technological University,
Singapore.

115

GODFREY SETHOLE, BUSI GOBA, JILL ADLER AND
RENUKA VITHAL

CHAPTER EIGHT

*Fine-Tuning a Language of Description for Mathematics Items which
Incorporate the Everyday*

INTRODUCTION

The new post-apartheid curriculum advocates for the teaching of mathematics which incorporates "relevant contexts for South African learners" (Department of Education, 2001, p.16). This call is neither strange nor new, it resonates with a standpoint and position held by a number of mathematics educators both within and beyond South Africa (e.g. Skovsmose, 1994; Volmink, 1993). Inclusion of the everyday in mathematics, it is argued, will help assist in promoting mathematics as a discipline which is not external to human activities.

Mathematics, however, is a subject which is clearly demarcated from others. It is, in Bernstein's (1996) terms, a highly classified discipline which does not rely entirely on the mode of reasoning and argumentation employed in the everyday. Thus, a mathematics task which incorporates the everyday requires both abandoning and summoning a certain dose of realism. Consider, for example, the following task: 75 learners have registered to go on a school trip; how many buses should the school hire if one bus carries a maximum of 30 people. In solving this task, learners are expected to summon the everyday consideration that buses do not come in fractions. Yet, they are also expected to let go of other everyday considerations, such as the possibility of some students withdrawing from the trip or being offered lifts.

Our position in this chapter is that the incorporation of the everyday in mathematics classrooms inevitably leads to some form of tension regarding the type of considerations to be summoned when engaging mathematics items that incorporate the everyday. This position has developed through our study of three different Grade 8 mathematics classes in South Africa, more detailed descriptions of which can be found in Sethole (2005) and Goba (2004). Because of its interest in "classroom situations" (Clarke, 1998, p. 1), the Learner's Perspective Study in South Africa (LPS-SA) enabled us to observe ways in which teachers and learners handle mathematical tasks. We noticed the different ways and considerations brought to bear by the classroom community in reflecting on the everyday. We

D. J. Clarke, C. Keitel & Y. Shimizu (Eds.), Mathematics Classrooms In Twelve Countries: The Insider's Perspective. 117–130. © 2006 Sense Publishers. All rights reserved.

thus needed to develop a language of description, informed by and reflective of these different ways in which the everyday aspects are treated by the mathematics classroom communities.

In this chapter we argue that the recruitment of the everyday across diverse mathematics classroom practices is apparently varied, and at the same time substantively uniform. This argument arises from our study of such classroom practices, and has been framed by the following two questions:

1. How do the classroom communities negotiate mathematical and everyday considerations when dealing with tasks that incorporate the everyday?
2. What language of description might we use to embrace the treatment of the everyday by the mathematics classroom communities?

We begin this chapter with a discussion of some pertinent methodological aspects of the study, briefly highlighting how the LPS-SA enables us to talk about engagements of the everyday. We then introduce a selection from the worksheets used in the mathematics classrooms and their categorisations, using the descriptions of Dowling (1998). The third section discusses how the worksheets were engaged, which in turn pushed at the categories we had, leading to a more elaborate language of description. We elaborate this 'new language' in the fourth section and then close our discussion in the fifth.

REMARKS ON METHODOLOGY

The LPS is a study which was aimed at examining learning and teaching in competently-taught grade 8 classrooms in different countries (Clarke, 1998, p.1). As Clarke points out, the study was motivated by the need to capture "classroom situations" which would make it possible to identify connections between teacher actions and student practices. In each country, as has been highlighted in other chapters in this book; three classrooms were to be studied for a minimum of at least ten consecutive lessons. Each lesson required the use of three video cameras: One camera focusing on the teacher, the other on the learners and the third on a focus group for that particular lesson. We were thus able not only to note the texts used by teachers in the classroom but, most importantly, how these were engaged by learners in the classroom.

In South Africa, the study coincided with the phasing in of the new curriculum at grade 8 level, Curriculum 2005, as it is popularly referred to. This curriculum was upfront with, amongst others, the incorporation of the everyday in mathematics. We thus anticipated that through our study, it would be possible to observe how this policy position is interpreted by the teachers and enacted at classroom level. Thirty-eight Grade 8 mathematics lessons in three different schools, located in the Kwazulu-Natal province, were observed and videotaped. For each lesson we (1) focused on the interactions of a particular group of learners (2) collected their written work and (3) interviewed them at the end of the lesson. In addition, teachers shared their impressions about each lesson by completing a questionnaire. At the end of the data collection process at each school, each teacher was interviewed about various aspects of their lessons[i]. For purposes of this chapter we focus on two of the three teachers, Bulelwa and Kevin[ii].

Both Bulelwa and Kevin are regarded by their peers and learners as good teachers. They hold senior positions in their respective schools and by South African standards they are amongst the better qualified teachers. Bulelwa holds a Bachelor's degree in Science with Mathematics and Statistics majors, a Higher Diploma in Education and a BEd (Honours). Though black, she obtained all her qualifications from what used to be a University for Indians in Durban. She teaches in Umhlanga High school which is situated in Umlazi township, an apartheid-created single race residential area for blacks. She has been a teacher for over ten years, though at the time of data collection, she had only been teaching at this particular school for just over one year. All the staff members and learners at this school were black. We observed nine of Bulelwa's mathematics lessons in a class of 38 learners. Kevin taught at Settlers High school situated in a predominantly white affluent suburban area. The majority of learners and staff members at this school were white. Kevin holds a four-year teaching diploma and had been teaching at Settlers for twelve years. He obtained his teaching diploma from a College of Education historically meant for whites. Kevin's class had 28 learners, three of whom were non-white. We observed fourteen of the grade 8 mathematics lessons at Settlers.

Bulelwa and Kevin thus had different backgrounds both in terms of their educational histories, race, gender, orientation to the new curriculum reforms and teaching settings. Both recruited the everyday in some of the lessons we observed, though in different ways. In the following section we discuss these items, highlighting their similarities and differences. We then draw on and then develop the language used by Dowling (1998) in his analyses of mathematical texts.

CATEGORISING THE ITEMS IN THE WORKSHEET

The expression, mathematics tasks or items which incorporate the everyday, conceals the different types and distinctions that might exist amongst these items. Dowling, however, offers a useful language of description by which differences amongst these items may be identified. Using this language, it becomes clearer that 'mathematics tasks which incorporate the everyday' is not a homogenous category. Dowling (1998) employs two aspects to describe tasks: mode of expression and the nature of context drawn or the content. Tasks which have a highly classified mode of expression are those which communicate information in "unambiguously mathematical" terms (Dowling, 1998, p.135). Such tasks can either draw from the mathematics context or the everyday; in which case he respectively labels them 'esoteric' and 'descriptive'. Other tasks employ a weakly classified mode of expression and thus communicate information using non-mathematical expressions. Likewise, these tasks may also either draw from the mathematics or the everyday contexts; he respectively labels them 'expressive' and 'public'. The four possible categories emerging from this discussion can be presented in the quadrant in Table 1.

Table 1: Categories produced from interplay between mode of expression and content

	Strong classification of content	Weak classification of content
Strong classification of mode of expression	Esoteric Domain	Descriptive Domain
Weak classification of mode of expression	Expressive Domain	Public Domain

PHASE ORGANISER: CULTURE & SOCIETY
PROGRAMME ORGNISER: NUMBER AND PATTERNS
ACTIVITY 7 (Number Patterns in Nature)

Mathematicians have studied number patterns for many years. It was discovered that there are links between mathematics and our natural environment and sometimes events occurring in our societies. For this reason an understanding of algebra is central to using mathematics in setting up models of real life situations. Study the tables given and answer the questions that follow.

Year	1960	2000	2040	2080	2120
World population growth	3 000 million	6 000 million	12 000 million		

Year	1997	1998	1999	2000	2001
World increase in the number of AIDS sufferers	16.7 million	33.4 million	66.8 million		

From the tables, you can see that the AIDS and population figures follow trends, which can be seen, from the number patterns. These patterns allow researchers to predict what these figures will be for the future.

(a) Describe the pattern of population increase every 40 years as shown in the first table.
(b) Describe the pattern of increasing number of AIDS sufferers as shown in the second table.
(c) Fill in the missing numbers in each table.
(d) Researchers believe the earth cannot support a population approaching 192 000 million people. If the population continues to double every 40 years, then in which year will it be 192 000 million? Explain how you worked out your answer.
(e) The world population in the year 2000 is said to be 6 000 million. In which year will the number of AIDS sufferers be greater than 6 000 million if the trend in the second table continues? Discuss what you think this means.
(f) How is HIV virus/AIDS transmitted? Discuss.
(g) What can we do as a society to break the pattern of increasing number of AIDS sufferers? (i.e. decrease the number of AIDS sufferers).
(h) What is the percentage of people suffering from AIDS related diseases around the world in the year 2000?

Figure 1. Bulelwa's Worksheet

The figure on the previous page (see Figure 1) is a worksheet that Bulelwa used during lessons number 7 and 8, in which AIDS was incorporated as a context. Throughout lessons 3 to 9, Bulelwa's main mathematical theme was *number patterns*. It was the first time though, that the everyday was recruited for this theme.

All the items but (f) and (g) in the worksheet make mathematical demands. All the items which make mathematical demands employ a weakly classified mode of expression and draw from familiar everyday contexts, AIDS and world population. Thus, the mode of expression used and the contexts drawn in are accessible, so to speak, to non-native speakers of mathematics. These tasks may be categorised as belonging to the public domain. The following table communicates the type of mathematical demands made by each item, the context drawn in and the category to which it belongs.

Table 2: Classification of Bulelwa's worksheet according to Dowling's descriptions

Item (s)	Mathematical demand	Context recruited	Category
(a) and (b)	Pattern recognition and pattern description	World population in (a) and AIDS in (b)	Public
(c)	Performing calculations and Completing a pattern.	World Population and AIDS	Public
(d) and (e)	Performing calculations and explaining the calculations	World population in (d) and AIDS in (e)	Public
(f) and (g)	No mathematical demand	AIDS	Non-mathematical
(h)	Calculating percentages	AIDS	Public

In setting the scene for these questions, the introductory phrase sets the contexts of AIDS and world population as examples of 'real life' or authentic situations that mathematicians deal with. It is not explicit that the figures in the tables are not real, but are recruited to enable a generation of mathematics patterns defined by the functions $f(x) = (2^{x-1997})$ *16.7* million where $x \geq 1997$ and $x \in N$ for the spread of AIDS (second table) and $f(n) = 3.2^{\frac{n-1960}{40}}$ million where $n \geq 1960$ and $n \in N$ for the rate at which population increases (first table). We will discuss the authentic-not authentic aspect further in the next section. For now, we pay attention to the worksheet used by Kevin at Settlers.

Kevin's worksheet

Kevin used a total of 24 word problems over the first five days of our observations. The overall theme for these five lessons was *using equations to solve word problems*. The mode of expression used in 23 of the 24 items was, as characterised by the use of mathematical symbols, strongly classified. Some of the contexts he recruited were mathematical or strongly classified (e.g. triangles) and others were everyday familiar contexts or weakly classified (cars). We only highlight the first six items which were dealt with in the first day:

1.1   Lucy is 7 years older than her sister and their combined age is 21. How old is Lucy?[iii]
1.2   John's age is p years. Write down in terms of p, Sue's age if she is 16 years older than John[iv].
1.3   A bar of chocolate costs three times as much as a toffee sweet. Four bars of chocolate and three sweets cost R6.00. Write down the price of a bar of chocolate if the price of a toffee sweet is x cents.
1.4   The length of a rectangle is 1 metre longer than its breath. The perimeter of the rectangle is 42 metres. If the breath of the rectangle is x, calculate its length.
1.5   Do question 1.3 again, but in this instance, let the length of the rectangle be x. Do you get the same dimensions?
1.6   The Dodgers Basketball team won 14 games more than it lost during the past season. If the team lost x games during the season, write down in terms of x, the number of games the team won.

In the following table, we have used Dowling's framework to show the mathematical demands made by the items, the contexts recruited and the different categories to which the items belong.

Table 3: Categorisation of Kevin's worksheet according to Dowling's descriptions

Item number	Mathematical demand	Context recruited	Category
1.0	Calculating Lucy's age	Learners' ages	Public
1.1	Calculation of Sue's age	People's ages	Descriptive
1.2	Calculation of a price of bar	Chocolate and Toffee	Descriptive
1.3	Calculation of dimensions of a rectangle	Rectangle	Esoteric
1.4	Calculation of dimensions of a rectangle	Rectangle	Esoteric
1.5	Calculating the number of games by Dodgers basketball team	Basketball team	Descriptive

We have drawn a distinction between items 1.0 and 1.1 on the basis of the different types of modes of expression they employ. However, we wish to highlight another significant difference emanating from the use of names Lucy on the one hand and Sue and John on the other. Lucy is a real learner in Kevin's class whose age is indeed 14. She has a little sister who is seven years younger. This, we view as an attempt by Kevin to illustrate the usefulness of mathematics in dealing with real authentic context. On the contrary, Sue and John are imagined beings, more so because their ages are also given in terms of a letter symbol 'p', an unlikely reference to a real person's age. Like in Bulelwa's case, we observe a shift, subconscious and unintentional though it may seem, between referencing the real and authentic on the one hand and the unreal and inauthentic on the other.

Dowling's categorisation assists us to distinguish between different mathematical items. However, we also note that some of the distinction between the contexts recruited in mathematics items goes beyond the categories of strongly classified context - weakly classified context; it also includes the extent to which the context is authentic or inauthentic. This aspect became evident as we reflected on how the worksheets were engaged by the teacher and learners. That is, the teacher' and learners' relationship to the context in a mathematical task adds a different dimension to the language of description. In the next section we tease out this observation by discussing how the classroom communities engaged the worksheet.

SUSPENSION OF AUTHENTICITY: DEVELOPING A LANGUAGE OF DESCRIPTION

When she introduced lesson 7, Bulewa mentioned AIDS as a real authentic context which is closely related to the learners' experiences. She pointed out that "we are still looking at number patterns" but trying to incorporate "real life situations". So close was AIDS to the learners' experiences, Bulelwa felt, that she said to learners:

B: I even said it is most unlikely that no one has ever heard of an AIDS victim in our communities or society, you must know someone that actually died of AIDS: Whether they are not a member of your family, your neighbour or in the communities.

Key to symbols used in transcripts in this chapter
[text]   Comments and annotations, often descriptions of non-verbal action.

As outlined in the previous section, Bulelwa also included items (f and g) which required learners to summon the everyday considerations, as can be seen from the following worksheet.

(f) How is the HIV virus/ AIDS transmitted? Explain have AIDS
   when you sleep around with girls /boys the HIV /AIDS
(1) is transmitted trought the comunity
(2) If you use a same Blook with person have AIDS can be transmitted
  one

(g) What can we do as a society to break the pattern of the increasing
number of AIDS sufferers? (i.e. decrease the number of AIDS sufferers)
(1) We can make a drama about aIDS and worn people
(2) We can show that with How danger aids its with posters
(3) We can organise an awereness day
(4) Be faithful to your partner.

Figure 2: Learner's written responses to items f and g

It is interesting though, that the learners were able to switch to different rules of engagement when dealing with item *b* for example. As evidenced by their responses below, learners recognised the item as mathematical.

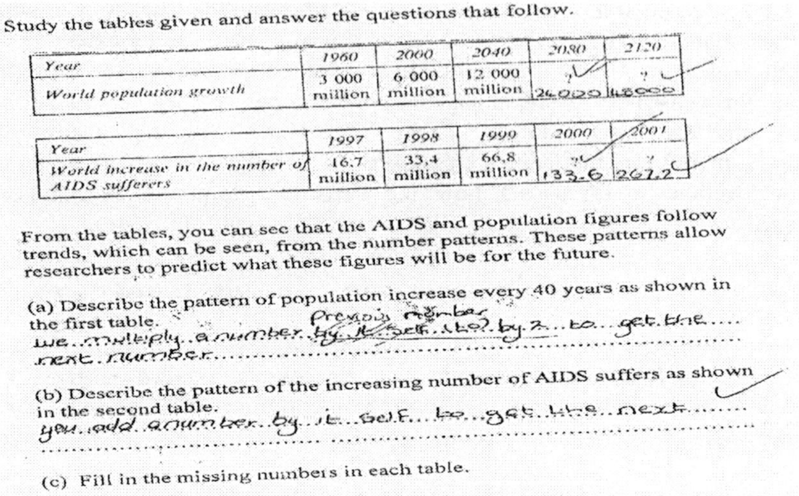

Study the tables given and answer the questions that follow.

Year	1960	2000	2040	2080	2120
World population growth	3 000 million	6 000 million	12 000 million	24 000	48 000

Year	1997	1998	1999	2000	2001
World increase in the number of AIDS sufferers	16.7 million	33,4 million	66,8 million	133.6	267.2

From the tables, you can see that the AIDS and population figures follow trends, which can be seen, from the number patterns. These patterns allow researchers to predict what these figures will be for the future.

(a) Describe the pattern of population increase every 40 years as shown in the first table.
Previous number we multiply a number by it self. it? by 2 to get the next number.

(b) Describe the pattern of the increasing number of AIDS suffers as shown in the second table.
you add a number by it self to get the next

(c) Fill in the missing numbers in each table.

Figure 3: Learner's assessed response to items a and b

As the teacher's ticks alongside the learner's responses (Figure 3) suggest, these responses met with Bulelwa's approval. Learners were expected, for example, to bracket off the possibility that as the number of years progressed, efforts to fight the spread of AIDS might begin to pay off, resulting in the reduction of the rate at which it spreads. The data in the table are not the real data and are more about eliciting a pattern rather than offering a description of AIDS. They were taken from a mathematics text as a mathematical number pattern and serve as a metaphor for the rapid increase rather than indicating the real rise in infection. The AIDS described by this data is not the real AIDS that learners may be familiar with, say from the media. Hence we call it inauthentic to distinguish it from the AIDS based on actual data and which calls for everyday considerations.

Kevin, like Bulelwa, alluded to the significance of incorporating real life situations. He introduced the first lesson as follows:

```
K:     Okay, what we doing now we doing equations and often you want
       to know how you can use something in, in maths in real life.
       So can you remember the problems. I was giving you, things
       like three CDs cost you three hundred and sixty rand what does
       one cost? Okay that's a real life situation you use in
       equations.
```

To illustrate this point, he used Lucy, one of the learners, as an example from which he formulated a question. Having established that Lucy is 14 years old and 7 years older than her sister (with the whole class listening), he formulated the following question:

– Lucy is seven years older than her sister, their total age add up to 21. Calculate Lucy's age.

In engaging this task, it was clear that both Kevin and the learners regarded it as mathematical. Firstly, when Justice explained, during the whole class discussion how he had used the equation $2x - 7 = 21$ and obtained $x = 14$, Kevin asked him to "finish it (the solution) off". Not making sense of what this phrase implied, Justice sought clarification.

```
J:    Pardon.
K:    You need to finish it off.
J:    I underlined fourteen.
K:    OK, I am just being pedantic have you actually answered the
      question.
J:    Err! [surprised]
K:    Good, so finish it off by writing the last statement. [Writes
      on board Lucy's age is 14]. Okay I'll be very surprised if
      they take a mark off by not putting it down there but let's
      not give them the opportunity to do. Did we all get that?
```

(lines 110 – 115, Settlers)

In this case Kevin suggests he is just being 'pedantic' and that it would be surprising if leaving out the everyday details would warrant taking off a mark. So whilst he initially declared the everyday as important, he also implied in this case that it did not matter much. Secondly, none of the learners went over to their classmate, Lucy, to ask for her age, as might happen in real life settings.

As in the case of AIDS in Bulelwa's class, we can distinguish between the real authentic Lucy and an imaginary inauthentic Lucy. Authentic Lucy was transformed into an inauthentic one during the formulation of a question by Kevin. He did this by asking learners to calculate her age even though Lucy had publicly said her age was fourteen. It is only in a mathematics class, and not in everyday settings, that this exercise might make sense.

We are now finding ways to distinguish mathematics items in terms of their authenticity. This distinction is not entirely new within the mathematics education domain. As Freudenthal (1973) noted:

> When speaking about mathematics fraught with relations, I stressed the relations with a *lived-through reality* rather than with a *dead mock reality* that has been invented with the only purpose of serving as an example of application (p.78, emphasis added).

We view these two Freudenthal-based categories as two opposite extremes. On the one extreme, using 'dead mock reality' to make reference to the everyday in a way which is highly unlikely or impossible. On the other extreme, 'lived-through experience' makes reference to genuine or not far-fetched use of the everyday. Authentic or inauthentic, a mathematics task may reference a context which is familiar or alien to learners. Moreover, there are shifts between the authentic and inauthentic as tasks take shape in classroom practice, as they come to be used for

125

the learning of mathematics in mathematical classroom communities. We bring in this dimension in the next section to extend our language of description.

ELABORATING A LANGUAGE OF DESCRIPTION

AIDS, in the last section, may be regarded as a context which resonates with learners' experiences because it occurs in areas where they stay and /or it takes place during the learners' lifetime. We regard such events as being '*near*' to the learners in terms of space (locality) and/or time (period of occurrence). Alternatively, a context may reference a scene or event which does not resonate with the learners' experiences either because it took place a long time ago or in a place situated physically far from where the learners reside. Such an event or context may be viewed as being '*far*' from the learners' experiences.

We have used the concept of 'near' similar to the way in which Royer (as cited in Billet, 1998, p. 8) uses 'near' as a qualification for knowledge transfer. He, for example, regards the ability of a university lecturer to teach with ease in another university as a case of 'near transfer' since it permits deployment of skills to a similar context. In a similar way, Royer uses 'far' as another qualification for knowledge transfer. Carrying on with an example of a 'university lecturer', Royer regards a requirement of a university lecturer to teach at a vocational college or primary school as far knowledge transfer. This is because in this case, there will be a deployment of a skill to a novel situation. In sum, the concept of 'near' is related to familiarity or similarity and 'far' is related to novelty or unfamiliarity.

Using the concepts of authenticity/inauthenticity and close/far to describe a context, the following four categories emerge (Sethole, 2005). For each category, we provide an explanation and an example of a task.

Table 4: Different categories of the everyday non-mathematical

	Authentic	Inauthentic
Near	Authentic and near	Inauthentic and near
Far	Authentic and far	Inauthentic and far

Authentic and near contexts

These refer to mathematical tasks in which the context is used genuinely or without major modifications and resonates with the learners' own experiences. This category is in line with what Freudenthal (1973) terms 'lived-through experiences'. An item which required learners to 'suggest ways in which AIDS can be prevented" falls within this category since it references a real context with which learners are familiar. A mathematical task using data based on actual levels of substance abuse in a neighbourhood in which learners live is authentic and near (Goba, 2004).

Authentic and far context

These are tasks in which the context, which does not resonate with learners' immediate experiences, is genuinely drawn in. Such contexts, for example, could be based on events that took place either long before the learners' lifetime or at a locality far away from where the learners stay or both. A mathematical task based on packaging of pills may be authentic because it reflects the reality of the activity but far as an adult activity (Goba, 2004).

Inauthentic and near context

These are tasks in which context, which resonates with learners' experiences, is not genuinely summoned. A task such as, 'The children ate p eggs', references the familiar (children and eggs) and therefore near contexts. However, it also introduces letter symbols which are not used in everyday settings, as a quantity for eggs. The earlier example of demonstrating a number pattern drawn from a mathematics text to describe AIDS sufferers is an inauthentic but near context.

Inauthentic and far contexts

These are tasks in which the context, which does not resonate with learners' experiences, is also not genuinely used. Developing number patterns or sequences in a task based on the context of on a drug company increasing its sale of drugs using fictitious numbers is both inauthentic and far (Goba 2004). A task which references a black president in the United States for South African learners can be viewed as belonging to this category from the learners' point of view. This is because the United States never had a black president (inauthenticity) and it is far for South African learners.

EMPLOYING THE ELABORATED LANGUAGE OF DESCRIPTION

Using these categories, it would seem that items f and g in Bulelwa's worksheet can be categorised as authentic and near since AIDS resonates with learners' experiences and is not trivialised by any form of inaccuracies. The rest of the items also reference AIDS or world population or both (i.e. near contexts) coupled with inaccurate data and thus not used genuinely. With regard to Kevin, all items but 1.0 are described in terms of letter symbols and therefore rendered inauthentic. The contexts used, however, resonate with learners' experiences.

Table 5: Recategorisation of items in Bulelwa's worksheet

	Authentic	Inauthentic
Near	(f), (g)	(a), (b), (d), (e), (h),(c)
Far	None	None

Table 6: Recategorisation of items in Kevin's worksheet

	Authentic	Inauthentic
Near	1.0	1.1, 1.2, 1.3, 1.4, 1.5
Far	None	None

The interesting and illuminating moment here is how this categorisation shows that despite visible differences in the tasks drawn in by both Bulelwa and Kevin, there are significant similarities in their tasks. Both use near contexts, though these are quite different kinds of contexts, with AIDS being far more provocative. Yet, in both classrooms inauthenticity comes to prevail as mathematics comes to the fore. Retaining authenticity increases complexity. The actual AIDS data might not produce neat patterns discernible to grade 8 learners.

The use of near contexts by Bulelwa and Kevin perhaps illustrates the intention to draw in contexts which resonate with learners' daily experiences. In order to enable access to mathematics, however, these contexts are coupled with either inaccurate data or with mathematical symbols. Perhaps the tension between wanting to enable access to mathematics on the one hand and using familiar contexts explain the high number of items falling within the category of near and inauthentic. The language of description elaborated in this article illustrates the non-simplistic nature of categorising contexts.

The call to incorporate the everyday into mathematics lessons seems to result in the mathematics communities having to "manage something rather complex" (Cooper & Dunne, 2000, p. 35): Creating a platform where authentic contexts can coexist with mathematical considerations. A better understanding of the nature of contexts used to embed mathematical tasks may also assist to illuminate how and why learners engage the tasks in the way they do. In what ways and for what learners does increased authenticity and nearness make the mathematics accessible, interesting and relevant or distract and block participation. Some learners resisted the closeness of authentic contexts like HIV/AIDS and substance abuse. There is also the question of what purposes and intentions are served in keeping near but inauthentic contexts that appear to dominate and how do learners mediate transitions from authentic to inauthentic contexts within any near or far setting. In Bulelwa's class most learners became preoccupied with the authentic and near items (f) and (g) and a special effort had to be made to get learners to engage the mathematics in the inauthentic but near items that involved working with the number patterns. How would a shift to authenticity and nearness have changed or impacted on these learner engagements? These are some further questions that the language of description developed here begins to raise.

NOTES

[i] Whilst differently resourced; all the three schools had access to telephone services and electricity and were thus not a representative sample of schools in the province, the majority of which have no access to these resources ((Financial Mail, 02/1999 page 25).

[ii] All the school names, teachers' names and learners' names are pseudonyms

[iii] *This item is not reflected in the worksheet. The teacher used it as a warm-up exercise. However, he did expect learners to solve it and later on he discussed the solution with the learners

[iv] Items 1.1, 1.4 and 1.5 had subquestions. We have not included all the subquestions because our aim is to highlight the substance of the questions.

REFERENCES

Bernstein, B. (1996) *Pedagogy, symbolic control, and identity: theory, research, critique.* London: Taylor and Francis.

Billet, S. (1998). Transfer and social practice. *Australian and New Zealand Journal of Vocational Education Research, 6*(1), 1-5.

Clarke, D. (1998) *Application to conduct research (summary):* Learner's perspective study.

Cooper, B. & Dunne, M. (2000). *Assessing children's mathematical knowledge: Social class, sex and problem-solving.* Buckingham, Open University Press.

Department of Education (2001). *National curriculum statement - Mathematics.* Republic of South Africa: Author.

Dowling, P.C. (1998). *The sociology of mathematics education: Mathematical myths/pedagogic texts.* London: Falmer Press.

Goba, B. (2004). *Grade 8 learners' experiences of mathematics in outcomes based research.* Unpublished master's thesis. University of Durban-Westville, South Africa.

Skovsmose, O. (1994). *Towards a philosophy of critical mathematics education.* The Nertherlands: Kluwer Academic Publishers.

Sethole, G. (2005). *Learners' perspectives on the incorporation of the everyday in mathematics. Unpublished doctoral thesis.* University of the Witwatersrand: Johannesburg.

Volmink, J. (1993). When we say curriculum change, how far are we prepared to go as a mathematics community? In C. Julie, D. Angelis & Z. Davis (Eds.), *Political dimensions of mathematics education 2* (pp. 12-129). Cape Town: Maskew Miller

Godfrey Sethole
Faculty of Education
Tshwane University of Technology
South Africa

Busi Goba
Faculty of Education
University of KwaZulu-Natal
South Africa

Jill Adler
Faculty of Humanities
University of the Witwatersrand
South Africa

GODFREY SETHOLE, BUSI GOBA, JILL ADLER AND RENUKA VITHAL

Renuka Vithal
Faculty of Education
University of KwaZulu-Natal
South Africa

SOLEDAD ASUNCION ULEP

CHAPTER NINE

'Ganas' – A Motivational Strategy: Its Influence on Learners

INTRODUCTION

What should a teacher of a mathematics class with 57 students do when they do not work on practice exercises the way she expects them to or worse some do not want to work on them at all? Ms. Santos, the teacher who was faced with this problem used a motivational strategy that was consistent with her belief of what students value. It was a strategy that was primarily shaped by her experiences as a trainer of contestants in student mathematics competitions. As something that they typically encountered in their mathematics class, the strategy had influenced the students in specific ways not all of which might be known to Ms. Santos. Documenting these ways had been made possible through the complementary accounts methodology (Clarke, 2001) of the Learner's Perspective Study (LPS). This chapter presents these influences and draws implications on teaching and learning mathematics from them.

BACKGROUND INFORMATION

The sample consisted of a grade 8 mathematics class with 34 girls and 23 boys of heterogeneous mathematics ability. The class ranked fifth among the 44 grade 8 classes in a public secondary school in an urban area in the Philippines. At the time the data were gathered, twenty-nine year-old Ms. Santos who had been teaching the subject Mathematics 2 in the school for 8 years handled the class. She was identified as the third most competent teacher by the mathematics supervisor of the city's public secondary school system. Class time was 40 minutes per day. The lessons were on geometry. Ms. Santos mainly used English in teaching but both she and the students often code-switched to Filipino, the national language.

Studies Related to Motivational Strategy

In a nationwide survey that assessed the training needs of Filipino secondary school mathematics teachers, 92% of the respondents indicated that they sometimes or always used motivational activities in their lessons (High School

D. J. Clarke, C. Keitel & Y. Shimizu (Eds.), Mathematics Classrooms In Twelve Countries: The Insider's Perspective. 131–149. © 2006 Sense Publishers. All rights reserved.

Mathematics Education Group, 1996). They considered these as important contributory factors to an effective mathematics teaching. In another survey on what high school mathematics teachers wanted to change in the way they teach mathematics, a grade 10 teacher who had taught for 10 years and had undergone many in-service training programs expressed that she needed "to improve the art of questioning and also motivation skills so that the students will not find mathematics boring" (Gallos, 1998, p. 62). Indeed, in the Philippines teachers seem to recognise the truth that Deitte and Howe (2003) say: "Student motivation is an essential component of successful student learning but it is also one of the most challenging areas for teachers" (p. 280). Their claim is consistent with what Sobel and Maletsky (1988) argue that proper motivation for the teaching of mathematics is important because "students except for the very few who seem to have a natural love for the subject, need to have their interest stimulated through suitable teaching techniques and procedures" (p. 27).

Motivation refers to any reason why a person behaves in a particular way in a given situation (Middleton & Spanias, 1999). In the words of Slavin (1997), motivation is "what gets you going, keeps you going and determines where you're trying to go" (p. 345). In this chapter, a motivational strategy refers to anything that the teacher deliberately does to motivate students to engage in a mathematical task or activity.

There are two types of academic motivation – intrinsic and extrinsic (Middleton & Spanias, 1999). In intrinsic motivation, a student works on a learning task for learning sake, that is, to gain competence. It is associated with desirable behaviours such as doing more than what a task requires and persevering at it in the face of difficulties. Intrinsically motivated individuals have learning goals (Slavin, 1997). In extrinsic motivation, a student engages in a learning activity because doing so will enable him/her to obtain extrinsic rewards (e.g. high grades) or favourable judgments of his/her competence (e.g. praises) or avoid negative judgments of the same (e.g. criticisms). Extrinsically motivated individuals have performance goals (Slavin, 1997). This difference in reason for getting involved in a learning task shows that motivation is part of what one considers as important (Ames, as cited in Middleton & Spanias, 1999).

Research studies have shown however, that extrinsic motivation was relatively ineffective in maintaining motivation for learning (Condry & Chambers, as cited in Nicholls, 1983; Lepper & Greene, as cited in Nicholls, 1983; Maehr, as cited in Nicholls, 1983). It could also hinder learning because completing the task or obtaining extrinsic incentives became the main concern of students rather than the task itself (Condry & Chambers, as cited in Nicholls, 1983).

Certain strategies are likely to foster learning goals and intrinsic motivation. Cobb et al. (1991) found out that second-grade students who were in classes that promoted learning through inquiry tended to believe that success in mathematics depended on attempting to understand it, explaining their ideas to others, and inventing their own solutions to problems. Students in traditional classes, however, tended to believe that such success depended on following the procedures used by the teacher or others. They also tended to develop performance goals. The study of

House (2003) revealed that the use of several teaching strategies such as cooperative learning was significantly related to student enjoyment in learning mathematics. Based on the results of their study, Schiefele and Csikszentmihalyi (1995) suggested that in order to increase their interest in mathematics, students have to be actively involved in mathematics laboratory or hands-on activities, small group work, and in solving problems in real world contexts. Moreover, they found it reasonable to assume that intrinsic motivation could be sustained to the extent that learning activities bring about a certain level of positive emotional experience.

There are also strategies that tend to promote performance goals and extrinsic motivation. In an experiment by Jagacinski and Nicholls (as cited in Nicholls, 1983), results showed that in a highly competitive situation, the students felt less able when the needed effort in engaging on a task was much. They viewed their competence in relation to that of the others and believed that their ability set the limit to what their effort could contribute to success. But in non-competitive conditions where learning goals were more valued, students did not have these views. These findings were consistent with those of the studies by Ames (as cited in Nicholls, 1983) that children emphasised ability and effort, respectively in interpreting their performance in competitive and non-competitive contexts. Nicholls (1983) argues that "competitive conditions produce unrealistic levels of aspirations and maladaptive levels of anxiety and helplessness" (p. 218). According to Middleton and Spanias (1999) it is important that students believe that ability is not fixed, that more effort could lead to greater mastery or success, and that occasional failure is part of learning. As such, they would tend to put in more effort and so achieve better in mathematics. In relation to this, persons who attribute success to ability and lack of success to lack of ability and so view success as unattainable have acquired the trait known as learned helplessness (Slavin, 1997).

In a competition, the success of one can mean the non-success of another. In considering the results of the study by Ames (as cited in Slavin, 1997), Slavin cautions that highly competitive grading or incentive systems should be avoided. Those who perceive their ability to be low are likely to give up in advance when they perceive that there is only one measure of success in class and only a few people can get it. The action research results of Francis (2003) showed that using a hands-on approach to learning positively affected students' engagement and motivation in numeracy. In contrast, using extrinsic rewards and creating a competitive classroom environment had no lasting positive effects on engagement in numeracy. An implication drawn from the study of Meyer et al. (1997) is that it is possible that typical classroom goals such as speed and accuracy may run counter with mathematics goals such as being able to justify one's method of solving a problem.

The studies cited show that a learning environment based on inquiry methods done in small groups can promote learning goals and intrinsic motivation while competitive learning environments such as one that emphasises speed and accuracy can promote performance goals and extrinsic motivation and may lead to learned helplessness.

THE TEACHER'S PROBLEM

The 10 lessons observed ranged from Lesson 6 to Lesson 15. In two of these lessons, Lessons 10 and 15, a test was administered. In the other 8 lessons, the activities consisted of introduction, lesson presentation, practice exercises, discussion and marking of answers to the exercises, quiz, and giving of assignment. The introduction was either a discussion of the answers to the assignment or a review of the previous lesson through a quiz. To introduce Lessons 6 and 8 respectively, a visualisation problem and a guessing activity were used. Ms. Santos presented the lessons using mainly exposition in a whole class setting in which she engaged the students in a rapid exchange of questions from her and answers from the students. According to Ms. Santos, when she presented a lesson particularly when the topic is new, the students were very attentive and participative. Partly, this might be due to their question and answer engagement. But when it was already time for practice exercises where they were to work and think on their own, they would not readily start working. It took them a long time to finish. And there were some who did not want to do them. She said that the students found the practice exercises boring. And she found the students slow.

Possible Reasons for the Teacher's Problem

During practice exercises, Ms. Santos must have considered going around the room to assist and monitor every student, a wish that she actually expressed during the interview. But because they were so many and the room was cramped, it was observed that she could only attend to very few students and they were mostly those who were sitting along the aisle. Since she could hardly get into the space between the rows of chairs it was seldom that she was able to assist the other students. It might be that with personal assistance particularly to the slow ones, she could have made students interested in doing the practice exercises. But with the large class size this was impossible.

Another possible reason why the students were not interested in the practice exercises was because they did not have the opportunity to work on their own prior to this part of the lesson. Since Ms. Santos only used exposition, there were no opportunities for them to explore and discuss mathematical ideas based on performing some activities or tasks. They were most of the time listening only to the teacher and responding to her questions.

'GANAS': THE TEACHER'S SOLUTION TO THE PROBLEM

To address the students' lack of interest in doing the practice exercises, Ms. Santos motivated them by using 'Ganas' which according to her was "giving additional points to the students who first finished the exercises with all answers correct." To signal, she would announce Ganas before an exercise and students would then take it to mean that additional points were at stake.

What Really Is Ganas?

Ganas is not a Filipino word but a Spanish word which means desire. In Filipino, words related to Ganas are "gana" and "nakakagana." Gana means appetite, eagerness, enthusiasm, or interest. Nakakagana means appetizing, interesting, or stimulating. Gana was used 3 times in the lessons.

Ganas was a word which Ms. Santos used to label the practice exercises for which the students can earn extra points besides their score if they were able to satisfy the specified requirements. Based on class observations, there were two kinds of Ganas. One kind (G1) involved giving additional points, also referred to as plus factor, to the students who got perfect or almost perfect scores in a practice exercise. The other kind (G2) involved giving additional points to a predetermined few number of students who first answered all the items in a practice exercise correctly. The essence of G2 is really competition. A student's success deprived others of being successful too just because they did not get as many correct answers as he/she did and/or they were not as fast as he/she.

How did Ganas Work?

G1 depended only on accuracy. There was no limit to the number of students who could earn a plus factor as long as they were able to get all or almost all the correct answers within the set time that the practice exercises should be completed.

In G2, it was not enough that students could only answer correctly or quickly. They needed to do both. For this type, students were instructed to stand up or raise their notebook where they wrote their answers, the moment they finished. The teacher counted each student as he/she stood up and stopped counting after the predetermined number had been reached. At times, not all those who were counted necessarily earned additional points. It was only after all or most of the students in class had finished working that the answers to the exercises were discussed, the students' works were marked, and their scores were determined. Only then would students know if they really qualified to get a plus factor. As such, there were students who were not included among those who were counted although they got perfect or nearly perfect scores because the quota had already been satisfied before they could stand up or raise their notebooks. Such was the case of Michaela, the best student in the class in Lesson 14.

When G2 was carried out, Ms. Santos asked the rest of the class to continue working after those who were included in the quota had been counted. However, it was observed that those who finished early had nothing to do anymore while the rest of the class was busy. This might well be the case also for those who were quick to finish when G1 was applied. As documented on the whole class video of Lesson 13, the potential recipients of additional points stood for a long time while waiting for what Ms. Santos would instruct them to do next. Later, when she asked them something about an exercise item, it distracted some students since they paid attention to them. And then much later, she eventually asked them to sit down while the others still worked.

How did Ms. Santos keep track of the plus factors that the students had already obtained? She gave out stickers, each of which was equivalent to two points. The students kept them and submitted them to her at the end of the grading period. She added the total extra points that a student had earned to the component of his/her mathematics grade where most of the students in all the classes that she was handling got the lowest.

The students who did well in the mathematics class were often the ones who qualified for the extra points. Having realised this, Ms. Santos banned those who already garnered a certain number of plus factors in a particular grading period from continuing to participate in Ganas for the rest of the grading period. However, those banned were still expected to answer the practice exercises.

What Led to the Decision to Use Ganas?

Ms. Santos' decision to use Ganas had been influenced by what she thought about what the students value. In the interview, when asked about what her students' answers might be if they were asked why it was important for them to learn the lesson, she said "in order to pass or get a high score if there is a quiz. Perhaps if the student is deep, it (the lesson) may be used in succeeding lessons. But I guess in general, to get a high score on a quiz or to pass Math 2 (grade 8 Mathematics)." Perhaps she thought that since the students value grades, then by giving them additional points if they work on the practice exercises quickly and correctly, this would make them want to work on them. Since scores were generated from practice exercises, giving additional points became a logical choice.

Qualifying for additional points in G2 required speed and accuracy. By requiring speed besides accuracy, Ms. Santos was able to get the students to start right away, kept them focused, and got them to finish soonest. But what she admitted as the possible reason for these requirements was her experience of being a trainer of student contestants in mathematics competitions for the past 8 months at the time of this research. In the trainings for and in the competitions themselves, the contestants were asked to solve problems with the time limit of 60, 30 or 10 seconds. Naturally, such time pressure was applied to contestants because speed and accuracy were the essence of competitions. But Ms. Santos required the same from all the students not all of whom were good in mathematics or liked it. That was why she must have found them slow. She recognised that such a requirement created pressure on students but she told them to "get used to pressure because everything in life involves pressure." However, it was observed that she extended the time a little longer when the students complained that they were not yet finished working on the practice exercises or she observed that they were having difficulties in answering them.

How Was Ganas Used?

Practice exercises were orally read by Ms. Santos or written on manila paper that was posted on the board. The nature of the items given in Ganas differed. Examples taken from the different lessons are presented below:

– Non-routine problem solving (Figure 1.):

I have six dots.

Move any 2 of the dots to reverse the direction of the figure.

Figure 1. Non-routine problem

– Identification: Tell what the following represents a) tip of a ballpen b) clothesline
– Concept Understanding (Figure 2.):

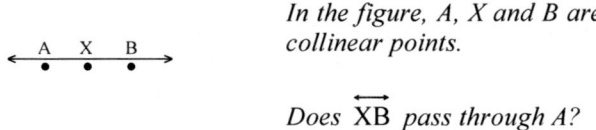

In the figure, A, X and B are collinear points.

Does $\overleftrightarrow{XB}$ pass through A?

Figure 2. Problem related to conceptual understanding

– Routine problem solving: Let the 2 angles ∠BAC and ∠BAD be coplanar and also let m∠BAC = 80 and m∠BAD = 60. If D is in the interior of ∠BAC, what is m∠CAD?

Table 1 that follows shows how Ganas was used in the lessons.

137

Table 1. Use of Ganas

Lesson Number	Number of Items in the Exercise	Nature of Ganas	Time Start	Time End	Duration m=min, s=sec	Frequency of Ganas
6	1	G2	2:14	4:02	1m 48s	
	3 ⎫	G1	17:57	20:12	2m 15s	4
	3 ⎬		21:49	24:42	2m 53s	
	5	G2	38:51	40:41	1m 50s	
7	None					0
8	3 ⎫	G1	1:21	2:30	1m 9s	2
	9 ⎬		18:15	28:40	10m 25s	
9	2	G1	20:35	24:56	4m 1s	1
10	Test					0
11	None					0
12	5	G2	32:37	33:49	1m 12s	1
				36:27	2m 38s	
13	3	G1	30:59	36:51	5m 52s	1
14	2	G2	12:23	13:34	1m 11s	1
				15:53	3m 19s	
15	Test					0

It can be gleaned from Table 1 that there were lessons where Ganas was given at least once (from 1 to 4 times). Ganas was not used when a test was administered. But there were also lessons where there was no test and yet Ganas was not used. Ganas was utilised a total of 10 times. The number of times G1 and G2 were employed was about the same. Ganas was given at different parts within a lesson such as in Lesson 6 and also across lessons as shown by the start and end times.

The practice exercises differed in the number of items. The time spent for them varied from about 2 minutes to 11 minutes. This means that the items must be answered rather fast. It also implies that the items were not those that required a lot of cognitive processes or high level thinking for they will require more time. The start was when Ms. Santos begun to give instructions and the end was when she asked the class to stop answering. For Lessons 12 and 14, the first end time was when the targeted number of students who were supposed to have first answered all the items correctly had been met.

The practice exercises that Ms. Santos gave were the same for all students regardless of their ability. And she had no provision for what the students who completed them quickly and correctly at that, would do while other students were still working on them.

THE INFLUENCE OF GANAS ON THE LEARNERS

According to Ms. Santos, Ganas had different effects on the students. She cited that the accuracy and speed of some had been enhanced. They had become used to thinking that they had to work as fast as they could and get all answers correct as much as possible. But there were others who still could not cope. When she announced that time was over, they would say "Ma'am wait." She recognised that at least they tried and somewhat answered. There were some others who until then were still forcing themselves to accept that they would participate in Ganas exercises. And there were still others who did not care about Ganas and that they would just answer the practice exercises. In short, among those who had gotten used to the requirements of Ganas, there were those who had satisfied them and those who had not. However, there were those who still had not accepted the requirements of Ganas and those who did not at all bother about participating in it.

Ms. Santos also commented that some felt insecure when the same students stood up and eventually earned extra points whenever there was Ganas. So she made sure that those who were supposed to be banned already, were really banned. Those who had not been banned were happy for this increased their chance to get additional points. But there were also those who were thankful and happy for getting banned.

Besides these accounts by the teacher, there were other influences observed in students such as those on the nature of interaction between the teacher and the students, the behaviour of students during practice exercises with or without Ganas, students' personal objectives in learning the lesson, students' attitude towards Ganas, students' reactions to unmet expectations, students' attitude related to banning, and how students related with each other on mathematical tasks.

Teacher and Students' Interaction Concerning Ganas

For the most part of any lesson without a test, the students mainly listened to Ms. Santos' explanation of the lesson and responded in chorus or individually if called to answer her questions. There were few instances when the students asked Ms. Santos questions and when they explained their answers. A different kind of interaction happened when Ganas was used.

In Lesson 6, Ms. Santos gave a practice exercise. At the end when she asked how many got a perfect score, several students raised their hand. She said that next time if they got a perfect score again, then that would be the time that she would give. Some students interpreted what she referred to that she would give as Ganas, said "no" in chorus, frowned, and with quite loud voices, they waved their hands in protest. They clamoured that she should consider the practice exercise as Ganas because they got high scores. She argued that they were many and added that in the first place, she did not say plus, to which they agreed. After the answers to the next set of exercises were discussed and the students' scores were determined, she asked who got perfect scores in both exercises. She distributed stickers to those who raised their hands. A few students pleaded that she should consider one less than the perfect score. She responded by asking them about what she said before the second exercise was given and their reply settled the matter. Then she announced that later, there would be another Ganas again.

G1 was used in the account in Lesson 6 presented earlier. Apparently, Ms. Santos wanted to give extra points only to those who consistently got high marks perhaps to challenge the students to always do well. Interestingly, the students attempted twice to make Ms. Santos change her mind so that they could earn Ganas points. But she remained firm by reminding them of what she initially told them.

After some more discussions, Ms. Santos announced, "Okay. Last. Perfect score plus 2." She asked if the students wanted it and they said "yes." First she said, 30 seconds. But the students complained, "That's too much." She responded, "It's easy, it's only yes or no." They consented to it and then asked her about what they were supposed to do. And she retorted "It's easy so I won't give you a time limit. So first 10 students."

G2 was used in the case just cited. Just like in the earlier situation mentioned, Ms. Santos and the students seemed to negotiate on whether to have Ganas and on how it should be done.

Student Behaviours With or Without Ganas

To the extent that it could be viewed in the whole class video, it was observed that in all the lessons where Ganas was used, the moment Ms. Santos announced Ganas the students seemed to hurry. The students were all focused - reading and trying to understand what was written on the manila paper, thinking about the answer and writing it on their notebook. No one turned his head to glance or talk with his classmates. Meanwhile, Ms. Santos walked back and forth along the aisle. At times, she stopped to look at the work of some students, and clarified some items with individuals. In Lesson 14, while the students were busy working on the practice exercises, she stated the rule that had to be applied to get the answer. Then she said that she had given them a clue. When the time was over, she told the students to stop writing and exchange notebook with their seatmate. She called on individuals to answer the items. While the answers were discussed and their work were scored, some students smiled or giggled when their answers were correct. Finally, Ms. Santos asked how many got specific scores, then gave a sticker to all who qualified for Ganas points.

In contrast, when the students worked on a practice exercise that was not Ganas, they did not seem to be in a hurry. Instead, they looked relaxed. They would look sideways, talked a bit with their seatmate or students near them, then wrote their answers on their notebook. They were not so focused as when there was Ganas. Such was the case in Lesson 13. But in that same lesson where there was Ganas, they behaved as described earlier. While students sometimes talked with their classmates only after they had completed their work when there was Ganas, they talked with their classmates even when they were still working on the practice exercises when there was no Ganas. So the students behaved differently depending on whether there was Ganas or not.

Students' Personal Objectives in Learning the Lesson

Ganas also influenced the students' personal objectives in learning the lesson. Table 2 below summarises the distribution of the 20 focus students who were interviewed based on what they thought about how good they were in mathematics (best - B, good - G, average - A, below average - BA, or weak - W) and what they expected or hoped to happen in the lesson in relation to whether the lesson for which they were interviewed had Ganas or not.

Four (20%) of the focus students explicitly mentioned getting the Ganas points or plus factor as what they expected to happen in the lesson. All of them were at least of average achievement, with the best student in mathematics included. This means that the better students aimed for the plus factor. Except for one, all the lessons for which they were interviewed had Ganas. A student wishing for a plus factor even when the lesson did not offer one suggested the intensity of her desire to have Ganas points. The answers in (2) might have indirect reference to Ganas while those in (1), (2), (3) and (4) might indicate that the students considered grades as important. And this finding is consistent with the perception of the teacher about what students value, cited earlier. The students in (5) were intrinsically motivated because despite the administration of Ganas in the lessons for which they were interviewed, what they wanted was to learn. They were in contrast with the students in (1), (2), (3) and perhaps (4) who were extrinsically motivated by high grades.

Table 2. Personal objectives of the students in learning the lesson

Objectives	Lesson		Number of Students
	With Ganas	*Without Ganas*	
(1) To answer Ganas or get plus factor	G, A, B	A	4
(2) To answer all the teacher's questions, to get perfect in all activities, to have things get better	G, 2A	BA	4
(3) To get high score in the test	0	2G	2
(4) To get correct answers to difficult test items	0	1A	1
(5) To learn, to understand the lesson	2G	0	2
(6) To answer the interview questions	W	0	1
(7) None	2A, BA	3A	6

The objective of Michaela, the student who had the highest grade in the mathematics class was to get the plus factor. Mathematics being her favourite subject, she answered the items in "Extra for Experts" in the textbook even without being assigned. She found them challenging and she liked them. Because of her high grade, she qualified to be a member of the Mathematics Club. She also attended the Program of Experts of Mathematics that was intended only for those who excelled in mathematics. Her objective may be understood in light of her account of her previous experience and belief concerning her performance in mathematics.

> ...maybe because of my experience in the elementary when I was dumb in Math. I said, I'll reach the top. So what I can do, I will really do so that I'll be the highest in Math. Because if you're the highest in Math, as if you're really smart because almost all students do not like Math.

Michaela's view matched what Scwartz (as cited in Gilroy, 2002) commented that if students are not good in mathematics, they feel a bigger sense of failure and they believe that they are not smart because society emphasises so much that mathematics is an intelligence indicator. Her view was also shared by a classmate, Kath who thought that a very good student in mathematics was one who when the teacher asked a question only he/she could answer.

The prestige or recognition that was associated with being the best in Mathematics motivated Michaela to do well in it. And for her, Ganas provided the opportunity for obtaining the desired recognition. Apparently, the extrinsic motivation was effective to the extent that it was consistent with the objective of a person.

Students' Attitude Towards Ganas

Ganas elicited different attitudes. In Lesson 7, after Ms. Santos announced "Ganas", Emil retorted "There is Ganas but I could get nothing." Possibly, he meant that it did not make any difference to him whether there was Ganas or not. He did not benefit from it anyway. According to Michaela, Emil was banned during the first grading period. In the interview after Lesson 8, Jenny commented that she did not always enjoy mathematics because "Sometimes the number of items is 1 to 5. Sometimes my score is 4 only. I could not get 5. It's only when you get 5 that you get a plus factor. Others were already getting 7. We still didn't have any." Both students expressed frustration and helplessness. In the case of Emil, it might have been because of his previous experience and also perhaps of his present situation while for Jenny it was because she could not satisfy the requirements of Ganas.

Kath in Lesson 6 when she got one less than the perfect score but additional points were given only to those who got a perfect score said "Disgusting. Huh...huh. I lost interest again. This one caused me to lose interest again." So what was supposed to motivate or stimulate interest actually had an opposite effect. Later, when another Ganas was given she resolved, "I'll make up. I'll make up." When she was not among those who first stood up, she said, "I lost again."

However, there were also students who responded positively to Ganas. According to Ren in Lesson 6 "we become more enthusiastic because we want to get it (plus factor). Her giving plus factor makes us more enthusiastic in solving problems."

As captured by the teacher's video camera, some students anticipated every practice exercise as Ganas. After Ms. Santos had done an exposition, they would automatically ask "Ma'am Ganas?" when she told them to get their notebooks. However, Ms. Santos wanted that they work equally well in all practice exercises that was why she asked, "will you only work when it is Ganas?"

Reactions to Unmet Expectations

Four focus students identified getting the plus factor as what they hoped or expected to happen during the lesson. But none of them got the additional points that they aimed for. In one case, a quiz was given instead of a Ganas exercise. In the other three, the students did not get the additional points although Ganas exercises were given. Nevertheless, each one said that it was alright.

Ana justified that "those who got a plus factor solved fast. If I'm in the mood, I solve fast. But not today, I'm tense." She said that anyway, she already had four plus factors last time. Apparently, Ana thought that she was still in control and her past performance supported this. So her not getting what she wanted was no cause for alarm.

Elena thought that there was no Ganas but in fact there was. She only got two correct answers out of three. She rationalised "it's okay. I didn't understand only one. Now I already understand it. Even if I didn't get Ganas, as long as I understood all of her lesson. At least for the test, I can answer. "

Michaela, could not get a plus factor even if there was Ganas because she was banned. In the interview, she said that it was alright because a student whom she regarded as more analytical than she, was also banned in the previous grading period so she was not the only one banned. But even if she was not banned, she was not among the first five students who stood up because "others were very fast. The problem was hard to explain. I could not picture out the figure so I found it hard to answer. But when I was already solving it, it was easy." Actually she got the correct answer. She expressed a feeling of loss for the additional points she missed.

Students' Attitude Towards Banning

Banning also elicited different attitudes from those who were banned and those who had not yet earned enough plus factors to be banned. Michaela did not want to be banned perhaps because she still wanted to earn points for which she was very capable of doing. In Lesson 8 when Ms. Santos informed her that she was banned, she exclaimed "unfair!" Ms. Santos retorted that it was not unfair and that she should learn how to give chance to others. In Lesson 14, when Michaela was interviewed, she said that it was alright even when she was banned. In a sense, she must have realised that it was not really unfair for Emil whom she considered better than her and so was also capable of getting extra points likewise experienced being banned. Those who heard what Ms. Santos uttered when Michaela exclaimed unfair said in chorus "Yes. We're pitiful. You should give to us." This was an admission that unless the better students give way, others have no means to get Ganas points also.

143

Still in Lesson 8, in contrast Kris was happy when Ms. Santos announced that he was also banned. Perhaps this was because he would temporarily not have to compete. The teacher asked, "Why, when you are banned, won't you study anymore?"

How Students Related With Each Other on Mathematical Tasks

Six out of the 20 focus students asked their brighter classmates to teach them when they did not understand the lesson or when they missed a class. In fact, Myrna claimed that she enjoyed their mathematics class because they helped each other. When they would have a test and some of them were confused, they would ask Michaela to give them examples for clarification. Elena also asked Michaela for help and she in turn was asked by her seatmate to teach her. So in this sense, there was sharing in the class.

A different atmosphere was created when there was Ganas. It involved students in close monitoring of the plus factors others had already obtained. Comparisons became inevitable due to competitions as exemplified in the excerpt below from Lesson 13. Ms. Santos asked the students to pass their papers to her according to the scores she called out.

```
43:28 Elena:    It's such a loss. I wasn't able to answer a while
                ago.
43:30 Tina:     Just one. I only answered one a while ago. Three is
                finished. We got two.
43:43 Elena:    Vida, Vida do you already have? Vida already has.
                How many did you already get? Two? You only got one
                so far? Me, I also got one already… What did we
                pass? Is it Ganas? It's not Ganas, right?
```

Key to symbols used in transcripts in this chapter
...   A short pause of one second or less.

The students were also vigilant in making sure that those who had to be banned were really disqualified. Obviously, this could increase their chance of getting extra points. In the portion of Lesson 14 presented below, Ana asked a seatmate how many plus factors she already had. She said that she would inform the teacher that the seatmate should also be banned. She accused the seatmate of claiming fewer points than she actually had. Michaela who was banned was also conscious of the plus factors of the seatmate.

```
16:59 Ana:       Six…six. You're cheating. Ma'am will also ban you,
                 I'll tell Ma'am to ban you also.
17:10 Michaela:  She has four only, hasn't she? Only four?
17:15 Michaela:  Yes. She's cheating. It looks like she already has
                 five.
```

So in Ganas exercises particularly G2, the students regarded each other as competitors. Having Ganas created an atmosphere of competition in the class.

144

DISCUSSIONS AND IMPLICATIONS

Ganas had become the preoccupation of some students. More than 50% of the focus students claimed that their objective in learning the lesson was related to getting high grades which Ganas could make possible. There was a student who had become enthusiastic in solving practice exercises because of the extra points that Ganas could give. Although the teacher had no prior announcement that the practice exercises that she would give were Ganas exercises, students tried to interpret that whenever she asked them to get their notebooks, this meant that they would do Ganas. And whenever they got high scores, they asked the teacher to consider the practice exercises as Ganas so that they could earn Ganas points. The goal of all these students in engaging in the mathematical task of doing practice exercises was to obtain an extrinsic incentive. Hence, they could be characterised as having a performance goal and were extrinsically motivated. Apparently, grade consciousness was pervasive in the class.

The Influence of Ganas as a Competitive Task

Ganas was at times a competition under time pressure. As such it kept the students focused when doing practice exercises. They appeared to attend seriously to the task. As a result of Ganas, the speed and accuracy of some were enhanced. Still some were slow but they attempted to work faster. These may be viewed as positive influences of Ganas. But once it was not given when it was expected, there appeared to be possible negative influences.

Almost every time that the teacher asked the students to bring out their notebook, they asked her if there was Ganas. To this, she responded, "will you only work when it is Ganas?" When she announced the names of students who were banned and they were glad about it, she also asked them "Why, when you are banned, won't you study anymore?" Both questions were expressions of her concern that the students might not work on the practice exercises anymore nor take them seriously if there was no Ganas or they could no longer compete for Ganas points. The brighter students were happy that they were banned perhaps because they were relieved from the pressure of competing. It was possible also that they no longer performed as well as they did when they were not yet banned. They were merely contented to get high scores without working fast enough even if they could for anyway, they had nothing to do while waiting for the rest of the class to finish. These student behaviours could be indications that Ganas was unable to maintain the motivation to learn. If so, this finding is consistent with what research had substantiated that extrinsic motivation was ineffective in sustaining motivation to learn (Condry & Chambers, as cited in Nicholls, 1983; Lepper & Greene, as cited in Nicholls, 1983; Maehr, as cited in Nicholls, 1983; Francis, 2003).

As a competition that required a combination of speed and accuracy, those who qualified for Ganas points, were mostly the brighter students. As such, some less able students had come to believe that unless the bright ones were prevented from participating in Ganas exercises, then they could not avail of what Ganas offered. They did not possess enough of what it took to qualify for Ganas points. An instance when they expressed this feeling of helplessness was when the best student in the mathematics class exclaimed "unfair" when the teacher announced that she was banned. The teacher countered that she should give a chance to others to which some of the students added, "Yes. We're pitiful. You should give to us." Another example that showed this negative feeling of helplessness was the teacher's account that they felt insecure when the same students were able to avail of the Ganas points. Moreover, it was possible that earning Ganas points had been elusive for some students. They could be some of those who did not bother to participate in the competition but might have answered the exercises just the same, as the teacher had observed. They, too, might have developed learned helplessness (Slavin, 1997; Middleton & Spanias, 1999). Still another focus student declared that she did not like mathematics when she did not get Ganas points. She complained that those who got perfect scores still got additional points because of Ganas while she whose score was just one less than the perfect score got no additional point. Her complaint expressed both her frustration and helplessness. Possibly she viewed that her ability could not match the demands of the mathematical task and more so that it was given in a competitive context. Because of this, she disliked mathematics. All these accounts are consistent with research findings that (1) competitive learning environments reinforce students' belief that success is attributable to ability, (2) consistently experiencing non-success in competitive situations leads to learned helplessness, and (3) weak students get discouraged when they are in such competitive conditions (Ames, as cited in Slavin, 1997; Ames as cited in Nicholls, 1983; Jagacinski & Nicholls, as cited in Nicholls, 1983).

Even the brighter students also had negative feelings relative to Ganas. One lost interest when after attempting to do better next time, she consistently could not get Ganas points in a lesson. Another felt a loss when she did not earn Ganas points because she was slow although her answers were all correct. One was frustrated because even when he was able to satisfy the requirements of Ganas, he was banned.

There was a natural tendency for students to help one another when they did not understand something. The weaker ones sought help from the brighter ones. The latter provided help possibly because by doing so they demonstrated that the providers were better than the recipients. It suited their performance goals. But this cooperative tendency was transformed into a competitive stance when Ganas was used. Those who were banned were on the lookout if others who were also supposed to be banned were really banned while the others compared their scores with fellow competitors. Apparently, the main concern of the students was how they fared compared to the others rather than on how they improved in mathematical knowledge and skills.

Alternative Teaching Strategies

In almost all the lessons, the students passively listened to the teacher's exposition and responded to her questions. As such, there were very rare opportunities for them to experience how to explore mathematical ideas on their own and to discuss these with others. This may explain why they did not have the inherent desire to work on the practice exercises. They were not used to thinking and working on their own. Using a variety of teaching strategies particularly those that involve students in exploring, generating, and communicating mathematical ideas can train them to be independent learners. Then during times of practice exercises when they are left to themselves, they will still be interested to work for they are already used to the independence that these required (Cobb et al., 1991; Schiefele & Csikszentmihalyi, 1995; House, 2003). They will do them not for the extra points but for the extra learning that they may get from them. The inherent enjoyment that is derived from discovering mathematical ideas rather than getting them simply by listening to the teacher can sustain students' interest even in practice exercises (House, 2003). Ironically, though they were mainly passive in learning, some students were very aggressive in insisting that the teacher consider the practice exercises where they got high scores to be Ganas exercises. Such a behaviour indicated that the students were really very concerned with getting high grades.

Using small group work may have merit in this class of 57 students since they were inclined to discuss with others when they needed help in learning. Weaker students may particularly benefit in cooperative rather than competitive classroom settings (Nicholls, 1983; Middleton & Spanias, 1999).

The practice exercise items need to present appropriate challenge to students of various ability levels otherwise the bright ones will find them boring and the weak ones will find them overwhelming. These items should include those that require higher order thinking skills so that they can deepen students' conceptual understanding and enhance their problem solving skills (Meyer, Turner, & Spencer, 1997). In Ganas exercises, students who finished ahead of the others were idle while those who were not as fast were busy. So the time that the fast learners saved for being able to work quickly was just wasted. Extra problems may be made available for them to solve so that precious time can be used to their advantage, that is, to further enhance their mathematical knowledge and skills. All students should take a common set of practice exercises that are reasonably within the ability level of the weak students and yet sufficiently challenging for them also. However, there should also be more challenging problems intended for the better students or fast learners only. In this scheme, no group will be disadvantaged. Everyone will grow or improve in knowledge and skills in relation to one's previous performance and not that of others. So there is no need to have competitions where the success of one deprives others of success, too.

CONCLUSIONS

This study has shown that a teacher used an extrinsic incentive, high grade through Ganas, to motivate her students to do practice exercises. This decision was based on her belief that the students value grades and on her experiences as a mathematics contestants' trainer. Ganas influenced the students in different ways. They engaged in the learning task not primarily for the opportunity that it offered for them to master and deepen their understanding of the mathematics lesson but to get a higher grade. So they viewed learning as a means to obtain an extrinsic reward. When Ganas was given as a competition, the weak students were disadvantaged for they could not cope with its demands of accuracy and speed. They felt frustrated and helpless especially as they compared themselves with the others who were more able. Hence, the very means which the teacher thought could motivate students to master mathematics could become a hindrance to their learning because of negative feelings (Schiefele & Csikszentmihalyi, 1995). Though they more often qualified for Ganas points, the brighter students were also disadvantaged because there were no provisions for them to do something more to further improve their knowledge and skills once they finished early or when they were already banned. While the more able students felt unfair that they were prevented from competing, the less able ones also felt unfair that their different abilities did not allow them to compete fairly.

The mechanism of Ganas to a certain extent sorted students based on ability. Those who were weak had come to accept their relatively low standing compared with the more able ones that they thought that without the latter ones being banned, they have no hope of getting higher grades. Bearing in mind the students who were extrinsically motivated, in the language of Clarke (2001), Ganas provided mutual affordances to both these students and the teacher. They both achieved their objectives. But considering those who were similarly motivated but whose achievement levels were not high, Ganas led to a frustrating experience. Though less obvious, Ganas also disadvantaged the better students for though they got the grades they aimed for, opportunities for them to learn more while others were still engaged in the practice exercises were not provided. Rather than using Ganas which is based on competition, it would be better to use motivational strategies that heavily involved students in their own learning so that their main concern will not be to outscore others but to progressively outlearn oneself.

REFERENCES

Clarke, D. (2001). Complementary accounts methodology. In D. Clarke (Ed.), *Perspectives on practice and meaning in mathematics and science classrooms* (pp. 13-32). Dordrecht: Kluwer Academic Publishers.

Cobb, P., Wood, T., Yackel, E., Nicholls, J., Wheatley, G., Trigatti, B., & Perlwitz, M. (1991). Assessment of a problem-centered second-grade mathematics project. *Journal for Research in Mathematics Education. 22*(1), 3-29.

Deitte, J., & Howe, R. M. (2003). Motivating students to study mathematics. *The Mathematics Teacher, 96*(4), 278-280.

Francis, J. (2003). *A study of motivation and engagement in numeracy.* Retrieved 24 July, 2003 from the World Wide Web: http://www.gu.edu.au/school/eps/ mteach-abstracts.html.

Gallos, F. (1998). *Evaluating action research as a model for school-based professional development of secondary mathematics teachers in the Philippines.* Unpublished doctoral thesis, Edith Cowan University: Australia.

Gilroy, M. (2002). Waking up students' math/science attitudes and achievement. *The Education Digest, 68*(4), 39-44.

High School Mathematics Education Group. (1996). *Report on the assessment of training needs of high school mathematics teachers.* Quezon City: National Institute for Science and Mathematics Education Development.

House, J. D. (2003). The motivational effects of specific instructional strategies and computer use for mathematics learning in Japan: Findings from the Third International Mathematics and Science Study (TIMSS). *International Journal of Instructional Media, 30*(1), 77-96.

Meyer, D., Turner, J., & Spencer, C. (1997). Challenge in a mathematics classroom: Students' motivation and strategies in project-based learning. *The Elementary School Journal, 97*(5), 501-521.

Middleton, J. (1995). A study of intrinsic motivation in the mathematics classroom: a personal constructs approach. *Journal for Research in Mathematics Education, 26*(3), 254-279.

Middleton, J. & Spanias, P. (1999). Motivation for achievement in mathematics: Findings, generalizations, and criticisms of the research. *Journal for Research in Mathematics Education, 30*(1), 65-88.

Nicholls, J. (1983). Conceptions of ability and achievement motivation: A theory and its implications for education. In S. Paris, G. Olson, & H. Stevenson (Eds.), *Learning and motivation in the classroom* (pp. 211-237). New Jersey: Lawrence Erlbaum Associates, Inc.

Schiefele, U., & Csikszentmihalyi, M. (1995). Motivation and ability as factors in mathematics experience and achievement. *Journal for Research in Mathematics Education*, 26(2), 163-181.

Slavin, R. (1997). *Educational psychology.* Boston: Allyn and Bacon.

Sobel, M., & Maletsky, E. (1988). *Teaching mathematics.* New Jersey: Prentice Hall.

Soledad Asuncion Ulep
National Institute for Science and Mathematics Education Development
University of the Philippines
The Philippines

149

CHAPTER TEN

Case Studies of Singapore Secondary Mathematics Classrooms:
The Instructional Approaches of Two Teachers

INTRODUCTION

Research on instructional approaches in terms of the classroom practices of Singapore mathematics teachers in the past has been relatively sporadic. Most of these studies were done in primary classrooms (e.g. Chang, Kaur, Koay & Lee, 2001 and Kay, 2003). One notable exception, which looked at the ways mathematics was taught in secondary classrooms, was the Kassel project (started in 1993 by England, Scotland and Germany) in which Singapore participated in 1995 for two years (Kaur and Yap, 1998). In this study, observations of 43 lessons involving Grades 8 and 9 classrooms suggested that the main teaching model appeared to be fairly traditional with teacher exposition as the main teaching approach and student activities confined mainly to individual seatwork.

More recently, with the setting up of the Centre for Research in Pedagogy and Practice (CRPP) at the National Institute of Education (NIE) in Singapore, there is an increasingly concerted and coordinated effort to collect data on the Singapore school system including its pedagogical practices. As part of the CRPP Core Research Program (Luke, Freebody, Lau & Gopinathan, 2005), Yeo and Zhu (2005) reported the classroom coding of 118 mathematics lessons, half each from Grade 5 and Grade 9 classes. Their analysis suggested that mathematics teaching in these classes was predominantly teacher-centred with emphasis on the memorisation and acquisition of basic concepts and routine procedural skills. They concluded that there was much room for the integration of higher-order thinking skills in mathematics teaching in Singapore.

While these studies are useful in providing us with the general features of mathematics instruction in Singapore classrooms, what is lacking is the documentation of details of the teaching practices at the micro level. For example, one of the previous studies mentioned above (Kassel project) noted that lessons were "highly structured" (Kaur, 1999) but there was no discussion on how the lessons were structured. Similarly, the study noted that teachers "presented knowledge to the pupils as a class by telling and explaining" (Kaur, 1999, p. 195), but there was little information on what the telling and explaining involved. This

D. J. Clarke, C. Keitel & Y. Shimizu (Eds.), Mathematics Classrooms In Twelve Countries: The Insider's Perspective. 151–165. © *2006 Sense Publishers. All rights reserved.*

chapter, through an exploratory analysis of two case studies, attempts to explicate the instructional approaches of the two teachers in two ways: (i) by showing how mathematical lessons were 'structured' in terms of the instructional activities that constitute the lessons and (ii) by describing the nature of the classroom talk in terms of the general content and interactional pattern. With just two case studies as the backdrop, it is not our intention to generalise such approaches across mathematics teaching in Singapore. Rather, we aim to provide a more in-depth analysis of the instructional approaches of Singapore mathematics teachers than attempted by previous studies.

Much research has been carried out on the activity structures of lessons (e.g., Mesiti, Clarke & Lobato, 2003; Shimizu, 2002; Stigler & Hiebert, 1997; Stodolsky, 1988). Understanding teaching processes in terms of the kinds of instructional activities and the ways these are sequenced in classrooms of different countries has been argued to be important for finding ways to improve students' learning and for developing sound education policies (Stigler & Hiebert, 1997). Besides instructional activities, sociocultural theories of learning also emphasise the role of social interaction as a mediator of students' learning (e.g., Vygotsky, 1978). Analysing the talk that takes place during the instructional activities is therefore important in understanding the kind of learning that is afforded to the students. How does classroom talk mediate learning? One perspective is to see "talk as social action" (Mercer, 1995, p. 67). Like all kinds of talk, it is through classroom talk that teachers strive to achieve their objectives (ibid.). The aims and purposes of the instructional activities that constitute the lessons can therefore be illuminated by examining the classroom talk. We hope, by identifying more features of the instructional approaches of Singapore mathematics teachers than previously attempted, our study would contribute to further examination of the potential successes and pitfalls in the teaching of mathematics in Singapore in relation to not just the intended curriculum but also in preparing our students for the global knowledge-based economy (see for example Gopinathan, 1999).

METHOD

Data

From the corpus of data collected in accordance with the protocol set out in the Learner's Perspective Study (Clarke, 2002a; 2002b), the video records of two ten-lesson sequences were the main source of data analysed. The sequences of lessons enable us to analyse for the overarching features of the lesson structure within a sequence rather than for the lesson script (Stigler and Hiebert, 1997) of individual lessons, which may vary depending on at which stage the lesson was within a unit (Shimizu, 2002). As the data collected from the third classroom have yet to be processed at the time of writing, analysis was confined to the data collected from the first two classrooms.

Both are Secondary Two Express (Grade 8) classes with a size of 37 for the class in School 1 (SG1) and 40 for the class in School 2 (SG2). Both teacher participants were females with many years of teaching experience (SG1: 21 years and SG2: 27 years). All lessons in both classrooms were about an hour each. The topics covered during the lesson sequences were Arithmetic Problems and Standard Form, Congruence and Similarity, Scales and Maps in SG1 and Factorisation of Algebraic Expressions, Algebraic Manipulation and Formulae and Simultaneous Equations in SG2.

Data Analysis

The constant comparison method (also known as the grounded theory approach) (Glaser & Strauss, 1967) was used to identify for activity segments that characterised the two lesson sequences that were analysed. As an exploratory study, activity segments – "the major division of the lessons", served as an appropriate unit of analysis for examining the structural patterns of lessons since it allows us "to describe the classroom activity as a whole" (Stodolsky, 1988, p.11). According to Stodolsky:

> In essence, an activity segment is a part of a lesson that has a focus or concern and starts and stops. A segment has a particular instructional format, participants, materials, and behavioral expectations and goals. It occupies a certain block of time in a lesson and occurs in a fixed physical setting. A segment's focus can be instructional or managerial (Stodolsky, 1988, p. 11).

For the purpose of this chapter, the activity segments were distinguished mainly by the instructional format that characterised them, although there exist other segment properties such as materials that differed among the various activity segments identified. Six categories of activity segments emerged through reiterative viewing of the video data. These mutually-exclusive segments were found to be able to account for most part of the 20 lessons from SG1 and SG2. Three categories were prominent in both classrooms. They are:
- whole-class demonstration [D], characterised by whole-class mathematics instruction that aimed to develop students' understanding of mathematical concepts and skills;
- seatwork [S] during which students were assigned questions to work on either individually or in groups at their desks;
- whole class review of student work [R], during which the teacher's primary focus was to review the work done by students or the task assigned to them.
The three other categories, two of which were exclusive to a single classroom, are:
- group quiz [Q] found only in SG2, during which a representative from each group (usually four groups at a time) was asked to compete with other representatives to solve a problem on the board;
- test [T] found only in SG1; and
- miscellaneous [M], a catch-all category during which the class was involved in mainly managerial and administrative activities.
Such a coding scheme closely corresponded with another study on Singapore primary mathematics classrooms which had also identified similar activity

segments (Ho & Hedberg, 2005). To code for the activity segments, *Studiocode*[TM], a video coding software was used which allowed us to see the patterns in which the various segments were sequenced in the lessons.

As suggested by previous studies on lesson structures, which have shown that the structure of lessons varied from lesson to lesson (Mesiti et al., 2003) and depended on the stage at which the lesson was located within a teaching unit made up of a lesson sequence of related content (Shimizu, 2002), the mathematical content is an important factor that accounts for the structural variations of lessons. As such, further analysis was carried out to identify the instructional objective that each segment was concerned with, based on the mathematical content that was covered. Table 1 below shows an abridged version of the analysis that combined both the lesson structure and the mathematical content.

Table 1: Analysis of lesson structure with mathematical content

Lesson no.	Activity segment code	Mathematical content	Instructional objective	Instructional cycle no.
1	[S]	Practice task: $2x + 4y - 3(x + 2y)^2$	Factorisation by grouping	1
1	[R]	Student wrote answers for practice task on board		
1	[D]	Worked examples: $x^2 - 9$, $y^2 - 1/16$, $9y^2 - 4z^2$	Factorisation of expression in the form of difference of two squares	2
1	[S]	Practice tasks: $a^2x^2 - 16y^2$, $50x^2 - 2p^2$		
1	[R]	Teacher and students worked out practice tasks on board		

FINDINGS

Pattern and Structure of Lessons

Coding of the video data revealed a fairly consistent pattern in the ways the activity segments were sequenced throughout each lesson sequence. Figure 1 shows the segment sequence for the 20 lessons from SG1 and SG2. Activity segments that served different instructional objectives were separated by a dotted vertical line. The most typical pattern (best exemplified in SG1-L01 and SG2-L02) found in both lesson sequences was composed of the three consecutive segments, whole class demonstration [D], seatwork [S] and whole class review of student work [R]. These three segments [DSR] together make up what we called an instructional cycle, one in which the mathematical problems used as worked examples during demonstration or as tasks for seatwork within the same cycle shared the same instructional objective. Each lesson was constituted of a varying number of instructional cycles.

Besides [DSR], there were other combinations of segments that constituted a cycle. In some cycles of SG1, the typical round of demonstration, seatwork and review [DSR] was followed by another one or two rounds of seatwork and review ([DSRSR] or [DSRSRSR]). In others but less typical ones, the instructional cycles were made up of a standalone segment (typically [R] as in the review of homework), or other combinations of segments (such as [DS], [DR], [SRS] or [SRSR]).

Similar patterns existed in SG2. The most distinctive difference was however the presence of the segment identified as group quiz [Q]. This particular segment gave greater variations to the composition of the instructional cycles present in SG2 though the most typical one was still [DSR]. Examples of combinations of segments found in SG2 include [DQ], [DS], [SR], [DRSR], [DSRS], and [DSRSRQ].

The instructional cycle in both classrooms shared several similar features. Firstly, it was not necessarily completed within one lesson but could be resumed in the next lesson. In one instance in SG1 and six instances in SG2 (shaded regions of Figure 1), the same instructional cycle continued into the following lesson. Another similar feature was that while the mathematical questions within the same instructional cycle were related to one another, they could also be related to the objective focused on during the demonstration segment in the preceding cycle. In both SG1 and SG2, the lessons invariably began and ended with the miscellaneous segments, moments which were characterised by the obligatory greetings ('Good morning/afternoon, class/[name of teacher]' and 'Thank you, class/[name of teacher]') and instructions for homework collection and other administrative matters.

The instructional models of both teachers can be described as cyclic because of the repetitive nature of the instructional cycles, and incremental as the mathematical knowledge was built up from one cycle to the next. Each cycle focused on a single instructional objective, or part of an objective. Such an instructional model, characterised by its highly episodic structure with each self-contained instructional cycle making up an episode, tended to mask the developmental nature of the spiral curriculum (cf. Alexander, 2000).

Besides the difference in the compositions of the instructional cycles between SG1 and SG2, another difference was the frequency with which the segments changed. The duration of the activity segments and hence instructional cycles in SG1 were generally shorter than in SG2 with the concomitant increase in the frequency of each category of segments (except for group quiz and miscellaneous) (see Table 2).

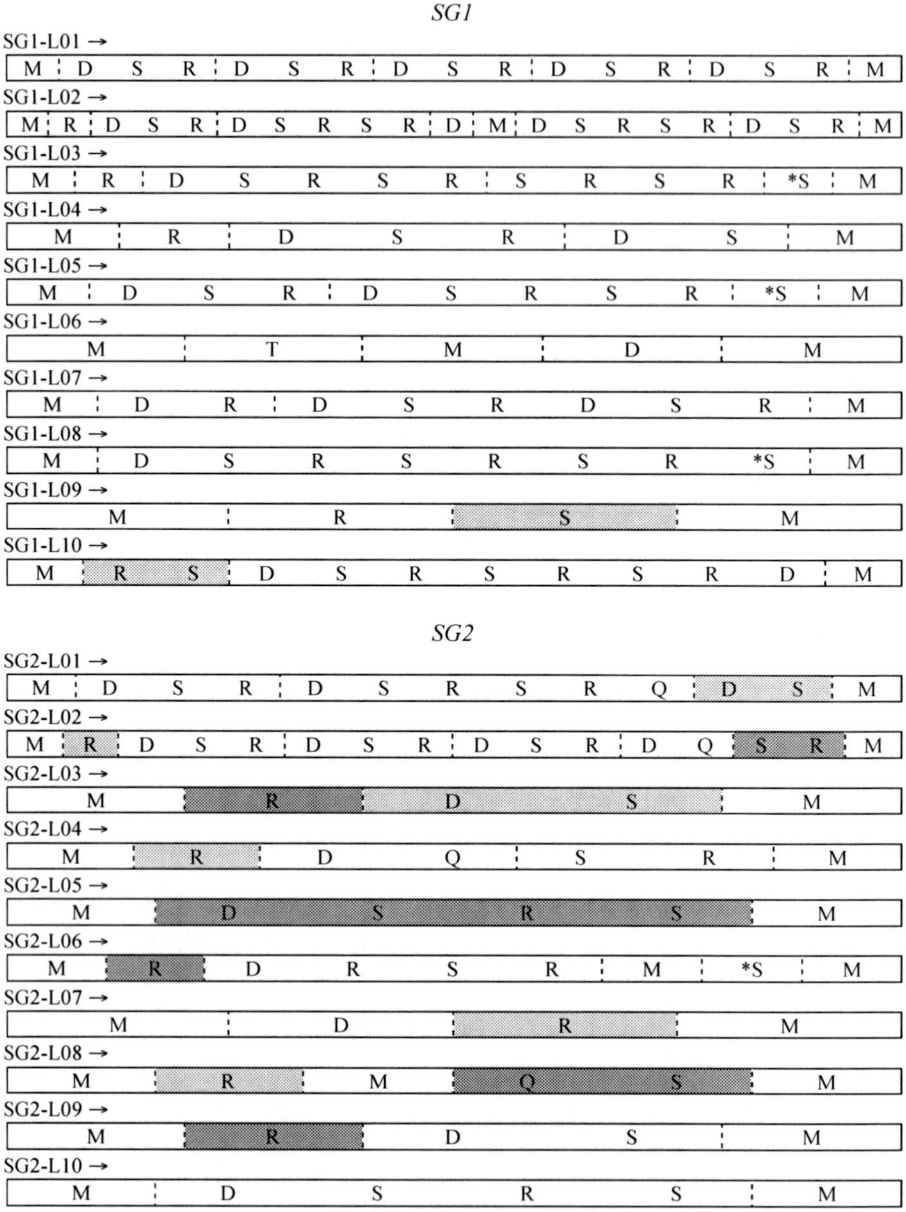

SG1

SG1-L01 →
| M | D | S | R | D | S | R | D | S | R | D | S | R | D | S | R | M |

SG1-L02 →
| M | R | D | S | R | D | S | R | S | R | D | M | D | S | R | S | R | D | S | R | M |

SG1-L03 →
| M | R | D | S | R | S | R | S | R | S | R | *S | M |

SG1-L04 →
| M | R | D | S | R | D | S | M |

SG1-L05 →
| M | D | S | R | D | S | R | S | R | *S | M |

SG1-L06 →
| M | T | M | D | M |

SG1-L07 →
| M | D | R | D | S | R | D | S | R | M |

SG1-L08 →
| M | D | S | R | S | R | S | R | *S | M |

SG1-L09 →
| M | R | S | M |

SG1-L10 →
| M | R | S | D | S | R | S | R | S | R | D | M |

SG2

SG2-L01 →
| M | D | S | R | D | S | R | S | R | Q | D | S | M |

SG2-L02 →
| M | R | D | S | R | D | S | R | D | S | R | D | Q | S | R | M |

SG2-L03 →
| M | R | D | S | M |

SG2-L04 →
| M | R | D | Q | S | R | M |

SG2-L05 →
| M | D | S | R | S | M |

SG2-L06 →
| M | R | D | R | S | R | M | *S | M |

SG2-L07 →
| M | D | R | M |

SG2-L08 →
| M | R | M | Q | S | M |

SG2-L09 →
| M | R | D | S | M |

SG2-L10 →
| M | D | S | R | S | M |

⋮ Represents the border between instructional cycles   * Time-filler
Shaded regions represent the same cycles across adjacent lessons.
Note: The lengths of segments do not reflect their duration within the lesson.

Figure 1: Structural patterns of SG1 and SG2 lesson sequences

Table 2: Duration of lesson segments

Activity Segments	SG1		SG2	
	Total duration	Total no. of instances	Total duration	Total no. of instances
Demonstration[D]	1 hr 51 min	22	2 hr 09 min	14
Seatwork [S]	3 hr 37 min	33	2 hr 33 min	18
Review of student work [R]	2 hr 15 min	33	2 hr 23 min	19
Group Quiz [Q]	-	-	1 hr 15 min	4
Test [T]	38 min	1	-	-
Miscellaneous [M]	32 min	22	34 min	22

Unlike the teacher in SG1 who usually restricted each instructional cycle to one type of questions, the teacher in SG2 tended to include several more types. Despite these differences, the total time taken for each segment in both classrooms over the ten-lesson sequence was fairly similar particularly with the demonstration [D], review of student work [R] and miscellaneous segments [M]. Although the total time taken for seatwork differed quite significantly, this difference was somewhat insignificant if this segment was combined with the group quiz segment, both of which can be considered as different forms of student activity.

Nature of Classroom Talk

To understand the instructional approaches further, we need to go beyond structural patterns of the lesson sequence. The following section describes the key features of the classroom talk through which the teachers realised their roles in not just the teaching of mathematics but also in engaging students to learn it. Given the exploratory nature of the study, description of the classrooms' discursive practices in this chapter, specifically that between teacher and students, is confined to the main themes that emerged from the video analysis and complemented by focused transcription of relevant episodes and field notes recorded during classroom observations. As the content focus and interactional pattern of the classroom talk were observed to be different across categories of activity segments, the following section describes the distinctive characteristics found in each category except those found in test and miscellaneous segments where the talk tended to be minimal or non-instructional.

Whole-class demonstration Occupying more than one fifth of the total class time in both classrooms, these were the segments during which both teachers played the most active role in expounding mathematical concepts and problem-solving skills mainly through the use of examples, both in the form of concepts and mathematical problems, as their teaching tool. The examples used were often carefully selected on the basis of systematic increase in complexity.

The discourse patterns adopted by both teachers appeared to be very similar. The most common interaction pattern in both classrooms was the initiation-response-feedback (IRF) discourse format (Sinclair & Coulthard, 1992). That is, the teacher asked a question, students responded and teacher provided feedback. Both teachers seldom nominated specific students to answer their questions. Instead students responded spontaneously to the teachers' frequent questioning. These questions could be those that elicited from students direct computational answers for the various steps during problem solving, or those that led students to think about what the next step of solving the mathematical problem should be. This is illustrated by the example below when the teacher in SG2 introduced the topic on manipulation of algebraic formulae:

SG2-L06

T:	So part A ah, given to you a formula and you have to make the letter in the brackets to be the subject. Instead of X to be the subject, you have to be A as the subject. That means A on the left side of the equal sign. So the rest will be on the right side of the equal sign, okay …[10s] What should you do? I want the A to be on the?
S:	Left.
T:	The left hand side of the equal sign. That means A must be the subject. A must be here. So what should you do from this given formula here? What's the first step that you will do?
S:	Times, times three
S:	Times three
T:	Times three. Where?
S:	Both sides.
T:	Both sides. Both sides times three …[7s] And then? And then? What should you do after you times both sides by three? Both sides of the equal sign by three. And then what should you do next?
S:	(Equals) three X. Three X (equals) ( )
T:	Three X
S:	Equals (to) two A plus fifteen Z
T:	Equals to?
S:	Two A
T:	That means you expand this one right?
S:	Yeah
. . . .	
T:	And then the next step?
S:	Divide ( )
T:	Yes, divide throughout by two. …[8s] So from X as the subject, now become A as the subject. So this is the answer. Okay?

Key to symbols used in transcripts in this chapter
//	Indicates the beginning of simultaneous speech.
…	A short pause of three seconds or less.
…[time]	Pause of more than three seconds
( )	Indecipherable utterance.
(text)	'Best guess' for speech that was difficult to hear.
[text]	Annotations, comments by transcriber or author, or descriptions of non-verbal actions.
….	Indicates that some words have been omitted from the transcript.

The discourses, which both teachers built around the examples, were different, particularly on the kinds of connections they made. In particular, the teacher in SG1 tended to select daily life examples that were related to the mathematical concepts and used them to show the relevance to students' everyday experiences. In the ten lessons analysed, there were six lessons in which the teacher related the mathematical content to everyday life examples. An example was when the teacher introduced the topic on congruency:

SG1-L04

T: Okay, some of you may be wondering why learn this ah? I mean
        why why learn this in mathematics. Actually congruency is
        important in our life. Alright, in fact it plays a role in
        your daily life. For example, let's say this windowpane ah,
        this glass is broken. Then you have to replace it right
        because the room is aircon right, you have to replace it. Now
        when you replace it, you must get exactly the same shape and
        size as the old broken piece, do you agree? So the old broken
        piece and the new one are actually congruent. Okay they're
        actually congruent. But when you talk about in daily life we
        don't say they're congruent. We say oh you must go and buy one
        of the same shape and size ah. You must replace one with the
        same shape and size.

On the other hand, such everyday examples were absent in SG2. Instead, the teacher tended to identify distinctive features of the mathematical problems and compare them with those that students had encountered previously as illustrated below:

SG2-L09

T: Simultaneous equations that mean you solve the two equations …
        simultaneously. Okay. In sec one you only solve one equation
        right? For example, given to you er X er two … alright X plus
        two equal to zero. Simple one. Or three X plus two equal to
        zero. Or some other complicated kind of linear equations. Only
        one unknown inside, okay. So sec one you solved the more or
        less er simple kind of equations. But now you are solving two
        equations together. Alright how are you going to solve it?

By doing so, the teacher was effectively making a different link from the one made by the SG1 teacher, one which involved linking the known and the new, that is, between what she has taught before with what she was now teaching. Such different kinds of shuttling between different forms of knowledge suggest different kinds of learning were being developed (Luke, Cazden, Lin & Freebody, 2005).

Student activities In SG1, the main student activities that students participated in were either individual or group seatwork while in SG2, students had in addition to the two types of activities, a competitive activity represented as group quiz where representatives from the groups were asked to solve questions on the board. Both classes spent about 41% of their class time on student activities alone. During seatwork, both teachers were actively engaged in between-desk instruction by moving from desk to desk either to monitor students' progress or to provide students with individual guidance. But there were distinctive differences in the ways the teachers enacted their instructional role as evidenced from their private interactions with the students.

In SG1, there were basically two scenarios. In the more common scenario, the teacher, while providing individual guidance, was also actively identifying common mistakes, misconceptions and difficulties that students encountered and bringing them to the attention of the whole class during the subsequent review segment. She also selected students' work for presentation to the class to illustrate the kinds of mistakes that students commonly made. In the second scenario, the teacher would provide one-to-one instruction to selected students who had shown difficulties with their homework at her desk while the rest of the class worked on the assigned task independently. The selected individuals often received specific feedback on their homework performance as well as guidance in solving those questions that they found difficult.

In SG2, the individual and group work were often highly competitive in nature as lollipops or group points were awarded to individuals or groups who were among the first to correctly complete the mathematical questions that the teacher assigned. As a consequence of the competition that characterised seatwork, the teacher, who apparently acted out of fairness, often avoided giving direct instruction to students who had encountered difficulty with the questions. She did provide individual guidance but only to a few selected students whom she appeared to believe were lagging behind the class. Her roles, unlike the SG1 teacher, were confined mainly to confirming the correctness of students' answers and encouraging more active discussion among group members.

Whole-class review of student work   This segment typically followed independent student work and in total occupied about one-quarter of the class time. Although it still appeared to be teacher-dominated in form, the main source of content for the discourse actually came from the students' work. During these segments, written input from the students was foregrounded in the discourse. One significant similarity in both classes was the ways in which the discourse was built around student work, particularly the emphasis placed by both teachers on learning from mistakes. The main thing both teachers stressed was not what the final answers should be but the kinds of mistakes students made while working through the steps needed to arrive at the final answer. Often both teachers did not stop at just locating and identifying the kinds of mistakes that were made but went on to hypothesise the possible reasons why students made those mistakes. Occasionally, comparison was made between the erroneous workings and the correct ones. The appropriate way of correcting the errors, the correct workings and sometimes, alternative approaches to solving problems were also provided by either the teacher or the students themselves. Comparatively, the provision of corrective feedback was significantly more substantial, specific and detailed than positive feedback as illustrated below:

SG1-L02

T:	Okay, now let's take a look at the solu er ... from one of your classmate ah, the answer for the first one, number six ah. Can you tell me what is missing?
S:	Bracket ... bracket
T:	Can you tell me what is missing?
S:	Bracket
T:	You look at her working. I believe they know what they are doing alright. But something is missing here. You look at the first line and the second line. Ten to the power of six for this term is missing. This is six point five two seven times ten to the power of six, then suddenly, it becomes like this. How can the first line be equal to the second line? ... What is missing? Can you tell me? Do you agree that the first line is not equal to the second line, do you agree?
Ss:	Yes
T:	Okay, so what has gone wrong?
Ss:	No bracket.
T:	Yes, where should you put the bracket? At the beginning and ... //after what number?
S:	//Before, before the power
Ss:	Before the power
T:	Before the power, yes. Alright, so what she did was actually she take out this common ... factor, alright, she took out ten to the power of six. Then she takes this number minus this number. [T put a tick besides the final answer.]

Apart from the review of student work done in class, the teacher in SG1 also allocated time for review of questions assigned for homework. This often constituted the single-segment instructional cycles that were found in the beginning of some lessons immediately after the miscellaneous segment (e.g., SG1-L02, L03, L04). During such time, the teacher checked with students whether they had any difficulty with their homework, sometimes going through one or two questions for which students had requested assistance, or going through some of the common mistakes students had made, which the teacher had identified while marking their homework. Review of homework was however non-existent in SG2.

Both teachers spent a considerable amount of time reviewing the test questions (as represented by the [R] segments in SG1-L09 and SG2-L07, L08). SG1 spent about 30 minutes while SG2 spent about an hour spread out over two lessons. What was significant was the amount of feedback that both classes received on the test itself. The focus of the teachers' discourses include: common mistakes made by the students, reasons for making the mistakes, how these mistakes could be avoided, how the marks were allocated, the different approaches used by students as well as the expected workings and their rationale. A typical example would be:

SG2-L07

T:	They mixed up the divide and the times. ... Alright, they do this part first. ... Okay, what's the difference? Can you see the mistake here? Can you see the mistake here? Because you know that five to the power of zero is one so divided by one times one-fifth. So the next step the pupils write er wrote it this way. Is it correct? ... Is it correct?

161

```
Ss:      No
T:       No. Where's the mistake? Where's the mistake?
S:       You must divide first.
T:       Yes, you must change the divide to times first. From left to
         right, according to the rules, should change the divide to
         times and then after that then times one-fifth. So the pupils
         doing this part first, become one-fifth and then you divide.
         So of course the final answer will be different, okay. So this
         is the mistake here. So be careful of the sign.
```

In fact, SG2 teacher had with her a piece of paper on which she recorded a whole list of mistakes that the class made for each question. She also included those mistakes made by her other grade eight classes. The rationale appeared to be that the more mistakes students learn about, the more they will be able to pre-empt and avoid those mistakes and hence the greater the likelihood they will be able to answer the questions correctly. Both teachers made sure students handed in their test corrections for marking. Interestingly, both teachers followed up on the review by making sure that students had additional practice on the topic that had been tested either through providing a worksheet (in SG1) or conducting a group quiz (in SG2). The questions asked were different but related to those which the students had done in the test.

DISCUSSION

There were many commonalities in the instructional approaches of the two highly task-orientated teachers. Specifically, both teachers relied mainly on the use of examples and mistakes made by students as their teaching tools in the development of students' mathematical understanding. Their highly structured instructional models as a whole also shared similar features, such as being cyclical and incremental in nature. Their instructional cycles, often focused on narrowly-defined content units, also shared fairly similar compositions, the most typical one being constituted by demonstration-seatwork-review of student work [DSR]. Interestingly, such a composition is like a macro-form of the IRF discourse pattern that was widely reported to characterise classroom discourse (Sinclair & Coulthard 1992). The pattern of presentation is that the teacher first 'initiates' the content and asks students to answer questions, students 'respond' during seatwork and finally the teacher provides 'feedback' during the review phase. However, as a macro-form of IRF, the [DSR] instructional cycle necessarily contained elements embedded within the individual segments that essentially extend and broaden the roles of the cycle beyond those that typically characterised the IRF structure at the discourse level.

The two teachers' instructional approaches also showed distinct differences. Differences abound in the compositional varieties of the instructional cycles and their duration, the kinds of shuttling between different forms of knowledge, and the teachers' roles during seatwork. These differences, particularly in the ways the teachers constructed their discourses – an important tool for learning from the sociocultural perspective (Vygotsky, 1978) – suggested differences in the kinds of learning that were afforded and constrained for the students. From our preliminary analysis of these two case studies, it appeared that the teachers' beliefs and knowledge, the mathematical content involved, the student profiles and the school cultural norms might account for some of these differences. Further analysis will be needed to examine the factors influencing these differences and how the differences impact on student learning.

One significant observation was that, despite the seemingly teacher-centred approach that characterised both teachers' instructional approaches, there were strong indications, particularly from the review segments, that the students' thinking was frequently taken into consideration during lessons and foregrounded in the discourse, albeit through the teacher. Such indications might not be apparent from one-off lesson observations but require careful analysis of the video data and classroom discourse. This suggests that previous studies of the instructional approaches of Singapore mathematics teachers (such as Kaur & Yap, 1998; Yeo & Zhu, 2005), that depended on classroom observations, might not be able to capture the extent to which teachers took into account students' contributions in their teaching. There is thus a possibility that the degree of 'teacher-centredness' may be overestimated. Moreover, teacher dominance need not be equated with student passiveness. Other studies have shown that the Western conception of teacher-centred versus student-centred learning might not be applicable in Asian contexts nor useful in explaining the 'Asian Learners Paradox' (Huang & Leung, 2004; Marton, Dall'Alba & Tse, 1996). Analyses taking into account the distribution of responsibility for knowledge generation may be one approach in which the degree of students' participation might be better represented (Clarke & Seah, 2005).

By showing structural patterns of the lessons, and the nature of the classroom talk, this chapter has attempted to explicate more features of the instructional approaches adopted by two Grade 8 mathematics teachers in Singapore classrooms than have previous studies. We hope that our analysis will lead to further research from which we can draw insight on how mathematics teaching in Singapore can be further improved. For example, the concept of instructional cycle can be used to analyse the structural patterns of other lesson sequences which may help us to better understand how the different ways in which instructional cycles are constituted may have an impact on students' learning. Given the corpus of data available that have not yet been analysed, such as the teacher and student interviews, it is also imperative that the perspectives on classroom practices of both teachers and learners, and in particular the importance they placed on them, are further examined and triangulated with what has been explored in this chapter.

ACKNOWLEDGEMENTS

This chapter was based on the funded project, CRP 3/04 BK, at the Centre for Research in Pedagogy and Practice, National Institute of Education, Nanyang Technological University, Singapore. Thanks are due to David Clarke and Anneliese Kramer-Dahl for their valuable insights as well as Kyungmee Park, Florenda L. Gallos, Catherine O'Keefe, Carmel Mesiti and Leong Yew Hoong for their feedback on this chapter.

163

REFERENCES

Alexander, R. (2000). *Culture and pedagogy: International comparisons in primary education*. Oxford: Blackwell.

Chang, A. S-C., Kaur, B., Koay, P. L., & Lee, N. H. (2001). An exploratory analysis of current pedagogical practices in primary mathematics classrooms. *The NIE Researcher, 1*(2), 7-8.

Clarke, D.J. (2002a). Developments in international comparative research in mathematics education: Problematising cultural explanations. In S. L. Y. Yam & S. Y. S. Lau (Eds.) *ICMI Comparative Study Conference 2002: Pre-Conference Proceedings* (pp. 7-16). Hong Kong: University of Hong Kong.

Clarke, D.J. (2002b, April). The learner's perspective study: Methodology as the enactment of a theory of practice. In D.J. Clarke (Chair), *International perspectives on mathematics classrooms*. Symposium conducted at the Annual Meeting of the American Educational Research Association, New Orleans.

Clarke, D., & Seah, L. H. (2005). Studying the distribution of responsibility for the generation of knowledge in mathematics classrooms in Hong Kong, Melbourne, San Diego and Shanghai. In H. L. Chick & J. L. Vincent (Eds.), *Proceedings of the 29th Conference of the International Group for the Psychology of Mathematics Education* (Vol.2, pp. 257-264). Melbourne, Australia: The University of Melbourne.

Glaser, B., & Strauss, A. (1967). *The discovery of grounded theory*. Chicago: Aldine.

Gopinathan, S. (1999). Preparing for the Next Rung: Economic restructuring and educational reform in Singapore. *Journal of Education and Work, 12*(3), 295-308.

Ho, K. F. & Hedberg, J. G. (2005). Teachers' pedagogies and their impact on students' mathematical problem solving. *Journal of Mathematical Behavior, 24*, 238-252.

Huang, R., & Leung, K. S. F. (2004). Cracking the paradox of Chinese learners: Looking into the mathematics classrooms in Hong Kong and Shanghai. In L. Fan, N. Y. Wong, J. Cai & S. Li (Eds.), *How Chinese learn mathematics: Perspectives from insiders* (pp. 348-381). Singapore: World Scientific Publishing.

Kaur, B. (1999). *Mathematics classrooms of Singapore schools: Implications for fostering thinking and creativity*. Paper presented at the 8th Southeast Asian Conference on Mathematics Education (Technical papers): Mathematics for the 21st Century. Philippines: Ateneo De Manila University.

Kaur, B., & Yap, S. F. (1998). *KASSEL Project Report (NIE-Exeter Joint Study)* (Technical report). Singapore: National Institute of Education.

Kay, H. G. (2003). *An exploratory study of the belief systems and instructional practices of four primary school mathematics teachers*. Unpublished MEd. thesis, Nanyang Technological University, Singapore.

Luke, A., Cazden, C., Lin, A., & Freebody, P. (2005). *A coding scheme for the analysis of classroom discourse in Singapore schools*. (Technical Report). Singapore: Centre for Research in Pedagogy and Practice, National Institute of Education.

Luke, A., Freebody, P., Lau, S., & Gopinathan, S. (2005). Towards research-based innovation and reform: Singapore schooling in transition. *Asia Pacific Journal of Education, 25*(1), 5-28.

Marton, F., G., Dall'Alba. G., & Tse, L. K. (1996). Memorizing and understanding: The keys to the paradox? In D. Watkins & J. Biggs (Eds.), *The Chinese learner: Cultural, psychological and contextual influences* (pp. 69-83). Hong Kong: Comparative Education Research Centre; Melbourne: The Australian Council for Education Research.

Mercer, N. (1995). *The guided construction of knowledge*. Clevedon: Multilingual Matters.

Mesiti, C., Clarke, D., & Lobato, J. (2003, April). The structure of mathematics lessons in the United States. In D. Clarke (Chair), *Mathematics lessons in Germany, Japan, the USA and Australia: Structure in diversity and diversity in structure*. Symposium conducted at the Annual Meeting of the American Educational Research Association, Chicago.

Shimizu, Y. (2002, May). *Capturing the structure of Japanese mathematics lessons: Some findings of the international comparative studies*. Paper presented at the ICMI-Second East Asia Regional

Conference on Mathematics Education and Ninth Southeast Asian Conference on Mathematics Education. National Institute of Education, Singapore.

Sinclair, J., & Coulthard, M. (1992). Towards an analysis of discourse. In M. Coulthard (Ed.), *Advances in spoken discourse analysis* (pp. 1-34). London: Routledge.

Stigler, J. W., & Hiebert, J. (1997). Understanding and improving classroom mathematics instruction. *Phi Delta Kappan, 79*(1), 14-21.

Stodolsky, S. S. (1988). *The subject matters: Classroom activity in math and social studies.* Chicago: The University of Chicago Press.

Vygotsky, L. S. (1978). *Mind in society: The development of higher psychological processes.* Massachusetts: Harvard University Press.

Yeo, S. M., & Zhu, Y. (2005, May/June). *Higher-order thinking in Singapore mathematics classroom.* Paper presented at the conference Redesigning Pedagogy: Research, Policy, Practice, Singapore.

Seah Lay Hoon
National Institute of Education,
Nanyang Technological University,
Singapore

Berinderjeet Kaur
National Institute of Education,
Nanyang Technological University,
Singapore

Low Hooi Kiam
National Institute of Education,
Nanyang Technological University,
Singapore

ASTRID BEGEHR

CHAPTER ELEVEN

Students' Verbal Actions in German Mathematics Classes

INTRODUCTION

This chapter analyses student and teacher participation in eighth-grade mathematics instruction in Germany. It examines how students participate in mathematics lessons and their scope of verbal action. Furthermore, the consequences for learning mathematics are discussed.

In class, students talk too little and teachers talk too much. Based on this highly debatable and apparently well-known assumption, the parts of talk of those participating in mathematics instruction are analysed and set into relation. In a triangular approach, comprehensive data is evaluated and linked to results from qualitative empirical classroom research. Qualitative and quantitative research methods – frequently presented as disparate approaches – are combined in order to enrich one another. Interpretative reconstructions of selected representative transcript passages extend the empirical results of the analysed German lessons.

FRAMEWORK AND GOALS OF THE STUDY

Database

The material from the Learner's Perspective Study (LPS) forms the basis of the empirical section of the present analysis. In addition to the videos, the slightly modified transcripts of this study were mainly used for the data analysis. Given the need to develop evaluation criteria before conducting quantitative analyses designed to provide answers to the research questions in connection with the students' verbal activity, the most important characteristics of the data collection will be introduced.

The LPS set itself the goal of further developing the video portion of the Third International Mathematics and Science Study (TIMSS) by placing the focus on the learners. Whereas the 100 lessons videographed within the framework of TIMSS had only one camera recording the class – and it, moreover, was generally aimed at the teacher – LPS made use of three cameras per class, which permitted the observation and recording of more events transpiring in the lesson. One camera monitored the class as a whole; another filmed the actions of two (on rare occasions three or four) selected students; while the third recorded the activities of the teacher in sound and pictures. The fact that the chosen, individualised learners ('focus students') had their own microphone set up in front of them placed their contribution to the lesson on an equal footing with that of the instructor, at least in terms of the structure of the research. As opposed to the video analysis of the TIMS Study, which was a longitudinal study comprising 100 individual eighth-grade German mathematics lessons recorded at 100 different schools, the LPS represents a convenience sample, for which ten consecutive lessons were recorded at three diverse German schools. This approach provided especially deep insight into everyday learning and teaching activity in the classroom. The observation of ten successive lessons makes possible a more profound understanding of everyday mathematics instruction with its diverse forms and accentuations of learning and teaching processes over a longer period than would be represented in a study of isolated single lessons.

Objectives of the Study

The content-based verbal activities of the students are a fundamental prerequisite – from the didactic-methodological perspective as well as on theoretical grounds – for learning processes and therefore a matter of vital concern for classroom instruction. In many teachers' manuals that are based on research results, the prevailing opinion is that students' learning processes are positively influenced when they are given the opportunity to participate actively and independently in classroom activities. The aim of independence does not deny the significance of social interaction among the students, but focuses on the need of the learners to work independently among themselves and on their own instead of being too dependent on the teachers' help. Regardless of which subject is being taught, the assumption is that the comprehension and acquisition of learning content by the students are facilitated when they perform the work of acquiring it on their own (Forytta & Linke, 1981; Petersen & Ritscher, 1996). Whereas the significance of making one's own contribution used to be viewed primarily in its function of clarifying thoughts and increasing motivation and memory performance (R. Tausch & A. Tausch, 1971), constructivist learning psychology sees independent activity as the fundamental prerequisite for all learning.

According to R. Tausch and A. Tausch (1971), the verbalisation of thought processes is the precondition for better understanding. If students are forced to reduce their contributions to classroom discourse to the interspersion of sentence fragments, however, this impedes the development of the complex connections that the teacher is trying to convey.

Ritz-Fröhlich (1976) articulates for teachers what she considers to be a generally recognised consensus, albeit one that is not accorded enough attention in teaching practice, as follows:

It is beyond dispute in contemporary didactics that learning processes are made more effective for the students to the extent that they are granted independence and the opportunity to participate actively in the learning process (p. 34, translated by author).

Jahner (1978) adds to this:

The assignment itself will certainly often originate from the teacher at the beginning, but one can at least attempt to work it out jointly with the students. [...] In the actual discussion phase (at the latest), the teacher must remain in the background as much as possible (p. 47, translated by author).

Abele also criticises the conventional form of lesson design in his 1988 article on the communication process in mathematics classes as a structural problem:

The discussion-initiating situations that allow the students to report on their own (mathematical) experiences or require individual students or groups of students to reach a decision occur far too infrequently in normal mathematics classes. Thus arises the methodological need for the structurally narrow discussion network, consisting of (teacher's) questions and (students') responses, to be expanded to at least include the possibility of several different ways of solving the problem (p. 26, translated by author).

Roeder and Schümer (1976) interpret the results of their empirical study as indicating that classroom interaction structures grant students only very few participation opportunities:

Students who have few opportunities to express themselves at all have virtually no chance of exerting influence on the structure of classroom discussion, meaning no chance of acting verbally in such a manner as to determine their classroom role for themselves (p. 77, translated by author).

As far as the official, documentable verbal participation of students in the classroom is concerned, there have been a few classic studies (Rice, 1893; Stevens, 1912; Corey, 1940; Bellack, Kliebard, Hyman, & Smith, 1966; Hoetker & Ahlbrand, 1969) and one recent representative German study based on the TIMSS video data (Begehr, 2004), which allows inferences about the current state of mathematics instruction. The methodological paradigm of studies concerned with classroom communication has shifted over the course of time in favour of qualitative approaches.

The primary objective of the study outlined in this chapter is thus to document the current status of students' verbal communication in such a way as to subject it to closer analysis in order to enable conclusions to be drawn. The LPS data appear to be exceptionally suitable for this purpose, since they place the students at the centre of attention, not only supplying them with a voice by the employment of additional microphones, but also interviewing them following the lessons. Furthermore, the study is intended to contribute to the search for explanatory approaches for the documented forms and qualities of learners' verbal activities.

METHODOLOGY AND METHODS

Preliminary Methodological Decisions

Research into students' verbal participation is based on the one hand on the category-derived interpretation of classroom communication, and on the other hand on interpretations developed via systematic reconstruction of classroom practice in discrete episodes of mathematics instruction.

It is a matter of dispute as to which approach meets the generally recognised requirements of scientific research and, moreover, is capable of achieving meaningful, comprehensible, and reliable results that can be applied in the classroom setting. The decision with regard to the methodological procedure employed in the present paper is based on the portrayal and consideration of the various research paradigms.

Quantitative Approach. The quantitative, empirical-analytical research approach sets itself the goal of explaining the phenomena that occur in the classroom on the basis of universally valid models. The aim is to obtain practical guidelines by deductive inference with the aid of specifically developed assessment criteria. Hypotheses concerning the object of study are derived during the course of the research from previously selected theories and subjected to empirical falsification tests.

One significant criticism aimed at the quantitative research approach relates to its allegedly negligible action relevance to pedagogical practice. Fault is found with the fact that quantitative studies overlook important individual criteria of the process under examination and are virtually unable, due to the emphasis on summary aspects, to provide concrete suggestions for the improvement of the curriculum in question.

> The customary use of quantitative data, for example, may provide much useful information on student achievement. But there is growing evidence that quantitative data cannot supply satisfactory answers to many of the qualitative questions of education today. It is no longer enough to say that Johnny can't read; what is now being asked is why he can't and what will make him learn (Ianni & Orr, 1979, p. 88).

Qualitative Approach. In contrast to the quantitative approach, qualitative research is frequently conducted in inductive and abductive fashion. A qualitative researcher strives to comprehend the phenomenon under study in its social and cultural context (Nussbaum, 1984). Data is preferably collected in its natural social environment.

As Rist (1982) explains, data collection and data analysis cannot be separated in qualitative research:

> To state that data analysis occurs concurrent with data collection is only to acknowledge that fieldwork is not simply the mechanistic collection of predefined data from predefined sets of respondents. Rather, the entire time

the researcher is in the field, there is a constant dialectic between collection and analysis, i.e., a constant assessment of what is known versus what is to be learned. [...] The imperative of concurrent analysis and data collection for a considerable period during a qualitative study negates the possibility of applying the natural science model of strictly sequential activity. Those who would first collect qualitative data and then apply an analytic framework to it have abused the approach (p. 445).

In German research on classroom mathematics didactics, one method, known as extensional interpretation, is particularly widespread. It calls for the evaluation of texts under study – mainly transcript of observed lessons – in a systematic process during which the transcripts are first broken down into so-called episodes. The structure that results has been most frequently attributed to the classroom interaction between teacher and students. This division into episodes is not unproblematic, since an interaction unit is itself composed of multiple elements and can only be interpreted with reference to the interaction structure in which it is embedded (Beck & Maier, 1994; Flick, 1995; Brandt, Krummheuer, & Naujok, 2001).

One criticism of qualitative approaches can already be found in the work of Campbell and Stanley (1963). They consider a strictly experimental methodology to be indispensable and are fundamentally opposed to qualitative classroom research in the form of case studies:

Much research in education today conforms to a design in which a single group is studied only once, subsequent to some agent or treatment presumed to cause change. [...] Such studies have such a total absence of control as to be of almost no scientific value. [...] It seems well-nigh unethical at the present time to allow, as theses or dissertations in education, case studies of this nature (pp. 176-177).

The lack of objectivity and validity in the acquired data – as opposed to quantitative research – is thus presented as a fundamental weakness of qualitative research. Qualitative research is based on uncontrollable processes such as "understanding" or "intuition." Representative results cannot be obtained due to the concentration on situations, biographies and subcultures. While qualitative research does possess the ability to identify and differentiate subjective structures of meaning and everyday routines, it is not capable of transcending the level of description due to its phenomenological awareness. Finally, the issue of how qualitative research can be made teachable has yet to be resolved, so that its practical utility for scientific study remains an open question (Nussbaum, 1984).

Triangulation. Against the background of the above-described dispute between representatives of the quantitative and the qualitative research approach, the research community is increasingly coming to realise that one does not necessarily have to decide between these opposing approaches when developing a research design; neither of them need be declared the one and only correct and tenable

171

approach. Both quantitative and qualitative methods represent fundamentally legitimate approaches, which are more or less appropriate depending on the questions posed, the object to be studied, and the range of interpretation (Klieme & Bos, 2000). Under certain conditions, it may be reasonable to combine the two approaches; in particular, it is appropriate to provide an interpretation of the statistical parameters identified in a quantitative study before it is completed (Wolf, 1995). The aim is to present additional information that makes the extremely reduced numerical information comprehensible and reestablishes its relationship to the context of everyday life (Fromm, 1990).

Triangulation can represent a cumulative validation of research results in which convergences and divergences between quantitative and qualitative findings are actively sought. In combining quantitative and qualitative methods, the researcher strives to identify the potential for complementary compensation of the weak and blind spots of the individual methods (Flick, 1995).

In their comparative portrayal of the use of teacher's questions in eighth-grade mathematics instruction in Germany, Japan, and the United States on the basis of the data of the TIMSS Video Study, Kawanaka and Stigler (1999) see the integration of quantitative and qualitative methods as essential to the success of achieving reliable and valid research results. For their own study, Kawanaka and Stigler reach the following conclusion:

> Clearly, we must go beyond simply counting the number of questions students are asked if we are to make sense of these inconsistent results. As a first step, we need a clearer picture of how teachers use questions in the context of actual classroom lessons and how questioning interacts with other factors in influencing how much students learn (p. 256).

Method of Analysis

With the above considerations of the benefits of triangular research serving as the basis for the exploitation of the data from the Learner's Perspective Study, it is necessary in this project as well to conduct cumulative quantitative and qualitative research. Firstly, it is possible to use the database of LPS, given the application of scientific research methods and the development of analytical criteria specifically for this project, to achieve significant quantitative results regarding student participation in mathematics classes. The results thus obtained are in and of themselves sufficient to allow conclusions to be drawn regarding the success or lack thereof of the widely propagated pedagogical efforts to get students more involved in classroom instruction, to enable more participation.

There is general consensus that the compilation and analysis of quantitative data alone is not sufficient when reasons for the events observed are being sought and practical suggestions are to be developed, such as an improvement of conditions viewed as inadequate. For this it is necessary – and with the availability of the LPS classroom videos also possible – to observe exemplary lesson units in their entirety, to analyse them according to mathematics-related and didactic aspects, and to provide criticism when necessary in order to be able to develop appropriate alternatives.

For analytical purposes, the videographed LPS lessons were transcribed. The transcripts contain all verbal statements of the teacher and students which were audible. These include all statements by the teacher and students that can be considered part of the official classroom discussion as well as additional statements by students that are unrelated to the classroom discourse but are made at a volume that makes it probable that they will be heard by the entire class (public talk). Furthermore, utterances were also incorporated into the transcript that occurred during so-called quiet work phases or were not intended for the class as a whole (private talk). It is precisely here that the student video, which filmed the focus students, recorded statements that otherwise would have been lost. The transcripts were designed to reflect in written form all the audible information necessary to identify and isolate the verbal communicative actions in the lesson studied. Of course, not all verbal statements of all students or the teacher could be recorded. Even though a camera was aimed at the whole class, it was not possible to catch all verbal statements by each individual. However, the camera and microphone pointed at the focus students enabled to record nearly all statements by these learners. As mentioned above, this fact placed their verbal contributions on an equal footing with that of the teacher, at least in terms of the structure of the research.

AN ANALYSIS

On the basis of the LPS dataset, with the aid of the video data, transcripts, questionnaires, and interviews, indicators can be developed that allow conclusions about students' verbal participation to be drawn. The video recordings comprise the verbal contributions of both the learners and the teacher. The transcripts shed light on the students who participate in the lesson with their verbal statements. They provide insight into the number of utterances by various students and display the sequence of the statements made by the two speaker groups – students and teacher. The share of classroom speech accruing to both groups reveals the extent to which students play a part in the lesson in comparison with the teacher. The values are based on the combined statements of the two speaker groups, measured in lines of transcript.

Along with the division of the speech shares of teacher and students, the length of the respective individual statements can provide evidence as to the possibilities for students to express their thoughts in longer statements. The length of the statements is thus an indicator of the amount of time the students have at their disposal to speak and of their willingness and ability to contribute to classroom discussion. A high degree of speech communication on the part of the teacher allows students only very little speech communication (R. Tausch & A. Tausch, 1971).

When students' mathematics-related statements are predominantly short, this could reflect the fact that these are primarily answers to narrowly phrased questions, for example. In their responses to such questions, students are not asked to express fundamental thoughts about mathematics. Questions that are intended to enhance an approach in which students are guided to an understanding subject at hand in small steps are generally phrased in such a way that students can answer them rapidly and succinctly with no need for deliberation. Longer statements, on the other hand, imply more complex and qualitatively more sophisticated student contributions. A long statement suggests that the student has developed and formulated her or his own thoughts and that the teacher considers it appropriate to grant the student time to express them.

Despite its obvious merit, measuring speech times to determine the length of the statements was not an option. The software employed recorded only the starting time and it was inaccurate to individually time each statement, since utterances can be very short. For a reliable determination of the length of student and teacher statements, therefore, the method already practised by Bellack et al. (1966) was applied; the length of the verbal statements is established on the basis of the number of lines of transcript. In order to achieve a comparative measure, the transcribed text has been formatted at uniform line length. The number of lines counted provides a sufficiently clear and reliable measure for the determination of the length of speech.

The research conducted with the TIMSS data revealed that in eighth-grade mathematics classes, in phases of full-class instruction (public talk), the students' share of speech as compared to that of the teacher lies in the ratio of approximately one to three (Begehr, 2004). This finding is the result of a cluster sample, that is, a random selection of the schools and classes (Stigler, Gonzales, Kawanaka, Knoll, & Serrano, 1999). The video recordings of eighth-grade mathematics classes were made in a random sample of the 153 TIMSS classes and schools. In Germany, 100 schools participated in this study with one class each. This value for a representative sample of eighth-grade students in the Federal Republic of Germany is now to be compared with the verbal activity of the German eighth-graders in the LPS project. Whereas TIMSS concerned itself with individual mathematics lessons, LPS provides an image of classroom practice over a longer period by studying ten consecutive classes on related material as part of one teaching sequence and by placing much greater emphasis on the learners.

Verbal Actions in Mathematics Lessons

In the following, the analysis of ten lessons from LPS is reported. Of these, one lesson German School 1 - Lesson 1 (G1-L01) is studied in particular. According to the teacher questionnaire, it represented the 21st of 30 lessons on the topic of "Integral Rational Terms and their Reformulation, Simplification." A total of approximately fifteen lessons were recorded at each school, from which the last ten consecutive ones were used as the basis for the study. The lesson presented here, the first of the ten studied, was the fifth of the total of fourteen lessons filmed, so that students and teacher had already had the opportunity to develop a certain level of familiarity in dealing with the three video cameras.

Quantitative Analysis. The analysis of this lesson produced quantitative results that were very similar to those from the TIMSS project. Furthermore, the lesson analysed here is an image of the median of the ten studied LPS lessons with respect to the parts of talk (see Figure 1). The share of speech of both speaker groups exhibits a striking similarity to the average values determined in the framework of the TIMS Study: Whereas the publicly audible statements of all students in the class (G1-L01) cover a total of 93 lines of transcript, the public statements made by the teacher cover 259 lines of transcript. This corresponds to a speech share of public talk for the students as compared to that of the teacher in the ratio of approximately 1 : 2.78, that is, 26.4 % and 73.6 %. Within the ten studied lessons, in average, the students' share of speech as compared to that of the teacher lies in the ratio of approximately one to three, that is, close to 27.3 % and 72.7 %.

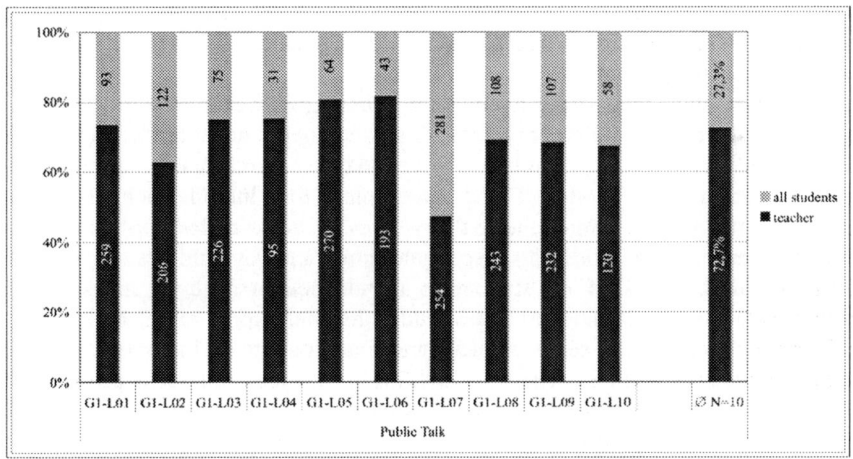

Figure 1. Distribution of verbal speech share in ten consecutive classes on related material as part of one teaching sequence (G1)

Although the numerical result could, at first glance, create the impression that the students' participation was quite limited, closer observation of the students' statements reveals that they did in fact contribute actively to the classroom discussion: Twenty-four of the twenty-six students present in this class participated in the classroom discussion. Two students do not appear in the transcript at all. Since there are a number of non-ascribable student utterances, even they, however, could have articulated themselves. Of the 172 statements made by students, comprising both public and private talk, 115 can be ascribed to individual students, whereas the speakers of 57 statements cannot be identified. If one thus examines only the attributable statements, which derive from 24 different students, and considers the fact that the students contributed to the class discussion with varying frequency – between one and fifteen times, with an average of slightly more than four statements per student – one receives the impression of rather lively and animated participation on the part of the students. Four students, two of whom are focus students, contributed over ten statements apiece. If the non-attributable statements are also taken into account, the average number of utterances rises to around seven per student. The teacher, on the other hand, spoke on 135 different occasions at this lesson. The fact that the statements under consideration here comprised both public and private talk must be taken into account. Only two students at the recorded lesson were technically – via the microphone – in a position to be heard as well as the teacher. It can therefore be assumed that a number of statements in the private talk of most of the students in the class were not taken into account. If one considers only the speakers equipped with microphones (i.e., the teacher and the two focus students, Felix and Torsten), the result is quite sobering. The 135 statements made by the teacher make up 325 lines of transcript, of which 259 lines of transcript consist of public talk and 67 lines of private talk. In contrast, Felix's fifteen statements – ten in public talk and five in private talk – make up six and three lines of transcript respectively. The eleven statements by Torsten, six in public and five in private talk, make up five and four lines of transcript respectively.

While the public talk of the lesson was quantitatively analysed in total (Figure 1), the complementary statements, that is, all statements not intended for the class as a whole (private talk), could only be analysed exemplarily through focus students like Felix and Torsten. These statements, which include such speech as the statements students exchange among themselves in the so-called application phase, are also of importance in order to investigate, in which way students articulate their mathematical thoughts. If all statements (total speech) of both speaker groups could be analysed, regardless of the intended listenership, it might be shown that Felix and Torsten are not representative and that students do have the opportunity to express their mathematical ideas. This, however, does not seem probable. The students' share of speech in the analysed lessons is lower than that of the teacher. The larger share of speech originates from one single speaker, the teacher, while the smaller share is composed of the statements of the entire class, which consists of a number of speakers. If one compares the speech shares of the individual speakers – that of the teacher with that of a single learner – the difference between the shares would thus turn out to be much more severe.

Qualitative Analysis. Description. The transcript excerpt is part of the roughly 26-minute (00:02:19:00 - 00:28:30:00) treatment of eight problems (see Figure 2) that the teacher had written on the board prior to the beginning of the lesson. It was preceded only by a greeting and some information regarding two class tests. In the second minute of the class, the teacher opened up the board and worked on the students to solve the algebraic terms. The following excerpt shows the subsequent discussion of the results.

Figure 2. Problems written on the board (G1-L01)

Transcript. The measured length of the individual statements can be found in the second column. Whereas the students generally only state their results, the teacher explains the way to reach the solution of the problem in question (G1-L01)

00:17:14:00	6 T	Okay I would like you all to now use a red pen for the corrections ... so that if you look over it again at home well wonders never cease you'll see immediately what mistakes you made. ... Now the first one. That was still quite easy. // ... Josefa.
00:17:30:30	1 Albert Me.	// Minus eleven.
00:17:33:23	1 Josefa	Minus eleven.
00:17:37:28	6 T	Yes and this here ... uh this is a plain and simple ... uh subtraction problem so what I don't need are the I can actually go without the brackets. //... Here in the second exercise what do I have to calculate first? What do I have to calculate first? Felix.
00:17:54:00	1 (Josefa)	// (I got something right.)
00:18:01:02	2 FELIX	Okay minus one times four and then again minus /
00:18:06:08	2 T	/ That would be a possibility that I again use this ... what would we have then if?
00:18:11:21	1 FELIX	Uh minus four plus seven.
00:18:15:25	5 T	Yes. If I take this okay this negative sign in front of the brackets into consideration ... regard it as if it the bracket is multiplied by minus one ... then I've got minus four plus seven and that equals?
00:18:30:00	1 Simonne	Three.
00:18:30:05	4 T	Three. But how could I have solved that in another way? Without me using this this regularity with the minus one. Could I have done this differently? Günther?
00:18:40:23	4 Günther	Yeah well first one could have done four

	minus seven. That would make minus three and due to the minus in front of the brackets (the sign turns around).
00:18:45:21 13 T	Correct. Yes that would be the second possibility. Yes that I work out // the brackets first four minus seven. That equals to … Olaf … minus three. Negative sign in front of minus minus three therefore equals to plus three. Correct. With the next one I can yet again… well I can't calculate the contents of the brackets first because well now we've got variables inside. How do I calculate the term minus open brackets X plus three Y? You have to be quite clear about that otherwise uh … you're going to have really big problems. How do I calculate that?
00:18:47:27 1 Albert Me.	// I didn't understand anything.

Key to symbols used in transcripts in this chapter:
...   A short pause of one second or less.
//   Marks the beginning of simultaneous/overlapping speech.
Ss   Indicates two or more students speaking simultaneously, saying the same words.
/   Indicates that one speaker cut in, interrupting another speaker before they had finished.
(text)   A plausible interpretation of speech that was difficult to decipher.

The teacher concluded the application phase by addressing the class and asking the students to take out a red pen. He thereby commenced the correction phase and with it the segment in which the results were to be compared. With the instruction "Now the first one." and the assessment of the degree of difficulty "That was still quite easy." (00:17:14:00), he asked the students to present their answers. Albert provided his answer without being called on, which the teacher ignored; he called on Josefa, who repeated the answer just given. He added the rationale for the result by providing a short account of the solution method: "Yes and this here … uh this is a plain and simple … uh subtraction problem so what I don't need are the I can actually go without the brackets." For the second problem "- (4 - 7)" he only asked for the first step and nominated Felix, who provided his first step as follows: "Okay minus one times four and then again minus /." He was promptly interrupted by the teacher, who said, "/ That would be a possibility that I again use this … what would we have then if," but was actually looking for a different way to solve the problem. Felix did not give his answer; Simonne was called upon to do so. The student replied, "Three." The teacher repeated the answer and posed a question related to the solution method he preferred: "But how could I have solved that in another way? Without me using this this regularity with the minus one. Could I have done this differently?" Günther first solved the subtraction within the parentheses (simplifies) and then considered the negative sign in front of the parentheses, thereby providing the solution method the teacher was seeking. The teacher seized on this answer, deeming it "Correct" and formulated the same idea once again: "Yes that would be the second possibility. Yes that I work out // the brackets first four minus seven. That equals to … Olaf … minus three. Negative sign in front of minus minus three therefore equals to plus three. Correct." He interrupted himself to discipline Olaf and was himself interrupted by Albert, who announced that he "didn't understand anything." The remaining ten minutes or so of the "reinforcement of the results" proceeded in similar fashion.

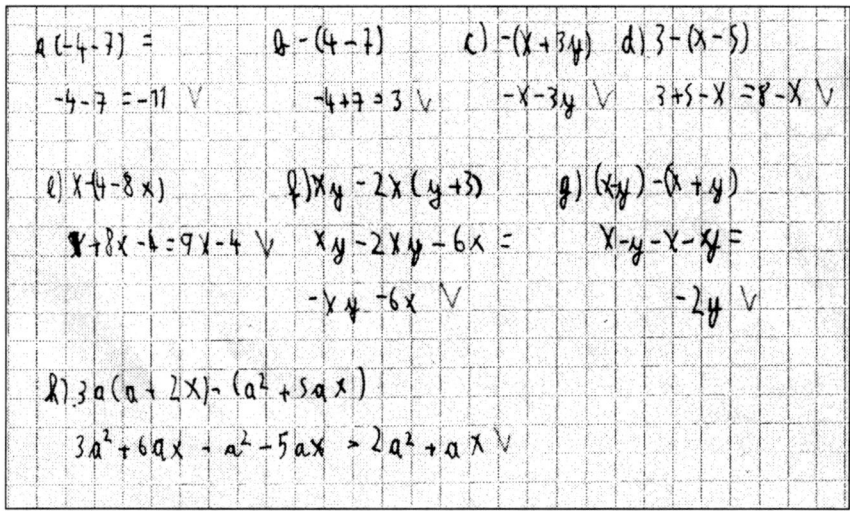

Figure 3. Felix's work on the problems (G1-L01)

Qualitative Analysis. Interpretation. Although the teacher declared in the questionnaire that this was a purely review lesson, described his most important goal as the recapitulation of knowledge by the students (restore the former level of ability/competencies in reformulation and simplification of algebraic terms), and moreover declared that the goal of the unit as a whole was development of a certain confidence vis-à-vis the reformulation of algebraic terms, he did not allow the students to present independently the results they came up with on their own or with their fellow students, nor to give a rationale for their solutions. His controlling approach served as an impediment to his students, as was already evident in the quantitative analysis and is now confirmed by the qualitative analysis. In the representative example discussed above, the teacher deprived the students of the opportunity to articulate their thoughts following the independent application phase. If he would have given them this chance, the students' share of speech would have been higher. However, once again, the parts of talk of the illustrative and representative transcript sequence can be portrayed through the characteristic ratio of approximately one to three. The teacher either did not recognise the need to give the learners this opportunity or did not believe that they were capable of putting forward their ideas without outside assistance, even though the students were well prepared for the problems, given that they had just finished working on them and were even able to solve them on their own. Felix, who was interrupted in his speech, since he first simplified the term by dissolving the negative sign in front of the brackets, solved all problems correctly. This can be inferred from his

179

notebook (Figure 3) as well as the transcript: Subsequent to this phase the teacher asked the students for a show of hands "Okay who has made no mistake?" (00:28:15:00) Felix raised his and stated "Moi." (00:28:13:14). Two-thirds of the lesson was spent on treating these eight terms.

CONCLUSION

Teachers "outtalk" the students, seemingly without being conscious of the fact. It might therefore appear reasonable to subject teachers to a confrontation with their own videographed lessons as part of their in-service training (and their basic teacher's education as well) in order to increase their awareness of this and other issues. Teacher talk is intended to support students in their efforts to master the subject matter. It thus appears even more important to point out that well-meant verbal support can have exactly the opposite effect when students are denied the opportunity to describe the content in their own independent statements and thereby to reflect on what they have learned. The verbal guidelines set by the teacher impeded students in their efforts to come to grips with the mathematical content and actually diverted attention from the subject matter at hand, since the learners only expressed themselves in disjointed fragments, more often than not in response to basic arithmetic problems. As a result, their ability to make their own discoveries and genuinely comprehend the material may fall by the wayside, as they are only called upon to designate small segments without being able to express them in the overall context. Teachers of mathematics classes must evidently learn to "let go", that is, not to guide their students along a narrow, predefined path, but to grant them the space, including the verbal space, to develop and express their own thoughts. Teachers will thus ultimately help their students more than by providing well-meant guiding instructions.

REFERENCES

Abele, A. (1988). Kommunikationsprozesse im Mathematikunterricht [Communication processes in the mathematics classroom]. *Mathematische Unterrichtspraxis, 9*(2), 23-30.

Beck, C., & Maier, H. (1994). Zu Methoden der Textinterpretation in der empirischen mathematikdidaktischen Forschung [About methods of text interpretation in the empirical research of mathematics education]. In H. Maier, & J. Voigt (Eds.) *Verstehen und Verständigung, Arbeiten zur interpretativen Unterrichtsforschung* (pp. 43-76), *IDM-Reihe, 19*, Köln: Aulis Verlag Deubner.

Begehr, A. (2004). *Teilnahme und Teilhabe am Mathematikunterricht: Eine Analyse von Schülerpartizipation.* [Passive, Active, and Interactive Participation in Mathematics Classrooms: An Analysis of Participation]. Berlin: Freie Universität Berlin.

Bellack, A. A., Kliebard, H. M., Hyman, R. T., & Smith, F. L., Jr. (1966). *The language of the classroom.* New York, New York: Teachers College Press, Columbia University.

Brandt, B., Krummheuer, G., Naujok, N. (2001). Zur Methodologie kontextbezogener Theoriebildung im Rahmen von interpretativer Grundschulforschung [On methodology of a context related formation of theories within the interpretative research in elementary education]. In S. von Aufschnaiter, & M. Welzel (Eds.), *Nutzung von Videodaten zur Untersuchung von Lehr-Lern-Prozessen: Aktuelle Methoden empirischer pädagogischer Forschung* (pp. 17-40). Münster: Waxmann.

Campbell, D. T., & Stanley, J. C. (1963). Experimental and quasi-experimental designs for research in teaching. In N. L. Gage (Ed.), *Handbook of Research in Teaching* (pp. 171-246). Chicago, Illinois: Rand McNally.

Corey, S. M. (1940). The teachers out-talk the pupils. *The School Review, A Journal of Secondary Education, 48*, 745-752.

Flick, U. (1995). *Qualitative Forschung. Theorie, Methoden, Anwendung in Psychologie und Sozialwissenschaften [Qualitative research. Theory, methods, application in psychology and social sciences].* Reinbek: Rowohlt.

Forytta, C., & Linke, J. (1981). *Ist Unterricht „gestörte" Kommunikation? Eine Untersuchung zum sprachlichen Handeln im Unterricht der Primarstufe, Band 2 [Is instruction "disrupted" communication? A study of linguistic acts in elementary school lessons].* München: Minerva Publikation.

Fromm, M. (1990). Zur Verbindung quantitativer und qualitativer Methoden [On the combination of quantitative and qualitative methods]. *Pädagogische Rundschau, 44*, 469-481.

Hoetker, J., & Ahlbrand, W. P., Jr. (1969). The persistence of the recitation. *The American Educational Research Journal, 6*(2), 145-167.

Ianni, F. A. J., & Orr, M. T. (1979). Toward a rapprochement of quantitative and qualitative methodologies. In T. D. Cook, & C. S. Reichardt (Eds.), Qualitative and Quantitative Methods in Evaluation Research (pp. 87-97). Beverly Hills, California: Sage Publications.

Jahner, H. (1978). *Methodik des mathematischen Unterrichts [Methodology of mathematical instruction].* Mannheim: Quelle & Meyer Verlag.

Kawanaka, T., & Stigler, J. W. (1999). Teachers' use of questions in eighth-grade mathematics classrooms in Germany, Japan, and the United States. *Mathematical Thinking and Learning, 1*(4), 255-278.

Klieme, E., & Bos, W. (2000). Mathematikleistung und mathematischer Unterricht in Deutschland und Japan, Triangulation qualitativer und quantitativer Analysen am Beispiel der TIMS-Studie [Mathematical performance and mathematics instruction in Germany and Japan, triangulation of qualitative and quantitative analyses exemplified by the TIMS-Study]. *Zeitschrift für Erziehungswissenschaft, 3*(3), 359-380.

Nussbaum, A. (1984). Quantitative und qualitative Unterrichtsforschung? Kontroverse Standpunkte und erste Schritte zu einer Annäherung [Quantitative and qualitative research in the classroom? Controversial viewpoints and first steps to convergence]. *Unterrichtswissenschaft, 12*(3), 218-231.

Petersen, J., & Ritscher, H. (1996). *Unterrichten lernen: Praxisbeispiele für die Lehrerbildung [Learning how to teach: Practical examples for the education of teachers].* Donauwörth: Auer Verlag.

Rice, J. M. (1893, [reprint]1969). *The public-school system of the United States.* New York, New York: Arno Press & The New York Times.

Rist, R. C. (1982). On the application of ethnographic inquiry to education: Procedures and possibilities. *Journal of Research in Science Teaching, 19*(6), 439-450.

Ritz-Fröhlich, G. (1976). *Verbale Interaktionsstrategien im Unterricht, Impuls – Denkanstoß – Frage [Verbal interaction strategies in the classroom, stimulus – thought-provoking impulse – question].* Ravensburg: Otto Maier Verlag.

Roeder, P. M., & Schümer, G. (1976). *Unterricht als Sprachlernsituation: Eine empirische Untersuchung über die Zusammenhänge der Interaktionsstrukturen mit der Schülersprache im Unterricht [Instruction as a language learning situation: An empirical study on the connections between structures of interaction and students' language in the classroom].* Düsseldorf: Pädagogischer Verlag Schwann.

Stevens, R. (1912, [reprint]1972). *The question as a measure of efficiency in instruction: A critical study of class-room practice.* New York, New York: Teachers College, Columbia University.

Stigler, J. W., Gonzales, P., Kawanaka, T., Knoll, S., & Serrano, A. (1999). *The TIMSS videotape classroom study: Methods and findings from an exploratory research project on eighth-grade mathematics instruction in Germany, Japan, and the United States - A Research and Development Report*. Washington, D. C.: U. S. Government Printing Office.

Tausch, R. & Tausch, A. (1971). *Erziehungspsychologie, Psychologische Prozesse in Erziehung und Unterrichtung [Psychology of education, psychological processes in education and instruction]*. Göttingen: Verlag für Psychologie.

Wolf, W. (1995). Qualitative versus quantitative Forschung [Qualitative versus quantitative research]. In E. König, & P. Zedler (Eds.), *Bilanz qualitativer Forschung, Band 1: Grundlagen qualitativer Forschung* (pp. 309-329), Weinheim: Deutscher Studien Verlag.

Astrid Begehr
Fachbereich Erziehungswissenschaft und Psychologie
Freie Universität Berlin
Germany

YOSHINORI SHIMIZU

CHAPTER TWELVE

Discrepancies in Perceptions of Mathematics Lessons Between the Teacher and the Students in a Japanese Classroom

INTRODUCTION

Classroom practice is a form of communal collaborative activity constituted as it is constructed through the participation of both teachers and learners and only understood and optimised through research that accords value and voice to all participants (Clarke, 2001). Given the fact that teaching and learning are interdependent activities within a common setting, classroom practices should be studied as such. The Learner's Perspective Study (LPS) has the potential in that it literally focuses on the perspectives of *learners* in mathematics classrooms as well as those of teachers. This chapter discusses the results of analysis of post-lesson video-stimulated interviews with the teacher and the students in an eighth-grade mathematics classroom in Tokyo that participated in the LPS.

The methodology employed in the LPS allows participants to identify those events that were significant to them. In the post-lesson video-stimulated interviews, which occurred on the same day as the relevant lesson, the teacher and the students were asked to identify and comment upon classroom events of their personal importance. The analysis of LPS data has revealed both patterns and variations in the ways in which the teacher and students perceived the lesson. The LPS then provides the researchers with the opportunity to explore the commonalities and differences in perceptions of mathematics lessons by teachers and students by means of juxtaposing their reconstructive accounts of the classroom.

The stimulated recall interview is one of the methods for exploring teachers' ideas and beliefs about teaching and learning (Clark & Peterson, 1986). A videotaped lesson as stimulus in the interview has been used in mathematics education research area. The entire videotaped lesson was, for example, used for examining American and Japanese teachers' ideas about what constitutes effective mathematics pedagogy (Jacobs & Morita, 2002). Also, video-stimulated recall interviews were conducted with the teacher and group of students to seek their interpretations of videotape excerpts as the events occurred in a classroom community of inquiry (Goos, 2004). Although video-stimulated interviews are used in examining teachers' and students' ideas and beliefs, earlier studies have no

D. J. Clarke, C. Keitel & Y. Shimizu (Eds.), Mathematics Classrooms In Twelve Countries: The Insider's Perspective. 183–194. © 2006 Sense Publishers. All rights reserved.

focus on contrasting perceptions between the teacher and the students of the same lesson they have just experienced.

In the classroom, teacher and student practice are supposed to be in a mutually supportive relationship. By contrasting their perceptions of particular lesson events, it is possible to identify the discrepancies between the teacher and the students in their perceptions of that practice. Furthermore, classmates can be more "significant others" who exerts a strong influence on individual students than the teacher in certain learning contexts (Bishop, Brew, Leder & Pearn, 1996). Through the analysis of perceptions of lesson events and associated values held by the teacher and the students, the influence of in learning mathematics can be explored.

The current approach to teacher and student perceptions of mathematics lessons is promising in bringing out the teaching and learning process that proceeds in a mutually supportive way for both sides. Japanese teachers, for instance, often try to organise an entire lesson around a few problems with a focus on the students' alternative solutions to them. In this "structured problem solving" approach, the "summing up" phase is indispensable to any lesson to be successful (Shimizu, 2006). Students' solutions are shared and pulled together during this phase in light of goals of the lesson. The teacher attaches different meanings to each phase of the lesson. How do the students perceive lesson events for that particular phase? While Japanese teachers may devote considerable effort to planning and structuring of their lessons around a climax of a lesson, these structures may not be perceived, or perceived differently by the students.

This chapter discusses issues and findings arising from the analysis of post-lesson video-stimulated interviews that juxtaposes perceptions of the event in the same lesson between the teacher and the students.

DATA AND METHODOLOGY

Data collection was conducted at three public junior high schools in Tokyo. The teachers, one female and two males, roughly represented the population balance of mathematics teachers at the secondary school level in Japan. The topic taught in each school corresponded to the three different content areas prescribed in the national curriculum guidelines: linear functions, plane geometry, and simultaneous linear equations.

The data of this study includes videotaped classroom data for ten consecutive mathematics lessons and transcripts of both lessons and post-lesson video-stimulated interviews with the teacher and students in each participating eighth-grade class as well as the data from the questionnaires completed by the participants and copies of relevant printed or written material.

The methodology employed in this study offered both the teachers and the students the opportunity in post-lesson video-stimulated interviews to 'parse' the lesson they had just experienced. That is, to identify for the interviewer those events in the lesson that the participant felt to be significant, the teacher and the students were given control of the video replay and asked to identify and comment upon classroom events of personal importance. Semi-structured post-lesson video-stimulated interviews occurred on the same day as the relevant lesson. The prompts for interviews include:

- *Prompt Four*: Here is the remote control for the video-player. Do you understand how it works? (Allow time for a short familiarisation with the control). I would like you to comment on the videotape for me. You do not need to comment on all of the lessons. Fast-forward the videotape until you find sections of the lesson that you think were important. Play these sections at normal speed and describe for me what you were doing, thinking and feeling during each of these videotape sequences. You can comment while the videotape is playing, but pause the tape if there is something that you want to talk about in detail.

- *Prompt Seven*: Would you describe that lesson as a good one for you? What has to happen for you to feel that a lesson was a 'good' lesson? Did you achieve your goals? What are the important things you should learn in a mathematics lesson?

During the interviews, interviewers were required to be explicit in specifying the time elapsed from the beginning of each lesson in the videotape, when the teacher and the students referred to an event as of significant to them. It is clearly possible that students identify as significant, lesson events that are quite different from those intended and/or identified by the teachers. Here is the potential to explore the differences and commonalities of their perceptions of mathematics lessons by juxtaposing teacher and student reconstructive accounts of the mathematics lessons that they have just experienced.

CONTRASTING PERCEPTIONS OF LESSON EVENTS BETWEEN THE TEACHER AND THE STUDENTS

The transcriptions of post-lesson video-stimulated interviews with the teacher and the students in one of the three eighth-grade mathematics classrooms were analysed by focusing on their responses to the particular prompts listed above. From a research perspective, and from a practical perspective as well, it is intriguing to explore the meaning of lesson events identified as felt to be significant by either the teacher or the students.

Table 1 shows the numbers of elements in each lesson identified as felt to be significant by the teacher and students in School J1. As Table 1 shows, the number of elements identified varied from one to twelve. Most students identified different numbers of elements as felt to be significant in two different lessons. Also, the numbers of identified elements both between the teacher and students and between two students in the same lesson were different. In the post-lesson interview of lesson J1-L05, for example, the teacher identified nine elements in the lesson to be significant, while each of two students interviewed identified eight and seven elements respectively. As for the lesson J1-L07, on the other hand, the teacher identified twelve elements in the lesson to be significant, while one of the two students identified eight elements and the other only three.

Table 1. Elements in the lessons felt to be significant: School J1

Lesson	Teacher	Student 1	Student 2
J1-L01	—	SHI: 2	TSU: 5
J1-L02	—	MA: 8	MO: 6
J1-L03	—	SHI: 5	TSU: 2
J1-L04	—	MA: 7	MO: 2
J1-L05	9	NI: 8	TA: 7
J1-L06	—	UN: 3	II: 4
J1-L07	12	NI: 8	TA: 3
J1-L08	—	UN: 1	II: 4
J1-L09	—	SU: 8	NO: 3
J1-L10	—	O: 4	IW: 4

Table 2 and Table 3 show the location of identified elements felt to be significant for each participant by the time elapsed (in minutes and seconds) from the beginning of each lesson. Although the number of elements identified as of significance in J1-L05 to both the teacher and two students were similar, their locations in the entire lesson were different except for a few elements. Only two elements were identical among three participants in lesson J1-L05. There were no elements identified as significant by all three of the participants in lesson J1-L07.

Table 2. Elements in the lessons felt to be significant: J1-L05

Teacher	Student 1	Student 2
3:50		
		6:00
9:29	9:23	
14:00	14:25	14:22
	16:00	
16:50		
		17:30
		24:44
	27:09	
28:00		
29:30		
32:26		
33:30		
	34:30	
	37:45	37:26
		40:00
42:50	42:55	43:02
	45:20	

Table 3. Elements in the lessons felt to be significant: J1-L07

Teacher	Student 1	Student 2
2:00		
	2:38	3:00
11:30		
15:00		
	17:25	
22:00		
	22:40	
		23:30
24:11		
	26:40	
$(**{:}**)^{a)}$		
30:00		
	31:17	
	31:40	
34:30		
	36:25	
37:00		37:00
40:00		
40:20	40:24	
42:00		

a) Unclear

The analysis of post-lessons video-stimulated interviews with the teacher and students reveals that their perceptions of mathematics lessons did differ significantly. The comparison of perceived events to be felt significant by the teacher and the students clearly shows that there are agreements and discrepancies between them on what is important in mathematics lessons and what is not. Here is the strength of the research methodology adapted by the LPS – to explore the differences and commonalities of their perceptions of mathematics lessons by juxtaposing teacher and student reconstructive accounts of the mathematics lessons that they have just experienced

The result of the analysis realises the issue of exploring the influences of such agreements and discrepancies in perceptions of lesson events between teacher and students on students' learning and teacher planning. The result also raises the need to attend the meanings constructed by the teacher and the students as the participants in the same lesson. If we take an approach of comparing specific components from which each lesson is constructed to examine the function of such components in the instructional sequence, we need to listen to the voice of the participants more closely. It is particularly interesting for the current study to examine the excerpts from transcriptions of post-lesson video-stimulated interviews with a focus on the event that was identified as of significance by *only* the teacher or the students as well as by *all* the participants.

A CLOSER LOOK AT PARTICIPANTS' PERCEPTION OF CLASSROOM EVENTS

When we look into our data more closely, we can see both conformities and discrepancies in perceptions of classroom events of personal importance between the teacher and the students. The following interview excerpts from lesson J1-L05 are examined for elements in the lesson felt to be significant by all three participants.

Ms. K (00:14:00)

```
01. Ms.K:   Can I just pause for a moment?
02. INT:    Okay. Here. Um, fourteen minutes, the scene of group
            discussion
03. Ms.K:   Well, this class is comparatively quiet, so I haven't let
            their desk in groups until today. But I thought it might be
            useful for students to complete the graph along with the
            discussion. They can help one another by thinking together,
            and it allows everyone to see other students' work. Maybe
            some of them can understand where they couldn't do by
            themselves
```

Key to symbols used in transcripts in this chapter

/	Indicates that one speaker cut in, interrupting another speaker before they had finished.
[text]	Comments and annotations, often descriptions of non-verbal action.
....	Indicates that a portion of the transcript has been omitted.

The teacher in the excerpt above emphasises the usefulness for the students to work and think together and mentions to the importance of understanding with the help of classmates. The two students also identified the same event as of personal significance and commented on the event as follows:

Student 1 (NI) (00:14:25)

```
04. NI:     Here. Transcript? [checks the rest at the bottom and top]
05. INT:    Here, right. Fourteen minutes twenty-five seconds. Where
            you started group work. Ok. Why here?
06. NI:     Discussing the graph and the chart in groups and finishing
            it up.
07. INT:    This scene, right? Ok, about this scene.
08. NI:     Here, I can share what I thought and what others thought,
            the answers I thought I had made a mistake on, could
            actually be right, not a mistake, my mistakes could
            actually be the answer. Huh?
09. INT:    What you thought was a mistake, might actually be correct.
10. NI:     Right. What I thought couldn't be incorrect, could actually
            be a mistake. I can compare my answers with other people
            and talk it over, so it's nice
```

Student 2 (TA) (00:14:22)

```
11. TA:     Here, in groups.
10. INT:    Yes, uh, fourteen minutes and twenty-two seconds, the scene
            where group learning has started. Ok. What did you think
            here?
....
11. INT:    Where you thought was important.
12. TA:     Yes.
13. INT:    Important in today's class.
```

```
14. TA:    In today's class, yes.
15. INT:   Um, what were you doing uh, then, together?
16. TA:    Oh, together, we first compared the part of the homework I
           did, and the part my friend did, and we discussed the parts
           that were different and shouldn't it be this, and such.
           Then, after that, there was a part we didn't understand and
           it was the same part, so we were asking each other how we
           were supposed to solve it, and so we asked the teacher.
```

Although Student 2 (TA) is not as explicit as both the teacher and Student 1 (NI) in explaining the reason why she thought the event important, the three of them appeared to share such a belief that they can help one another by thinking together, and to understand what they could not complete from each other. In this event, then, it is safe to say that the teacher's intention and what the students wanted to do at that moment were not different.

The discrepancies in perceptions of classroom events of personal importance between the teacher and the students became clear when the students realised that the teacher did not understand their solution to the problem. The two students who had worked together made similar comments as follows.

Student 1 (NI) (00:37:45)

```
17. NI:    Here, right here.
18. INT:   Uh, thirty-seven minutes forty-five seconds. Uh, where she
           went on to the explanation about the graph, NI, right?
19. NI:    The teacher, what the teacher was thinking, and what me and
           Ta-san were thinking were different, so, we were both
           trying to explain but the teacher didn't really understand
           us, and finally, after a while, she understood.
20. INT:   The teacher, she wrote down a graph but it was different,
           right? From what you and TA got, and there was a gap
           between what you were thinking and what she was thinking,
           right?
```

Student 2 (TA) (00:37:26)

```
21. TA:    Yes.
22. INT:   The scene where Ni comments on the graph. Thirty-seven
           minutes twenty-six seconds.
23. TA:    Uh, this, the teacher mistook what I said or something and
           Ni-san pointed out her mistake, but the teacher made
           mistakes once in a while so it would be nice if someone
           could say the correct answers in situations like this
24. INT:   Ohh. So, you felt to the teacher, the teacher that your
           ideas were being interpreted differently, so you were
           trying somehow to say this to her. Uh, actually she said
           this for you.
25. TA:    Said this for me. Yes.
....
26. INT:   That's important. So, then, Ta-san, anything you felt, or
           rather, thought?
27. TA:    Feel
28. INT:   If you got across what you wanted to say.
29. TA:    Oh, what I wanted to get across? I was explaining but Ms. K
           misunderstood again so what I wanted to say didn't really
           get across to her but there's a scene after this where I
           explain something else again.
```

```
30. INT:   Yes.
31. TA:    And, there, uh, I think Ms. Ka understood the correct
           answer.
32. INT:   Yes, where you went up to the chalkboard. The scene where
           you're explaining, the teacher understood you correctly.
33. TA:    Yes.
```

On the other hand, Ms. K did not realise until the final minutes of the lesson, that she understood differently the explanation made by two students during the instruction at desks (Kikan-Shido) and the whole class discussion. All three participants identified the element around sixteen to seventeen minutes from the start where the teacher was doing Kikan-Shido at their desks, as significant. At this moment, the teacher intended to assess what these two students were doing and she gave them a hint. The students commented in the interviews on the importance of Kikan-Shido in the current lesson and appreciated the teacher's help at that moment.

In the later part of the lesson, however, the students realised that the teacher understood their explanation differently. Two students identified as significant the event where Student 1 presented (S1: 37:45, S2: 37:26, see Table 2) and Student 2 presented (S1: 34:30, see Table 2) to explain again their response to the teacher's question in the whole class discussion. Ms. K, on the other hand, identified only one of the two events (Ms. K: 33:30) based on her original intention of having 'Yamaba', the highlight of the lesson, around the interpretation of the "seven point five matter", which means to examine what is happening on the graph, which was derived from the problem situation, at the point of $x = 7.5$.

In the post-lesson interview, she answered the question, "what were your goals in today's lesson?" as follows. This comment directly connected to the Yamaba in the lesson from her perspective.

```
34. Ms.K:  All right, um, talking about this lesson plan, it was one
           of our goals to learn about the change among diagram,
           graph, and formula. Also, to think about the domain of
           change. And, um, we had the point that how we determine an
           independent variable against an dependent variable. So, as
           our plan is along all of these factors, I know students
           were confused at several points. But for today, we could
           complete drawing a graph, and made a consideration when
           seven point five was given, in other words, we thought
           about the area between the dot and another dot. It was an
           introduction to regard a graph as an aggregation of dots.
           So, I wanted to make today's lesson as an entrance to the
           examination of general graph that will be shown in the
           lessons from now on. And also, there was another thing that
           today's class was aiming at unifying a graph and rate of
           change, not studying each one separately. So, um, the main
           point of the lesson was to be able to look for the answer
           of y to the number seven point five, I mean to the change
           of zero point five.
```

The main point of the lesson to her was "to look for the answer of y to the number seven point five" as she mentioned. She intended to highlight two different interpretations on for the answer of y to the number seven point five based on her anticipation of students' alternative approaches. She answered the question, "what do you think is a good lesson?" as follows.

35. Ms.K: Um, it depends on what kind of teaching materials you are using, but if we use this type of teaching materials about function, I would say, "the class having multiple perspectives in considering one phenomenon".
36. INT: Multiple perspectives from the students?
37. Ms.K: Yes.

So, she wanted to have multiple perspectives from the students on the interpretation of the "seven point five matter". Thus, Yamaba, the highlight of lesson, from the teacher's perspective was to be the whole class discussion of students' alternative solutions to the "seven point five matter". For the two students interviewed, however, the event in the lesson felt to be significant was more specifically focused on teacher's understanding of their own solution. The teacher reflected on her understanding of students' understanding as follows.

Ms. K (00:33:30)

38. INT: Okay. Um, let me just say the count. Thirty-three minutes and thirty seconds, the third female student, TA, is doing the presentation.
39. Ms.K: Um, here, I had already decided that I would center that seven point five matter, but I still wanted to make sure that every student had understood about another diagram. So, I let them do the presentation. But, my perception and their perception of perimeter were different. Um, I mean, students understood that part better than me. Um, I have just taught another class now, and I noticed that my point of view and students' point of view were different. I always explain the question by looking at the perimeter, but for those students like TA, they take the colored part as y. I think it is interesting as they take an independent variable as fifteen, fourteen, and thirteen when they drew a graph after this scene. That is, they took an independent variable in decreasing direction, so their graph was shaped in lower right direction. I thought it was very different from mine, but actually it was not a big problem. So, I felt that I could just leave it on because I understood it as a different way of making sense.
40. INT: Your perception of the wrong answer and the actual perception of those students had were somewhat different.
41. Ms.K: That's right.
42. INT: Okay. So, did you notice that through this presenter, TA's presentation?
43. Ms.K: Um, no/
44. INT: No?
45. Ms.K: No, well, I had this feeling,
46. INT: You felt it was rather different.
47. Ms.K: Yeah, I felt so. And, it became clear when zero was given, I mean, when they spoke up their opinion toward my question that why it would become twenty-six when one was given. That was the time I thought, well, I had better treat it in

```
        the next class. Um, I thought I could use their thoughts
        for next lesson of rate of change.
```

The excerpts from transcriptions of the post-lesson interviews suggest that students occasionally perceive the lesson differently from their teacher. As was mentioned earlier, one of the characteristics of Japanese teachers' planning of lessons is the deliberate structuring of the lesson around a climax within the entire lesson. The students in Japanese classrooms can be unaware of the occurrence of these climactic points or their intended significance. The teacher, in turn, can be unaware of students' perspective on what really matters to them.

DISCUSSION

The analysis reported in this chapter clearly showed that students identified significant lesson events differently from those intended and identified by the teacher. The analysis also revealed, by juxtaposing their perceptions of mathematics lessons, that there were both discrepancies and agreements between the teacher and the students in the event with the perceived significance. In other words, while they shared perceptions of the personal importance of particular lesson event, 'group work', for example, the students can attribute different meanings to the same event from those constructed by the teacher. The students in the classroom can be unaware of the occurrence of the event that is intended by the teacher as of significance to their learning. The teacher can also be unaware of the important moments in relation to her students' activity during the lesson for her further planning of lessons.

When we look into the interview data closely, the difference can be understood as discrepancies in perceptions of classroom events between the teacher and the students. The analysis described in this chapter raises the question about to what extent are teacher and students practices conceived in a mutually supportive relationship. The teacher, who realised that her interpretation of students' solution was different from those by the students, thought that she had better treat their solution in the next class *during* her teaching. In this sense, discrepancies in perceptions of classroom events between the teacher and the students can be building blocks for other lessons in an instructional sequence.

The analysis also reveals the richness and potential of the collected data, as well as strength of the methodology, in the LPS. The methodology employed in the LPS allows participants to identify those events that were significant to them. The analysis of the data revealed variations in the ways in which the teacher and students perceived the lesson. The LPS certainly provides the researchers with the opportunity to explore the commonalities and differences in perceptions of mathematics lessons by teachers and students by means of juxtaposing their reconstructive accounts of the classroom.

Issues raised by the results should be explored in a further analysis. A further analysis of the LPS data is needed to clarify the discrepancies and associated meanings. One arena to look at is the development of classroom norms in relations to the perceptions of the importance of particular classroom events (Yackel & Cobb, 1996). Students in mathematics classroom are learning not only mathematics itself but also how to learn mathematics. The recognition of the importance of a particular classroom event, summing up at the end of lessons, for example, derives from learning about learning mathematics.

Needless to say, there are both similarities and differences among classroom practices in different countries. The findings from the TIMSS 1999 Video Study included indicators of how talk was shared between teachers and students, the total number of words spoken by teachers and students during public interaction, for instance. The result shows that there is a strong similarity among participating countries (Hiebert et al., 2003, p.109). The data from the LPS can help us find differences in participants' perceptions of mathematics lesson that are otherwise not apparent.

It is of interest in this study whether the learner practices observed in one country show consistency of form and purpose such as to suggest a culturally-specific character. Whether or not such identifiable learner characteristics exist as cultural traits, this study is predicated on a belief that international comparative studies are likely to reveal patterns of practice less evident in studies limited to a single country or community. They might share the beliefs about learning mathematics in the classroom and be constructing different meanings associated to the same events.

CONCLUDING REMARKS

In this chapter issues arising from the analysis of post-lesson video-stimulated interviews with the teacher and students in an eighth-grade mathematics classrooms were discussed. The analysis revealed that students identified significant classroom events differently from those identified by the teacher. By juxtaposing their perceptions of mathematics lessons, discrepancies and agreements between the teachers and the students appeared. Then, the analysis raises the question to the extent in which teacher and learner practices are in a mutually supportive relationship. The results indicate that teacher and learner practices appear to be both conflicting and mutually sustaining.

The analysis in this chapter demonstrated the richness and potentials of the collected data, as well as strength of the methodology, in the Learner's Perspective Study. A further study of participants' perceptions of mathematics lessons is needed which explains the process of meaning constructions in the sequence of lessons and their impact on changes of perceptions.

ACKNOWLEDGEMENT

The research reported in this chapter was partially funded by 2004-2006 Grant-in-Aid for Scientific Research (B) of Ministry of Education, Science, Sports and Culture (Grant No. 16300249).

REFERENCES

Bishop, A. J., Brew, C. Leder, G., & Pearn, C. (1996). The influences of significant others on student attitudes to mathematics learning. In L.Puig & A. Gutierrez (Eds.), *Proceedings of the 20th Conference of the International Group for the Psychology of Mathematics Education*, Vol. 2. (pp. 89-96). Valencia, Spain.

Clark, C. M., & Peterson, P. L. (1986). Teachers' thought processes. In M.C. Wittrock (Ed.), *Handbook of research on teaching* (3rd ed., pp. 255-296). New York: Macmillan.

Clarke, D. (2001). Teaching/Learning (pp. 291-320). In D. Clarke (Ed.), *Perspective on practice and meanings in mathematics and science classrooms*. Dordrecht, Netherlands: Kluwer Academic Press.

Clarke, D. (2004). Learner's perspective study: Developing meaning from complementary accounts of practice. In M. J. Høines & A. B. Fuglestad (Eds.), *Proceedings of 28th Conference of the International Group for the Psychology of Mathematics Education* (Vol. 1, pp. 212-216). Bergen, Norway: Bergen University College.

Goos, M. (2004). Learning mathematics in a classroom community of inquiry. *Journal for Research in Mathematics Education, 35*(4), 258-291.

Hiebert, J., Gallimore, R., Garnier, H., Givvin, K. B., Hollingsworth, H., Jacobs, J., Chiu, A.M.Y., Wearne, D., Smith, M., Kersting, N., Manaster, A., Tseng, E., Etterbeek, W., Manaster, C., Gonzales, P., & Stigler, J. (2003). *Teaching mathematics in seven countries: Results from the TIMSS 1999 Video Study*. U.S. Department of Education. Washington, DC: National Center for Education Statistics.

Jacobs, J, & Morita. E. (2002). Japanese and American teachers' evaluations of videotaped mathematics lesons. *Journal for Research in Mathematics Education, 33*(3), 154-175.

Shimizu, Y. (2006). How do you conclude today's lesson? The form and functions of 'matome' in mathematics lessons. In Chapter 6 of D. Clarke, J. Emanuelsson, E. Jablonka & I. A. C. Mok (Eds.), *Making connections: Comparing mathematics classrooms around the world.*. Rotterdam: Sense Publishers.

Yackel, E. & Cobb, P. (1996). Sociomathematical norms, argumentation, and autonomy in mathematics. *Journal for Research in Mathematics Education, 27*(4), 458-477.

Yoshinori Shimizu
Graduate School of Comprehensive Human Sciences
University of Tsukuba
Japan

FLORENDA LOTA GALLOS

CHAPTER THIRTEEN

Students' Private Discourse in a Philippine Classroom: An Alternative to the Teacher's Classroom Discourse?

INTRODUCTION

In the Philippines, typical mathematics teaching/learning is dominated by extensive teacher-directed explaining and questioning in a whole group setting where teacher's questions required short and definite answers (Pascua, 1993). Students claimed that it is through listening attentively to the teacher and then providing answers to the mainly closed-type questions where mathematics is best learned (Gallos & Ulep, 2002). It is apparent that the typical classroom situation is that of a teacher doing most of the talking and the students are supposed to just listen and talk only when the teacher asks a question. Yet, it was shown in the Learner's Perspective Study (LPS) that students talked privately despite the situation that they were limited to public talks.

There are recommendations to allow students to talk privately in mathematics classrooms. The National Council of Teachers of Mathematics (NCTM, 1991) mentioned that students privately talking to their seatmates should be part of a classroom discourse. They claimed that it is important to know the nature of classroom discourse for it is a major influence on what students learn about mathematics. However, studies about students' private conversations in the Philippine mathematics classrooms are hard to find. One possible reason could be that findings were not published, if ever there are some around. Another reason could be that such study could pose difficulty on capturing these private conversations obliviously, thus researchers would not like to pursue such study. In the LPS, with the help of technology, the private conversations of students in a particular mathematics classroom were captured.

LOOKING BACK

Mathematical discourse involves students and teachers to be engaged in tasks that require ways of representing, talking and agreeing and disagreeing their mathematical ideas, either in oral or written form (NCTM, 1991). D'Ambrosio and Prevost (1995) described it as a process of engaging the members of the classroom

D. J. Clarke, C. Keitel & Y. Shimizu (Eds.), Mathematics Classrooms In Twelve Countries: The Insider's Perspective. 195–208. © 2006 Sense Publishers. All rights reserved.

community, which are the students and teachers in talking with one another. It is expected that when students engaged in such tasks, they could share their ideas with their classmates or initiate questions intended for their teacher and their classmates that could lead to better understanding of mathematics.

It was suggested that all discourse can be marked by two different texts functions: to convey meaning and to generate meaning (Lotman cited in Knuth & Peressini, 2000). Wertsch (1991) referred to the first function as *univocal* where the listener receives the message that the speaker intends to send. He referred to the other function as *dialogic* that is characterised by the give-and-take communication in which the listener initially receives the message sent by the speaker. At this moment, univocal discourse stops and dialogic discourse has just begun. In the dialogic discourse, the dialogue is being used as the "thinking device" (Lotman cited in Knuth & Peressini, 2000).

"Life is dialogical by its very nature. To live means to engage in dialogue, to question, to listen, to answer to agree, etc." (Bakhtin, as cited in Dysthe, 1996). This statement implies that the nature of life is for everyone to be engaged in a dialogue. And this dialogue can viewed as public, private or personal (Mifsud & Johnson, 2000). *Public* dialogue pertains to civic or community communication, *private* dialogue to interpersonal communication and *personal* dialogue to intrapersonal communication.

In Clarke's (2001a) statement, frequent communication among students is considered a key attribute of effective classrooms. The team of Backhouse, Haggary, Pirie and Stratton (1992) has a similar statement for they claimed that learners quite often sort out ideas by talking things through with the person sitting next to them. They added that talking things through does not work all the time, but when it does it is an effective means of learning for it is the learner's own achievement. According to Adler (1999) talk is considered a resource for learning mathematics in school. She pointed out that for talk to be a resource for mathematics learning, students must be able to see it and use it. In the study of Manouchehri and Enderson (1999) of a seventh grade classroom where mathematical discourse was being promoted, several findings were discussed. A significant mark was noted on the degree of ownership the students took of the situation that led to further dialogues that were productive in many ways.

In the initial data of the LPS, 12 of the 17 students interviewed about the best way to learn mathematics said that it is through listening attentively to the teacher (Gallos & Ulep, 2002). This learning style of students appeared to match exposition, the teacher's main teaching strategy. It is indeed a typical classroom culture in the country that the teacher does most of the talking and the students are supposed to just listen and talk only when the teacher asked a question. Thus, it can be said that in the classroom the students are not supposed to converse with their seatmates. However, in the studies of Wallace (1995) and Tenedero (2000) Filipino students were found to be visual and kinesthetic learners and least as auditory learners. Moreover, LPS data showed that students did talk to their seatmates while mathematics classes were ongoing. Thus, it would be important to have a close look at these students' private conversations in a Philippine classroom and try to find out what patterns of ideas can be derived from them. In this way students' learning styles and their understanding or misconceptions in mathematics can be unveiled, and hence can be a valuable input in the improvement of teaching and learning of mathematics.

HOW DATA WERE GATHERED

Data were drawn from video-recorded observations and interviews collected from an intact class in one of three schools in the Philippines which were sampled for the LPS. This particular class who was handled by a competent teacher, Ms. Santos, has 57 students and is the top 5 of 44 sections in an urban public high school. The class was observed for 10 consecutive school days (15 days including the familiarisation) and the lessons were recorded using three video cameras. The first camera focused on the teacher, the second one on two students (referred to as the focus students for that day) and the third one on the whole class. Video-simulated interviews with the focus students followed right after each lesson. The teacher video-simulated interview was done once a week. This documentation procedure was based on the Complementary Accounts Methodology (Clarke, 2001b). It captured the conversations of the focus students as well as those students near them, including private discourses, on video. Aside from the video-recorded lessons and tape-recorded interviews, the observer's notes, mixer's notes, copies of focus students' work and the teacher's lesson plans were also collected.

REFLECTIONS ON PRIVATE CONVERSATIONS

Analysis of data taken from the mixed video lessons, students' interviews, students' work, lesson plans and field notes could show a picture of the classroom scene. However, for this particular study, the discussion will focus on students' private conversations that are mathematical and between their seatmates.

It emerged that those patterns on students' private conversations vary, depending on the activities for that particular lesson and on students' participation. For example, in Lesson 12, the activity required students to measure angles using the protractor, thus, there were students' conversations recorded, while that in Lesson 15, minimal student talk was recorded for a test was administered in this session. With regards to the nature of the students themselves, an example of this was when the teacher used exposition as a teaching strategy in two different lessons. This difference in the nature of students' participation happened in Lesson 6 where the focus students talked a lot whereas in Lesson 8 just one conversation was recorded, which was mathematical, univocal and public.

The trend of the students' mathematical conversations was found to be mainly between their seatmates, which were dialogic and private in nature. This trend was derived by analysing students' talks recorded in a table, part of which is shown on the next page.

Table 1. Record of Time of Students' Mathematical and Private Conversations

Lesson No.	Name	Dialogic-Private		
		with Teacher	*with Seatmate*	*with Others*
6	Kath	17:00 - 28:06	2:11 - 2:20	
			3:34 - 3:35	
			4:17 - 4:18	
			6:09 - 6:10	
			7:05 - 7:21	
			10:19 - 10:32	
			12:36 - 12:37	
			20:08 - 20:09	
			21:44 - 21:48	
			22:44 - 22:45	
			24:37 - 24:39	
			25:05 - 25:18	
			27:33 - 27:47	
			31:09 - 31:24	
	Rene		7:20 - 7:22	
			11:34 - 11:36	
			13:57 - 13:58	
			20:11 - 20:12	
			24:36 - 24:45	
			25:19 - 25:20	
			27:21 - 27:26	
			31:08 - 31:24	
			32:58 - 33:13	
			33:04 - 33:08	

Further analysis of data showed more patterns regarding students' private and mathematical conversations. It emerged that there were private conversations that can be classified as those used by students as a means to understand mathematics using their own language, as an alternative to the public dialogue and those that interfere with the public dialogue.

Private Conversations as Means to Understand Mathematics in an Unhindered Language

It emerged that when the students talk privately to understand mathematics they appeared to be comfortable using their own language. Their unhindered private conversations were in the Filipino language, with a mix of some English words and phrases. For instance, the following excerpts are taken from Lesson 10:

```
0:36 Mich:      Nakapag-review ka na? (Have you reviewed?)
```

0:37	Abby:	Hindi. Kasi wala ng <u>time</u>. *(No. Because there's no more time.)*
18:59	Mich:	Paano? Ia-<u>add</u> yung <u>signs</u>? *(How? Shall I add the signs?)*
19:01	Abby:	Yung, yung ano, <u>ten minus</u>, ah, <u>ten minus three equal seven</u>. Tingnan mo, <u>ten minus three equals to seven. Plus ten, seventeen</u>. *(The, the what, ten minus, ah, ten minus three equal seven. Look, ten minus three equals to seven. Plus ten, seventeen.)*

Key to symbols used in transcripts in this chapter
(text) English translation of previous text.
[text] Comments and annotations, often descriptions of non-verbal action.
<u>text</u> Underlined text indicates use of English.

Aside from conversing in a mixed language, students used a language that is perhaps used by those of their age. Statements or phrases such as "Gets mo?" (*Did you get it?*), "Pa'no ba, ssht hoy!" (*How, ssht hey!*) and some swear words were not recorded as used when talking to the teacher.

Further analysis of the dialogic, private and mathematical conversations resulted in more patterns. Some mathematical conversations were found to interfere with the public dialogues and could impede students' learning of mathematics, while others can be considered as an alternative to the public dialogue that could enhance students' learning of mathematics. The discussions that follow are some situations that captured students' private conversations. Again, these conversations were in Filipino language with a mix of some English words and phrases but the translation in the English language was used in the transcripts that follow.

Private Conversations as an Alternative to Public Dialogue

The students' private conversations could be considered as an alternative to the public dialogue with the teacher. Firstly, students were apparently able to seek help from their seatmates in doing some tasks more easily. They also discussed mathematical concepts and clarified their ideas more easily. Lastly, they appeared more comfortable discussing with their seatmates for they can freely use any language they like.

Seeking for explanations or procedures. It appeared that the most common mathematics-related conversations of students pertain to seeking explanations or procedures. For example, in Lesson 10, the topic was on finding the coordinate of a point on the number line, where the item was like this: "Solve for x so that PQ = QR. Point P is x, point Q is -8 and point R is -2". Abby was trying to get an explanation from Mich how the answer was derived. Here is a transcript of their private conversation and the figure on the next page is Abby's solution:

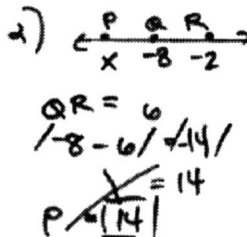

Figure 1. Abby's solution to the item

16:39	Abby:	Bakit fourteen? *(Why fourteen?)*
16:40	Mich:	Fourteen ang sagot mo, fourteen din ang sagot ni Barena. Mali mali nga ako dyan. Nakakainis! *(Fourteen, that's your answer, fourteen is also the answer of Barena. I'm wrong there. Annoying!)*
16:47	Abby:	Uh-hu-hu, bakit mali? Negative fourteen. Kasi sabi plus five daw eh. *(Oh, hu, hu,[an expression as if crying] why wrong? Negative fourteen. Because she said it's plus five.)*
16:53	Kath:	Fourteen! Ano ka ba? *(Fourteen! What's the matter with you?)*
16:54	Mich:	Ano ka ba? *(What's the matter with you?)*
16:55	Abby:	Sabi nya five. *(She said five.)*
16:57	Mich:	Fourteen.
16:58	Abby:	Sabi nya five. *(She said five.)*
17:00	Mich:	Fourteen!
17:01	Kath:	Di ba six 'to? Six na yung distance. [Counts using her fingers] Nine, ten, eleven, twelve, thirteen, fourteen. *(That's six, right? Six is already the distance. Nine, ten, eleven, twelve, thirteen, fourteen.)*
17:26	Mich:	Ahh, [then nods and paused], nakakainis, nalito ako dyan. *(Ahh, annoying, I was confused there.)*

It was obvious here that Abby wanted an explanation from Mich by asking a question. However, rather than explaining to Abby why five as an answer was wrong, she referred to an answer of another student and just insisted on it. The argument only ended when Kath explained to them how the answer was arrived at. The nod can be interpreted as indicating the explanation provided was clear. Here, it was apparent that Abby did not bother to look back on the reasonableness of her answer. As shown in her illustration, an answer of five or a positive number for this case, is not an acceptable answer.

In another session (Lesson 12), the two focus students were discussing the procedure for using the protractor. Before this activity on measuring angles, the teacher asked the students to raise their hands if they do not know how to use the protractor and no one raised their hands. But take note of their private conversations at this time:

```
13:07 Ken:      Hindi ganyan. H'wag dyan, hindi mo naiintindihan.
                Number one 'to, di ba ito thirty five? Thirty five
                di ba? (Not like that. Not there, you didn't
                understand. This is number one, it is thirty five,
                isn't it? Thirty five, isn't it?)
13:17 Nico:     [Asks Ken a question, inaudible]
28:05 Ken:      APN, angle APN. Angle P?
28:06 Danny:    [Nods]
28:12 Ken:      Pwedeng ito tapos paganon. (It can be this, then
                like that.)
28:15 Danny:    Tange! Paganto tapos ito at ito. Tange, siya na ang
                pinaka-vertex niya eh. (Stupid! Like this then this
                and this. Stupid, that's already its vertex.)
28:22 Ken:      Sabi ni ma'am hindi pwede. Kunwari maraming ganyan
                hindi pwede yan. (Ma'am said that can't be. Suppose
                there are several like that, that can't be.)
28:28 Danny:    Pwede yan! (That can be!)
```

It is clear from this transcript that the focus students had difficulty using the protractor. Ken was trying to help Danny by demonstrating it. Again, the nod could be interpreted that the explanation was accepted. Similar to what happened in Lesson 10, when an argument arose between them, it appeared that the reasoning is not mainly based on mathematics but on the authority, which is the teacher. Thus, despite the apparent practice of students to talk privately as an alternative to the teacher's public talk, the authority of the teacher was still obvious.

Here, what seemed to be the pattern is that, students seek help from another student in learning the mathematics lessons through private talks. In some instances they were able to provide help, yet, whenever they come across an argument requiring mathematical reasoning, opportunities to use this skill was not the immediate resource. They seemed to have difficulty providing the mathematical reasons, rather, they would just state the answers and insist they were correct, sometimes with swear words, or with a plan to refer the matter to the teacher or at times with other classmates.

Conversations in relation to understanding mathematics. Some students would initiate a conversation to understand mathematics. This could be in statements about what's in their minds or in a form of a question. For example, in Lesson 14, Ms. Santos was trying to draw out from the students a conclusion about segments PQ and PR when Q is a midpoint of segment PR. The conversation of the focus students was this:

```
38:14 Angel:     Com, ano yun? (Com, what's that?)
38:15 Michaela:  Coplanar?
38:16 Angel:     [Nods]
38:17 Michaela:  Hindi. (No.)
38:18 Angel:     Non coplanar?
38:19 Michaela:  Hindi. (No.)
38:20 Angel:     Oh?
38:27 Angel:     Dun sa plane. (There on the plane.)
38:44 Angel:     Di ba coplanar yung dun sa plane? (Isn't it that
                 it's coplanar, that one on the plane?)
```

```
38:46 Michaela:    Iba yun. (That's different.)
38:47 Angel:       I see.
```

Here, it appeared that these students tried to recall the concept being asked by guessing without understanding the given question. Their conversation on this part ended when Angel seemed to agree with Michaela's statement that did not even provide an explanation why a coplanar is different from what was being discussed at that time. At this stage, the concepts of midpoint, plane and coplanar appeared to be unclear to these students. Angel's understanding of midpoint was captured in the interview conducted right after the lesson. What she meant by midpoint is that any points between two given points on a line segment. In contrast, her seatmate Michaela appeared to have fully understood the concept of midpoint and distance of points on a line segment for she was able to offer to the interviewer an alternative solution to an exercise about midpoints. However, Michaela's conversations with Angel did not show much discussion on how the former could help the latter enhance her understanding of mathematics.

In Lesson 9, the first part of the lesson was a discussion of answers to their assignment. When the item was on this, "If $\overline{BD} \cong \overline{DC}$ and $|BD| + |DC| = |BC|$; then what conclusion can you make?" the focus students talked about it and part of that transcript was this:

```
1:15 Leo:    Line BD. If line BD?
1:18 Nol:    Ha? (What?)
1:19 Leo:    If line BD?
1:21 Nol:    If line BD if, eh equals, ano ba to! (If line BD if,
             ah equals, what's this!)
```

They both appeared to clarify their thoughts by asking questions but ended up not pursuing this. In the interview, it turned out that each of them did not earn a mark for this item on their assignment for each of them gave an answer of "The distance of segment BD and DC are equal". It was apparent that it would be difficult for them to pursue with the discussion on this item for both of them appeared to have difficulty answering this item. Nol's reasons as to why segments BD and BC are equal appeared to be confusing. At first his reason was that because *"these are the same"* then after more probing questions he shifted his answer to the concept of midpoint, then reasoned again that the distance *"as if these are near each other"* and finally said that *"same distance".* He even said that he could not even remember the right answer for this item. In the interview with the other focus student, it turned out that he also did not get this item right. Actually, the focus students have the same answers to all the items so each got a total mark of one out of four. Thus, based on this assignment results and the interviews, engaging in more dialogic conversation to clarify their mathematics ideas might be difficult.

Although there were attempts to clarify their thoughts through private talks, it appeared that its use as an alternative to public dialogue is minimal. Students hardly pursue the conversations to further clarify their ideas. Thus, what follows is a close look at the private conversations of students that may interfere with their learning of mathematics.

Private Conversations that Interfere with the Public Dialogue

While students' private conversations can be considered as an alternative to the public dialogue with the teacher that learning may take place, it may also interfere with the public dialogue and may hinder learning of mathematics. Some of the situations are as follows:

Conversations that interfere with teacher's input. It was documented that as the public dialogue with Ms. Santos (Ms. S) was ongoing, private talks among some students were also ongoing. This happened in Lesson 13 where the topic was on Angle Addition Postulate. In one stage of the lesson the class was supposed to discuss measures of angles where some of the measures were given based on this illustration:

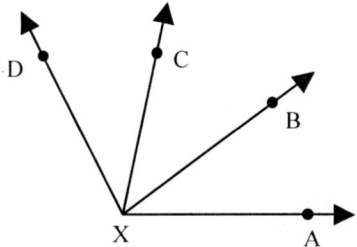

Figure 2. The item applying the Angle Addition Postulate

It was obvious that both public and private dialogues were ongoing while they were working on the above activity. Here is part of the transcript: (In all the two-column transcripts that follow, those at the left are public dialogues, either by the teacher or students, while those at the right are mainly private dialogues of focus students and their seatmates, with some occasional participation in public dialogues.)

	23:47 Kris: Di ba ganito lang 'to? Tapos AXC. Ito lang,
23:54 Ms. S: Okay, kasi as far as I know kapag may measure sya before or an 'm' before, wala, hindi na nilalagay yung degree. Kaya kung answer lang you	ito lang ba 'to? Ito lang? *(Just like this, right? Then AXC. Just this, just this? Just this?)*
	23:55 Mary: Ito. *(This.)*
	23:57 Kris: Ito lang. *(Just this.)*
	23:58 Mary: Ano ba yon? *(What is*

need here the unit
of measurement.
Nasa usapan natin
kanina. *(Okay,
because as far as I
know if there is a
measure before or
an 'm' before,
none, you need not
write the degree.
So if answer only
you need here the
unit of
measurement. It was
our agreement a
while ago).* We will
accept even without
the unit of
measurement. Okay,
but not kung exam
nyo na. *(Okay, but
not in your exam).*
Example in the long
test, we will not
accept it. Okay na?
(Is it okay now?)
Okay, number two.
Number two.

		it?)
23:59	Kris:	Halimbawa AXC. *(For example AXC.)*
24:00	Mary:	Ito, tapos ? *(This, then?)*
24:04	Kris:	Ito na yun! Minus BCX. Pa'no na? Ano ilalagay mo dito? *(This is it! Minus BCX. Then how? What will you put here?)*
24:14	Mary:	Ito. *(This.)*
24:15	Kris:	[Nods]

It was obvious that although students learn from each other, they may not be able to hear teacher's input and thus have lost the chance of learning those concepts and skills being discussed by the teacher. For this case, the teacher was emphasising when to include the degree measure and was also making a modification on what was agreed upon. However, as reflected in the transcript, the focus students appeared busy in their private talks and probably did not hear about this. It was also noticed in their conversations that Kris was naming the angle as BCX when it should be BXC or CXB. Thus, private talks may also lead to students' misunderstanding of concepts as will be elaborated in the next section.

Conversations that led to students' misconceptions. Students' talking to one another to seek further explanations, to clarify their ideas and to learn more about mathematical concepts could also lead to misconceptions rather than learning mathematics. This happened in Lesson 12 where Danny and Ken were discussing about how to measure angles and the teacher was asking about the relationship of the measure of angle one and the measure of angle four.

13:14 Ms. S: Okay, uhm [just an
expression], the
question is, what
can you say about
the measure,
measure of angle
one and the measure

of angle four?	13:21 Ken: Anong sign 'to? [pointing to the ≅ sign]. *(What sign is this?)*
	13:27 Danny: Ano, *(What,)* [paused] congruent or equal. [Paused] Equal [paused], equal.
13:40 Ms. S: O game, sige, Rochelle. *(Okay game, come on, Rochelle.)*	
13:42 R: Congruent.	
13:43 Ms. S: So congruent. Okay? If we speak of measurement, okay, how will you relate one form the other? So they are equal.	

As shown here, it was apparent that Ms. Santos wanted to emphasise the difference between the "≅" and "=" signs. However, instead of listening to the teacher Ken sought the help of Danny. But Danny's reply to Ken was not accurate. The next public dialogue of Ms. Santos could have corrected Danny's statement; however, it appeared that the focus students were still conversing privately as reflected in the transcripts that follow.

13:55 Ms. S: If we are talking about the measurement. But if we will be going to take the angle itself, then how are you going to describe the angle with the same measurement? They are what? Congruent. Naalala nyo yung sa segment? *(Do you still remember about segment?)* Okay, we take the distance, how do we describe the two? Equal, with regards to the measurement, they are equal. But if we take the segment itself, how are we going to describe?	13:55 Danny: Number two, one three five.
	14:02 Ken: Ano? *(What?)*
	14:05 Danny: Paano naging fifty five? *(How did it become fifty five?)*
	14:06 Ken: Number two? Tange, hindi pa ganyan. Pa ganito, o. Tapos pa ganyan. *(Stupid, not like that. Like this, see. Then like that.)*
14:22 Ms. S: Okay, the segments have the same distance and the segments are congruent. So that is the difference between	14:22 Ken: Congruent

```
the use of your equal
sign and the congruent
symbol.
```

The teacher's repetitious public dialogue and Ken's moving out of the private dialogue once in a while and participating in the public dialogue may have clarified his confusion of the mathematical sign. Thus, in the end he was able to provide the right answer.

In summary, these findings showed that despite the fact that the teacher used mainly exposition as the teaching strategy, students still talked with one another during mathematics classes. The most common mathematics-related conversations were those seeking explanations or procedures. The initiation of conversations sometimes started by asking questions from their seatmates. It was noted that their conversations were in the Filipino language with a mix of English words and phrases.

It was apparent from these results that students' conversations could either interfere with the class discussions that may obstruct learning of mathematics or serve as an alternative to public dialogue in order to enhance students' understanding of mathematics. Thus, allowing students to have private discourses in a mathematics classroom can be said to have its advantages and disadvantages.

IMPLICATIONS FOR PRACTICE

The teacher's lesson presentations were mainly using exposition where she did most of the talking and the students were supposed to listen. But there were still private talks that focused on mathematics that were recorded. It was apparent that students could have gained something out of these conversations. In the study of Pugalee (2001), he claimed that it was important for students, especially for students in the secondary level, to structure their thoughts, to express themselves, to listen to the ideas of others, and to think about the listeners when they expressed themselves. It is implied here that students should be given more opportunities to communicate mathematically with their classmates. Thus, the teacher could plan for mathematical tasks that would encourage students for more mathematical conversations and opportunities for him/her to listen to these conversations unobtrusively. In this way students' understanding or misconceptions can be heard and be discussed in the class.

The attempts of students to discuss mathematical ideas with their seatmates were found to be limited, for not much elaboration or arguments were pursued, rather the authority of the teacher was always insisted upon. It appears that further encouragement with students to talk more about mathematics in class especially with their seatmates is needed. Yackel, Cobb, Wood and Merkel (1990) found their experience in the classroom where mathematical discourse was given emphasis as realistic and attainable. Moreover, not only did it develop the skills of communication, reasoning and making connections but also resulted in an exciting adventure for the learning mathematics for both the teacher and the students. Again, this could happen when students are provided with activities that offer more opportunities to talk about mathematical ideas and to be active questioners, problem creators and problem solvers (Vace, 1993).

Despite the Department of Education language policy that Mathematics be taught in English, students were obviously comfortable using their own language especially in their private conversations. Thus, studies could be done to find out the effects of providing students with mathematical tasks that encourage them to talk more and use the language they feel comfortable to use in their private conversations in relation to the skills we want to develop among our students. Students' skills on problem solving, reasoning, and communication and in applying situations to other areas of the discipline and in real life could be looked into.

In summary, there emerged a pattern implying a change in the role of the teacher in the classroom in order to benefit more from the students' talks. According to Silver and Smith (1996) they envisioned a new role for a mathematics teacher that is bound to issues of communication. They elaborated that "in traditional mathematics instruction, the role of the teacher is essentially to transmit knowledge to, and validate answers for students, who are expected to learn alone and in silence. In contrast, according to the reform vision of mathematics classrooms, the role of the teacher is diversified to include posing worthwhile and engaging mathematical tasks; managing the intellectual activity in the classroom including the discourse; and helping students to understand mathematical ideas and to monitor their own understanding. Students are expected to engage in doing mathematics while participating actively in a discourse community." (p.20). And since the Filipino students were found to be visual and kinesthetic learners and least as auditory learners (Wallace, 1995 & Tenedero, 2000), it could be worthwhile for our teachers to consider this shift in the role and see its effects on them and on the learners as well. Teachers could lessen the time for using exposition and spend time using other teaching strategies such as group problem solving and practical work. Teachers might undertake some sort of action research for this purpose in their respective mathematics classes.

REFERENCES

Adler, J. (1999). The dilemma of transparency: Seeing and seeing through talk in the mathematics classroom. *Journal for Research in Mathematics Education, 30*(1), 47-58.

Backhouse, J., Haggary, L., Pirie, S., & Stratton, J. (1992). *Improving the learning of mathematics.* London: Cassell.

Clarke, D. (2001a). Teaching/Learning. In D. Clarke (Ed.), *Perspective on practice and meanings in mathematics and science classrooms* (pp. 291-320). Dordrecht, The Netherlands: Kluwer Academic Press.

Clarke, D. (2001b). The complementary accounts methodology. In D. Clarke (Ed.), *Perspective on practice and meanings in mathematics and science classrooms* (pp. 13-32). Dordrecht, The Netherlands: Kluwer Academic Press.

D'Ambrosio, B., & Prevost, F. (1995). Highlighting the humanistic dimensions of mathematics activity through classroom discourse. *The Mathematics Teacher, 88*(9), 770-771.

Dysthe, O. (1996). The multivoiced classroom: Interactions of writing and classroom discourse. *Written Communication, 13*(3), 385-425.

Gallos, F. & Ulep, S. (2002, December). *Students' emergent understanding of mathematics- Its links with teaching style and learning styles.* Paper presented at the Annual Conference of the Learner's Perspective Study, Melbourne, Australia.

Knuth, E. & Peressini, D. (2000). Unpacking the nature of discourse in mathematics classrooms. *Mathematics Teaching in the Middle School, 6*(5), 320-325.

Manouchehri, A. & Enderson, M. (1999). Promoting mathematical discourse: Learning from classroom examples. *Mathematics Teaching in the Middle School, 4*(4), 216-222.

Mifsud, M. L. & Johnson, S. (2000). Dialogic, dialectic, and rhetoric: Exploring human dialogue across the discipline. *The Southern Communication Journal, 65*(2/3), 91-104.

National Council of Teachers of Mathematics. (1991). *Professional standards for teaching mathematics.* Reston, VA: Author.

Pascua, L. B. (1993). Secondary mathematics education in the Philippines today. In G. Bell (Ed.), *Asian perspectives on mathematics education* (pp. 160-181). Rivers Mathematical Association, Lismore, Northern UK.

Pugalee, D. (2001). Using communication to develop students' mathematical literacy. *Mathematics Teaching in the Middle School, 6*(5), 296-299.

Silver, E. & Smith, M. (1996). Building discourse communities in mathematics classrooms: A worthwhile but challenging journey. In P. Elliot & M. Kenney (Eds.), *Communication in mathematics, K-12 and beyond (1996 Yearbook)* (pp. 20-28). Reston, Virginia: The National Council of Teachers of Mathematics, Inc.

Tenedero, H. (2002). Learning styles: A teacher's perspective. *Manila Bulletin,* p. B6.

Vace, N. (1993). Teaching and learning mathematics through discussion. *Arithmetic Teacher, 41*(4), 225-227.

Wallace, J. (1995). Learning styles in the Philippines. *Education, 115*(4), 552.

Wertsch, J. (1991). *Voices of the mind. A sociological approach to mediated action.* London: Harvester Wheatsheaf.

Yackel, E., Cobb, P., Wood, T., & Merkel, G. (1990). Experience, problem solving, and discourse as central aspects of constructivism. *Arithmetic Teacher, 38*(4), 34-35.

Florenda Lota Gallos
National Institute for Science and Mathematics Education Development
University of the Philippines
The Philippines

MICHAEL N. FRIED AND MIRIAM AMIT

CHAPTER FOURTEEN

The Israeli Classroom: A Meeting Place for Dichotomies

INTRODUCTION

It would be ideal if we could characterise the Israeli classrooms examined in the Learner's Perspective Study (LPS) by a set of typical, clear, and consistent classroom practices. But reality is rarely so simple. This is not to say that there are *no* such clear and consistent practices; rather the study of the Israeli case suggests that its classroom practices are marked by certain underlying tensions. These tensions can be formulated as (by no means exclusive) dichotomies: private versus public, collaboration versus authority, improvisation versus scripted activity, responsiveness versus directedness and so on. How these come into play within the classrooms varies of course; but most often they form a kind of silent backdrop against which actual practice – practice *in practice* – is always uneasily positioning itself.

Our general aims for the present chapter are to describe two dichotomies, private versus public and collaboration versus authority, to show the circumstances under which they arise in an Israeli classroom, to spell out their mutual relationship – for we believe that these two dichotomies *are* deeply related – and, finally, to link them to some other broader issues in mathematics education.

The chapter is structured as follows. In the first part, we give some additional information about methodology to the extent we go beyond the general LPS methodology set out in detail earlier in this book. In the second part of the chapter, we discuss mathematics notebooks in the Israeli classrooms, both from the teacher's and the students' point of view. This provides the context for discussing the first dichotomy, that between the public and the private domain. Next, we move on to the question of whom students turn to for help and support – the question of authority. The second dichotomy, that between relations of authority and relations of collaboration, is naturally highlighted in this part. Both of these dichotomies have already been discussed by us in detail elsewhere (Fried & Amit, 2003; Amit & Fried, 2005). However, they were treated then as separate issues; here they are placed side by side. Accordingly, in the concluding part of this chapter, we discuss how these two dichotomies are connected with one another and with more general dichotomies in education.

D. J. Clarke, C. Keitel & Y. Shimizu (Eds.), Mathematics Classrooms In Twelve Countries: The
Insider's Perspective. 209–220. © 2006 Sense Publishers. All rights reserved.

METHODOLOGICAL REMARKS

As we mentioned above, our methodology for the most part followed the general pattern set out in the original research design for the LPS (Clarke, 2000; Clarke 2001):

- Classroom sessions were videotaped using an integrated system of three video cameras – one on the class as a whole, one on the teacher, and one on a "focus group" of two or three students.
- Students in the focus groups were interviewed following each lesson, and the interviews were videotaped.
- Ten to fifteen consecutive lessons in each class were observed and videotaped, so that, among other things, almost every student, at one point or another, was a member of a focus group.
- The teachers were also interviewed once a week.
- All interviewees viewed and reacted to the videotape of the lesson during the interview.
- The researchers were present in every lesson, conducted all the interviews, took field notes, and collected relevant class material – including classroom notebooks, which was especially important for the research we describe below.

In our particular case, we also adopted a routine wherein following each filming session, we, our administrative assistant, and the technical crew engaged in a period of group reflection, in which we compared notes, discussed issues, and raised questions concerning that day's lesson. Regarding the interviews, basic questionnaires were constructed both for the student and teacher interviews. This was done, however, only to provide a rough guideline for the interviews; in practice, we allowed the interview protocol to remain quite flexible so that we could freely pursue particular classroom events. Our interview methodology, in this respect, was along the lines of Ginsburg (1997).

In general, our research style, consistent with the spirit of the LPS, tended to have a non-traditional ethnographic flavour (such as that described by Eisenhart, 1988). The open and exploratory character of the research methodology meant that specific teaching, learning, and classroom practices were not assumed in advance, nor even their existence; it allowed, accordingly, themes to emerge out of classroom situations naturally rather than be imposed on them. Without a methodology such as this, it is unlikely that we would have come to the dichotomies in the Israeli classrooms to be discussed in this chapter. For example, the issue of authority and collaboration arose according to the following nexus: One of the general research questions set out by Clarke (2000) asked whether teacher and learner practices are conflicting or mutually sustaining. With this in mind, we asked the students during the interviews to describe the circumstances in which they request help from their teacher, and, from there, whether they request help from other people as well. We put the question this way because we observed in the classroom sessions (as can be clearly seen in the video tapes) that although the students sat in pairs and were meant to work together, they rarely worked cooperatively; rather, they seemed to ignore their co-workers and turn to the teacher, or sometimes to a 'knowing' member of the class instead. These observations and our subsequent interviews engendered a whole new set of questions concerning students and authority, questions such as: Who is an authority for students? What is the extent of the authority of various people? How pervasive is the influence of authority in students' mathematical lives? What effect does students' relationship with authority have on their mathematical practice? Thus, questions, which were *not* among the initial research questions, were allowed *to become* research questions for us.

Two classrooms formed the basis for the specific research presented in this chapter. The first was taught by a dedicated and experienced teacher, whom we call Danit. Danit teaches in a comprehensive high school. Her eighth grade class is heterogeneous and comprises 38 students, mostly native-born Israelis, but also new immigrants from the former Soviet Union and one new immigrant from Ethiopia. Danit taught eighth grade mathematics many times in the past and knew the material well; she also seemed well informed about theoretical educational ideas, such as constructivism, which are important background ideas for the Israeli mathematics program. The 15 lessons observed in Danit's classroom belonged to a unit on systems of linear equations.

The second classroom, which was in a different school, was taught by a teacher whom we call Sasha. Sasha is a new immigrant from the former Soviet Union with several years' experience teaching in Israeli schools and much experience teaching in Russian schools. Like many teachers from the former Soviet Union (and unlike most Israeli teachers), his mathematical background is particularly strong, having completed advanced studies in applied mathematics. His eighth grade class is a high-level class and comprises 30 students. The ten lessons observed in Sasha's class all concentrated on geometry.

Danit and Sasha are, admittedly, very different teachers, not the least because of Sasha's background as a new immigrant. The latter involves complexities whose exact influence is not always easy to extricate; indeed, we have recently begun research directed specifically towards issues connected with new immigrant teachers within the Israeli system (for example, Amit & Burde, 2005). Nevertheless, we believe that Israeli schools are what they are partly because there are teachers both like Danit *and* Sasha; classroom practice or tensions in practice do not flow from one teacher alone.

NOTEBOOKS: PUBLIC-PRIVATE

Danit and Sasha, as we have just said, are very different teachers. Yet, in both classrooms the ways they and their students understood the character, function, and place of student notebooks in the mathematics classroom were strikingly similar. Let us begin with Danit.

Danit almost always began her algebra lessons with a frontal lesson lasting roughly ten minutes, after which students were assigned to work on exercises at their desks for the remaining 35 minutes of the class period. On days when there were two consecutive lessons, the students' work period was often extended through to the second lesson period. The students worked in groups of two or three while Danit went from group to group, checking the students' work, asking questions, and helping with difficulties.

We observed that at the beginning of each class period, when Danit presented new material, gave examples, and went over exercises from the previous lesson, the students' notebooks almost always lay closed on the students' desks. This raised obvious questions for us as researchers. Did the students not find Danit's examples and lessons worth recording? Did Danit not think that the students ought to write down her remarks and examples so that they might have them at their direct disposal to think about and refer to later? In several of the interviews, we asked the students about this. They replied that Danit lets them to choose whether they use their notebooks or not. In fact, in one of the interviews the students claim (though not unambiguously) that Danit is more liberal than other teachers in this regard:

Transcript 1: IS1-L01 Student Interview

I	I have another question. At the beginning of the lesson and during the lesson itself when [Danit] was speaking, did you write something down in your notebook?
Gila	Umm, no.
I	Did you open your notebook?
Gila	No, not exactly, because Danit always says you don't have to copy.
Sarah	/ [as if quoting Danit] "If you want to copy, if you think it will strengthen you, then you can copy what I write. I leave it to you."
Gila	In general, there are teachers that, for example, say "Copy now"
Sarah	"Copy!"
Gila	She [Danit] says it can be in the notebook.
Sarah	/ If it will strengthen you.

Key to symbols used in transcripts in this chapter

/	Indicates that one speaker cut in, interrupting another speaker before they had finished.
[text]	Comments and annotations, often descriptions of non-verbal action.
text	Italicised text indicates emphatic speech.
–	Indicates where the speaker breaks off what he/she is saying in the middle of a phrase.

On the face of it, then, students' notebooks are considered by Danit as the students' own property, and students can use their notebooks as they wish. In fact, however, the use of notebooks among the students bore no individual stamp; it was determined by Danit. Indeed, despite what Sarah and Gila said above, we observed several cases in which she *explicitly* told the students what to write in their notebooks. Moreover, she sometimes asked to inspect the students' notebooks. If the students know that Danit will eventually inspect their work, surely they must bear in mind how Danit might react if nothing were written, or if what was written down has been written badly, or wrongly.

Summarising, we can say that far from being a matter of students' choice, three basic rules for the use of notebooks guided students in Danit's class:
- During frontal lessons, students only listen; their notebooks are closed.
- When exercises are given, notebooks have to be opened immediately; all the exercises are to be recorded neatly (the students should use rulers, for example, as Danit indicated in one lesson).
- The teacher will occasionally inspect the notebooks.

Sasha's geometry lessons were evenly divided between those completely directed by the teacher in front of the class and those in which the students worked collaboratively on problems at their seats. The former were carefully planned and presented. Sasha's students, unlike Danit's, did open their notebooks at the start of the class and copied Sasha's lessons as they appeared on the blackboard; of course, they also used their notebooks when they worked on problems. Like Danit, however, Sasha also determined how the students' notebook should be used and how they should appear. In one lesson, for example, he said explicitly, "I want your notebooks to reflect what is here [on the board]." Also like Danit, Sasha's students know that their notebooks are always open to inspection. In a dramatic incident it was made clear to us that students' notebooks in Sasha's class could be inspected not only by Sasha at any time, but by the rest of the class as well: dissatisfied with how an exercise in one student's notebook appeared, Sasha picked up the notebook and held it open for the whole class to see, explaining all the while how exercises *should* appear in the students' notebooks:

Transcript 2: IS2-L12

```
Sasha    I ask you, please, sketch [the diagrams] in your notebooks.
         [Moves from the board and begins to look over the students'
         work]
         Oh, look, here [picks up the students' notebook and holds it
         up].
         Don't try and save space in your notebook.
         Look [pointing to the notebook], the diagram is in one place,
         the givens over here on the side — it's hard to read, to
         understand, to check — for everything, it's hard.
         Please [pointing back to the board], everything should be
         instead the way I asked [you to do] — the diagram, say, to the
         left, the givens next to it, the proof line by line below.
```

So, except for Sasha's expectation that material from the frontal lessons be copied in the students' notebooks, practices connected with notebooks in Sasha's classroom coincided with those in Danit's. These can be summarised as follows:
- Notebooks are to be kept open during the frontal lessons as well as the work periods.
- Notebooks must be neat and orderly (for example, diagrams must be drawn with care).
- Students must always be ready to show their notebooks to the teacher and even to the whole class.

The knowledge that the students' notebooks are to be open for inspection means that they are a public matter – they are not to be a record of the students' private thoughts about what they are learning: desultory reflections; false starts; mistaken

MICHAEL N. FRIED AND MIRIAM AMIT

conclusions and their, perhaps embarrassing, corrections. Where one can find these things, that is, preliminary work for exercises and calculations, is not in students' notebooks, but, to their teachers' continual annoyance, on the students' desks! Students learn, that the notebook must contain finished work that can be confidently held open to view. The notebook is, in this respect, a kind of rehearsal for that part of school life in which the students' work is inspected most closely: the examination. Indeed, in the students' interviews they often emphasised that they use their notebooks to record the sort of exercises that will appear on examinations rather than the explanations behind them. Here too, it is likely that the teacher is the source of this emphasis. Sasha, for example, remarked in one of our interviews with him: "A [good] exam paper goes with the notebook."

The public character of the students' notebooks is, interestingly enough, mirrored in the character of Danit's own lesson book. Her lesson books, which she unhesitatingly agreed to show us, consisted of pages of solved problems and numbered exercises. We could not see any stated goal, any exposition of the material, any distinction between major and minor examples, nor any lesson structure. She told us that she brings her lesson book to class and allows, even encourages, the students to compare their notes with hers (as we later confirmed in the lessons following this interview). We asked her if she hopes their notebooks will look like hers. She said, "Yes, yes." Elsewhere she also said:

Transcript 3: IS1-L09 Teacher Interview

Danit   *The students know that they can approach my notebook and compare and check exactly at what stage they've made a mistake.* If they don't understand where they've made a mistake, that's another story; then they wait for me so I can explain where their mistake is, that is, where it comes from. If they understand that it's just a careless error, then they can go on without waiting for me, without standing around. That's it.

I   That's interesting since most teachers see their notebooks as something maybe not exactly private, but still something only for the teacher.

Danit   No, that doesn't bother me. On the contrary, if I forget to take out [my] notebook, they [i.e. the students] remind me. And there've been students who wanted to photocopy [the notebook]; I said — they thought this was a joke, and said "Can we photocopy, can we?!" — I said, "Take it!" [smiles]. *I have no problem with this — what I have [in my notebook] is exactly what they need to have in their notebook. It is exactly the same thing. There are no secrets here.*

Thus, there is a notable consistency between teacher and student practice, firstly, a notebook is the place to record exercises only and secondly, a notebook may be inspected at any time, that a notebook is always public.

Standing opposed to the public character of mathematics notebooks, then, is, as we have suggested above, the private domain of the students, the place in which they make their own decisions and their own mistakes and reflect on what they have learned and how they learn. When Danit *says* that the students can choose what and when they write in their notebooks, she is, in effect, appealing to this private domain. Sasha, as we have noted, does not allow his students this freedom. Yet Sasha, accustomed as he is in seeing the notebook as public, does not aim that his students be adept only at public performance: in the interviews with him, he refers many times to students' thinking and, particularly, to their 'seeing' geometry. He wants geometry to belong to the students' *own* world of mind and eye. This too seems to mark an acceptance of the students' private domain. So, in both these classrooms, we discern a tension between the desire that mathematics become part of students' own worlds, their private domains, and the treatment of the notebook – which, because it is truly the students' own, is a natural locus for the private domain – as an utterly public object. This, then, is the first dichotomy: public vs private.

We shall say a little more about this in the fourth part below. But for now, we should also note that division between what is public and what is private has much to do with students' autonomy or agency. With that in mind, we want to turn to the question of authority.

AUTHORITY VERSUS COLLABORATION

Asked whom they approach for help when they run into difficulties, students provided always the following sources of help: their teacher, their friends, their parents, or their siblings. Among these, the students' teacher and friends were the dominant sources for help mentioned in the interviews. When asked to whom they turn first, some students said the teacher, and some, their friends. The reason given for turning to friends first was almost always that the "teacher is too busy and can't get to everyone." Often, however, it seemed to us from our observations of the class that students turned to their friends first simply because their friends were near, as they were usually students sitting at the same desk.

Teachers, friends, parents and siblings form a web of sources of assistance; when one source is unavailable or unable to help, one turns to another. For example, if Yara in Sasha's class cannot get help from Sasha, for one reason or another, she turns to one of her friends:

Transcript 4: IS2-L04 Student Interview

```
I       And if your friend doesn't know?
Yara    If my friend doesn't know, I ask someone else — or my father.
```

The web forms a hierarchy according to the degree of authority possessed by the sources, where by the latter we mean the degree to which a person's statements are taken unchallenged (this sense, in a way, is already built into the word 'source'; indeed, the Greek word for 'source', *arche*, also means 'sovereignty', and in the plural, *hai archai*, 'the authorities'). Conversely, turning to an authority means turning to a person for an answer or for instructions, not, by contrast, for a discussion. Moreover, we use the word 'authority' rather than, say, 'expertise' because the reason that a person's statements are not to be challenged is, as we shall soon see, not always dependent on the degree of the person's knowledge, though it may be perceived that way. And, in this hierarchy, there is no question, the teacher comes first.

The predominance of the teacher's authority was obvious in all of the student interviews, both in Danit's and Sasha's class. For example, at one point in our interview with two students in Danit's class, Michael and Saul, we asked whether a graphical method or algebraic method of finding the solution to a system of equations was more reliable. Here is the exchange:

Transcript 5: IS1-L11 Student Interview

```
Michael   If I get an answer for one and a different answer for the
          other, then you've got to check. If I get the same answer,
          then I'll believe it's correct. But if there's, maybe, still
          some doubt in my mind, I ask Danit.
I         What does Danit have that other people don't?
Michael   She's a teacher, she can help; if you make a mistake, she
          corrects it!
I         And if she errs?
Michael   She doesn't err.
Saul      She studies everything at home before she comes to class.
Michael   Otherwise she couldn't correct — she's a teacher!
I         But she did make a mistake at the board.
Saul      She got mixed up because she substituted wrong.
Michael   Those are nonsense things she gets mixed up about, but real
          things [gestures to show the weightiness of the things he
          has in mind] — if two exercises are supposed to get the same
          answer or not, it doesn't seem to me she'd get mixed up
          about that.
```

What is striking in this exchange is how Michael and Saul are willing to see Danit as nearly infallible, and how far they are willing to defend her authority, even when she is seen to make a mistake. The students view her, apparently, not only as one who knows more than they do, but also as a strong figure with powers they lack; she possesses not only expert authority, but also, to use Weber's (1947) term, charismatic authority, that is, authority whose power is supernatural and which commands devotion more than mere obedience. Thus, when Michael says, "She's a teacher, she can help; if you make a mistake, she corrects it!" he sounds as if he is speaking of a healer, a miracle worker, rather than of his eighth grade mathematics teacher. Similarly, when we asked Sylvia and Shari, also Danit's students, what exactly do they expect from the teacher, when they ask her for help, Sylvia said simply "That she will explain to us better," to which Shari added immediately, "When she comes over to me, when she explains to me, suddenly I understand *better*." Even Saul, who finds a rational reason for Danit's 'never' erring, namely, that she studies everything at home, does not seem willing to see any mistakes that she does make to be anything but trivial. Consistent with this image of Danit was the importance the students seemed to place on the mere fact of Danit's coming to help them when they worked on exercises. When we asked what the climax of the lesson was, Elana, in the same interview in which Sylvia and Shari participated, answered, "When I was having trouble with the book and I called [Danit]." In a different interview, another girl in Danit's class, Gila, answered the same question in precisely the same way. Conversely, on two different occasions we came across a student in Danit's class who also seemed to have trouble with the exercises, but did not ask Danit for help. When we asked why they did not ask for help from the teacher, we received the same response both times: "The teacher doesn't want to help me." Such a statement frames the attention the students receive from the teacher as something depending on the teacher's whim. The teacher becomes a dictator, in this interpretation, though, surely, for most students, a beneficent one who willingly helps them when they really need help. Nevertheless, conceiving the teacher as a creature of whim is to conceive the teacher as a creature with terrific power.

The extent of teachers' authority from the students' perspective was significant enough; however, we were more surprised to discover how easily students were willing to see *other* people as authorities to the same degree. For example, we were interested in how students understood the requirement of "showing their work," whether this was only a requirement of students or of mathematics itself. So, we asked whether a salesperson explaining to customers how much they should pay given such and such a discount should be required to show his/her work. To this, Ben, again from Danit's class, replied: "No, I can rely on him...I can rely on him – for sure lots of people come to him – there must be those who know percentages and things, and they rely on him, so I can rely on him too." It is worth noting here that the Hebrew word Ben used for "rely" is *somech* which is closely related to the word *somchute* meaning literally "authority."

In the students' world, authorities are ubiquitous; sources of assistance are invariably sites of authority. And the most striking fact is that this applies also to the students' friends and classmates. As mentioned above, friends in the class are a dominant source of help. But when the students turn to their friends, they tend to turn to them only for answers. And, as we saw with Sasha's student Yara, when one friend does not know, she turns to another. In one interview in Danit's class, we asked a student why he did not ask his friend for help at a certain point during the lesson. He replied, "I knew Yuri wouldn't know the answer..." Thus, when students are perceived by their fellow students as knowing the answer to some question *they are treated for that instant as an authority*, that is, the answer is accepted and not discussed. When students are not perceived as knowing the answer, they are usually not asked. In fact, the classroom videos showed quite often (though less so in the geometry classes) that students sit together, occasionally speak together, but do not really work together, even though they are not necessarily encouraged to work individually.

To understand the significance of this tendency of students to treat one another as authorities *ad hoc*, we have first of all to see what alternative stands opposed to it. This need not be considered hypothetically, for in Sasha's class we found an exception to the tendency. During his lesson, Sasha gave a geometry problem to the class; we watched as two girls, Yana and Ronit, solved the problem in a truly collaborative spirit. Ronit showed her diagram to Yana; Yana commented and pointed to her own diagram; they discussed the problem together, and, finally, came to a solution. Yana and Ronit happened to be our focus group for that lesson, and throughout the interview we saw how different their behaviour was from other students': they consulted with one another, raised possibilities on their own, revised opinions, and seemed to arrive at common conclusions. In other words, rather than treating one another as possible authorities, that is, only as possible sources of answers, Yana and Ronit treated one another as intelligent interlocutors who could work together to make progress on the question at hand. We should stress that this was, indeed, behaviour different not only from that of students in Danit's class, but also from that of other students in Sasha's class. For instance, at one point in our interview with Yara, we asked if she could draw a triangle having two acute exterior angles; she said that she could and proceeded to draw a diagram, which, obviously, could not be correct. When we asked Panina, the second girl in the focus group, whether Yara's diagram was okay, she assented immediately and with no further remark.

In both Danit's and Sasha's classroom collaboration is valued and encouraged. In both classrooms, students sit two or three at a table, and their tablemates are, in the work periods, meant to be their workmates. Of course, as Johnson and Johnson (1989) stress, collaboration is more than "Simply placing students in groups and telling them to work together" (p. 237). This is true, but here we have a definite force working *against* collaboration, for where students are accustomed to treating one another as authorities no true dialogue takes place between them, and where there is no true dialogue there can be no true collaborative learning. Thus, we have the second of our dichotomies: relations of authority versus collaboration.

CONCLUDING THOUGHTS: SOME CONNECTIONS

Now, in our introduction we said that the authority-collaboration and the public-private dichotomies were interrelated, and we suggested where the link lies, namely, in the ideas of agency (as discussed in, for example, Boaler, 2003) and autonomy. The connection between these notions and authority is obvious, for where there is authority there is agency. The connection with the other side of that dichotomy, with collaboration, is a little less obvious. However, as we argued at length in Amit & Fried (2005), the collaborative relationship can be seen to grow out of an authoritative one if, following Benne (1970) (with echoes also from Lave and Wegner, 1991), the conception of authority is redirected away from domination and obedience and towards negotiation and consent. In this light, the authority-collaboration dichotomy reflects a process in which agency is distributed, little by little, among the members of a community (see Amit & Fried, 2005 for details of this process).

But what can we say about the public-private dichotomy? This dichotomy is quite complex and even problematic (see Fried & Amit, 2003; Gal, 2002; Benn & Gaus, 1983). The connection with agency and autonomy comes clear, however, when one considers the dichotomy in its typical socio-political setting. Here, one speaks about public and private *domains* and does so with such terms as 'access' and 'accountability' (see Benn & Gaus, 1983). Public activities, then, are those activities with regards to which one is accountable to teachers, peers, or co-workers. In the public domain, therefore, one is bound to common practices and to the necessity of formal communication; one's own agency is derived from that of a greater community, but one is also, thereby, empowered by it. Private activities, on the other hand, are those with regards to which one is not accountable to teachers, peers, or co-workers. In the private domain, therefore, one is free from the expectations and constraints of common practice. Private activities, as one writer puts it, take place in "a zone of immunity" (Duby, 1985, p.10); here autonomy flourishes and agency is most fully one's own. Needless to say, once one mentions 'accountability', one must consider to whom or to what one is accountable, that is, what authority is at work.

With respect to mathematics learning, both dichotomies refer to the questions of where and by whom students' knowledge is formed: whether knowledge is formed in the quiet of one's own thoughts, in one's own room, as it were, or within the public arena governed by agreed upon or culturally conditioned forms of discourse and reasoning; whether knowledge comes from an external source or whether one is the agent of one's own knowledge, perhaps together with other such agents. These questions can be located within other dichotomies. For example, Sawyer (2004) speaks of improvisatory versus scripted teaching, where the former refers to teaching which is responsive and which bends itself to students' thinking and learning. A grand and overarching dichotomy, however, is that which Cobb (1994) discusses, namely, that between constructive and sociocultural perspectives on learning. This dichotomy also concerns where and by whom knowledge is constructed: inside or outside, by oneself or through others. Cobb makes a point of saying, moreover, "...that the apparent conflict between constructivist and sociocultural perspectives is not merely a matter of theoretical contemplation. Instead, it finds expression in tensions endemic to the act of teaching" (p.13). His article then goes on to try and resolve the conflict by a pragmatic argument whereby the different perspectives can coexist because they are used to justify distinct aspects of teaching. Perhaps, though, such dichotomies ought not be viewed as problems to be resolved in the first place, but as defining characteristics of classrooms. If so, this would mean, naturally, that simple categories of classroom practice would be of only limited use. Our findings from the Israeli classroom, in this regard, tend to support Cobb's *initial* remark that the dichotomy is the focus of a *tension* in classroom practice. Indeed, it has been our contention that the way in which that tension comes to the fore in practice is the truest way of describing the Israeli classroom.

REFERENCES

Amit, M., & Burda, I. (2005). "Why do we learn mathematics? Because it organizes our minds" An encounter between Former Soviet Union and Israeli Cultures of mathematics education. Plenary talk given at the annual conference of the *Commission Internationale pour l'Étude et l'Amélioration de l'Enseignement des Mathématiques (CIEAEM 57)*, Italy, July 23-29.

Amit, M., & Fried, M. N. (2005). Authority and authority relations in mathematics education: A view from an 8th grade classroom. *Educational Studies in Mathematics, 58*, pp.145-168.

Benn, S. I., & Gaus, G. F. (Eds.). (1983). *Public and private in social life*. New York: St. Martin's Press.

Benne, K. D. (1970). Authority in education. *Harvard Educational Review, 40*, 385-410

Boaler, J. (2003). Studying and capturing the complexity of practice – The case of the 'Dance of agency'. In N. A. Pateman, B. J. Dougherty, & J. T. Zilliox (Eds.), *Proceedings of the 27th Annual Conference of PME27 and PME-NA25*, vol.1, pp. 3-16. Honolulu, Hawaii: CRDG, College of Education

Clarke, D. (2000). *Learner's perspective study: Research design.* Unpublished.

Clarke, D. (2001). *Perspectives on practice and meaning in mathematics and science classrooms.* Dordrecht: Kluwer Academic Publishers.

Cobb, P. (1994). Where is the mind? Constructivist and sociocultural perspectives on mathematical development. *Educational Researcher, 23(7)*, 13-20.

Duby, G. (1985). Preface to P. Veyne (Ed.), *Histoire de la vie Privée* [History of private life], Vol. I, *De l'Empire romain à l'an mil.* Paris: Edition du Seuil.

Eisenhart, M. A. (1988). The ethnographic research tradition and mathematics education research. *Journal for Research in Mathematics Education, 16*, 99-114.

Fried, M. N., & Amit, M. (2003). Some reflections on mathematics classroom notebooks and their relationship to the public and private nature of student practices. *Educational Studies in Mathematics, 53*, pp. 91-112.

Gal, S. (2002). A semiotics of the public/private distinction, *Differences: A Journal of Feminist Cultural Studies, 13*, 77-95.

Ginsburg, H. P. (1997). *Entering the child's mind: The cognitive clinical interview in psychological research and practice.* New York: Cambridge University Press.

Johnson, D. W., & Johnson, R. T. (1989). Cooperative learning in mathematics education. In P. R. Trafton, & A. P. Shulte (Eds.), *New directions for elementary school mathematics* (pp. 234-245). NCTM.

Lave, J., & Wenger E. (1991). *Situated learning: Legitimate peripheral participation.* Cambridge: Cambridge University Press.

Sawyer, R. K. (2004). Creative teaching: Collaborative discussion as disciplined improvisation. *Educational Researcher, 33(2)*, 12-20.

Weber, M. (1947). *The theory of social and economic organization.* Henderson, A. R. & Parsons, T. (Trans.). London: William Hodge ad Company Limited.

Michael N. Fried
Program for Science and Technology Education
Ben-Gurion University of the Negev
Israel

Miriam Amit
Program for Science and Technology Education
Ben-Gurion University of the Negev
Israel

GAYE WILLIAMS

CHAPTER FIFTEEN

Autonomous Looking-In to Support Creative Mathematical Thinking:
Capitalising on Activity in Australian LPS Classrooms

INTRODUCTION

The lesson was almost finished when Leon suddenly realised something which changed the direction and intensity of his focus:

> Towards the end of the lesson, I started understanding it and I don't know but for some reason my goal sort of changed a little bit. ... it changed from *finishing* the work to actually understanding the work [L12-SI-Leon][1].

During the next lesson, Leon softly exclaimed 'Oh!' as what he had puzzled over suddenly became clear. Leon explained in his interview after the lesson: "It just sorta clicks into your head and- and you think oh! I know this now". When Leon recalled what had occurred, he described features that fitted 'flow' conditions:

> [I][ii] really didn't understand ... it was a bit of a challenge ... when I finally did understand it- it really made me feel good about myself [L13-SI-Leon].

Flow is a state of high positive affect during creative activity (Csikszentmihalyi, 1992). It occurs when people work to overcome a 'spontaneously' set challenge that is almost out of reach and develop new skills during the process. Leon spontaneously set himself an *intellectual* challenge: to develop *new conceptual understanding* thus setting up the conditions for flow specific to mathematical exploration ('discovering complexity') that I had previously identified (Williams, 2002a).

Discovering complexity involves a student or group of students spontaneously setting up an intellectual challenge and developing novel conceptual understanding as they overcome that challenge. My use of the term 'spontaneous' is consistent with its use by Steffe and Thompson:

> We do not use spontaneous in the context of learning to indicate the absence of elements with which the student interacts. Rather we use the term to refer to the non-causality of teaching actions, to the self regulation of the students when interacting. (Steffe & Thompson, 2000, p. 291)

D. J. Clarke, C. Keitel & Y. Shimizu (Eds.), Mathematics Classrooms In Twelve Countries: The Insider's Perspective. 221–236. © 2006 Sense Publishers. All rights reserved.

This chapter explores the process of discovering complexity where complex mathematical thinking and high positive affect co-exist during the creative development of novel (to the student) mathematical concepts. In particular, this chapter examines 'Looking-In activity' (that Leon undertook just prior to his goal change) and its role in this process.

Leon was a Year 8 student in Australian School 1 in the Learner's Perspective Study (LPS) who used the term 'Looking-In' to describe his own activity. I found this same activity common to two of the other three students who were identified discovering complexity in other Australian LPS classrooms. Some students engaged in Looking-In activity on more than one occasion. Looking-In (Williams, 2004) is an idiosyncratic student activity that supports creative mathematical thinking. It can occur when a student does not possess appropriate 'cognitive artifacts' to progress their exploration. Cognitive artefacts include mathematical ideas, and concepts assembled by the student during the process of abstracting (Hershkowitz, Schwarz, & Dreyfus, 2001). When Looking-In, the student focuses on a dynamic visual display generated by another class member and extracts mathematical ideas and/or concepts implicit within that display.

By studying Leon's Looking-In activity, and the conditions under which it occurred, the nature of Looking-In is elaborated and a theoretical language is developed to discuss the process of discovering complexity (creative mathematical thinking) in which this activity can be embedded. This study of creative mathematical thinking contributes to the body of knowledge about how students can develop relational understanding (Skemp, 1976): learning resulting from the interconnecting of concepts to develop new conceptual understanding rather than the learning of a fragmented set of rules that are applied without awareness of their meanings (instrumental understanding, Skemp, 1976).

This study informs teachers and teacher educators about pedagogy that supports the development of relational understanding, thus increases students' 'mathematical literacy' (Kilpatrick, 2002). Mathematical literacy is the ability to use mathematics flexibly in unfamiliar situations and to recognize mathematics in unfamiliar contexts during work and life in general. To be able to do this, students need to 'recognize' the usefulness of previously developed mathematical ideas and concepts in unfamiliar contexts, and/or recognize mathematics embedded within contexts. 'Recognizing' is one of the three 'observable cognitive elements' found to occur during the process of abstracting (Hershkowitz, Schwarz, & Dreyfus, 2001). Recognizing, and the other two observable cognitive elements 'building-with' and 'constructing' are illustrated through Leon's activity in this chapter. Recent focus on the importance of mathematical literacy has led to a valuing of creative student thinking in curriculum documents internationally (e.g., South Africa, Department of Education, 1997; Australia, http://vels.vcaa.vic.edu.au; USA, Kilpatrick, 2002).

STUDY DESIGN

Through the post-lesson interview process, students who explored mathematical complexities to generate new knowledge were identified and social influences upon their thinking were made explicit through their discussion of the lesson video. The probes used to clarify, elaborate, and/or extend student responses in the Australian interviews focused on student reconstruction of their thinking during the lesson and student identification of what they thought had influenced their thinking. The interviews in conjunction with the lesson video were used to identify the periods of time over which student cognitive activity was spontaneous (Williams, 2004) and those intervals of time were reanalysed to identify intervals of creative mathematical activity. The following analysis elaborates terms in this paragraph.

Ericsson and Simon (1980) have shown that verbal reports can provide valid data when attention is given to research design. There were features of the LPS interview probes in Australia that fitted with Ericsson and Simon's (1980) findings about how to generate high quality verbal data associated with cognitive activity. The interviews focused on: (a) using salient stimuli (mixed image lesson video of the student [centre screen] and the teacher [in the corner]) to stimulate student reconstruction); (b) allowing the student to focus the content of the interview; and (c) encouraging students to focus on lesson activity and their own thinking rather than the interviewer asking general questions. Ericsson and Simon state that where the researcher asks specific questions that include constructs the subject has not previously reported, the subject is more likely to "generate answers without consulting memory traces" (Ericsson & Simon, 1980, p. 217). On the other hand, if a subject spontaneously "described one or more specific sub-goals, and these were both relevant to the problem and consistent with other evidence of the solution process, ..." (Ericsson & Simon, 1980, p. 217) there was stronger evidence that the reported activity occurred. The multi-source data collection techniques in the LPS supported such analysis.

To study social and personal influences upon creative student thinking, simultaneous analysis of cognitive, social, and affective elements of the process of abstracting was undertaken in conjunction with analysis of student responses to successes and failures (resilience or optimism, Seligman, 1995). Seligman identified the engineering of flow situations as a way to increase student perception of their ability to overcome perceived failures. Connections between flow and optimism thus have pedagogical implications. There are three dimensions to optimism (Permanent-Temporary, Pervasive-Specific, Personal-External). Optimistic children see success as permanent, pervasive, and personal and failures as temporary, specific, and external. It was expected that optimism would be associated with student inclination to explore unfamiliar mathematics because problem solving in mathematics can be perceived as the encountering of failures along the pathway towards success. Discourse analysis (Säljö, 1999) of post-lesson interviews was undertaken to find indicators of optimism, and analysis of student cognitive activity to see how optimism was enacted . This enabled study of the theorised link between optimism and flow situations. The dimensions of optimism are elaborated through Leon's activity in this chapter.

SITE AND SUBJECTS

Australian School 1, the government school Leon attended in Melbourne, contained students from many cultural backgrounds (the 22 came from at least 12 cultural backgrounds). The families had generally been in Australia for more than one generation or had arrived in Australia more than two years prior to this research study. Leon's teacher (Mrs Milano) generally commenced her lessons with a whole class activity followed by pair-work, which was sometimes of an exploratory nature. Where exploratory activity was included, it was generally at the end of a topic for the purpose of reinforcing new ideas. During pair-work, Mrs Milano moved around the classroom assisting individual students or groups of students particularly when they requested her attention.

Leon reported in his interviews (after Lesson 12 and Lesson 13) that mathematics learning was a process of adapting what was previously learnt to new situations. This was an indicator of optimism along the Permanent-Temporary dimension showing that he perceived 'not knowing' a solution method ('Failure') as temporary (Seligman, 1995). I have previously provided detailed analysis of Leon's optimistic orientation (Williams, 2003). The way he described himself as learning mathematics did not place a heavy reliance upon assistance from the teacher because he perceived success as resulting from his own effort ('Success' as 'Personal'). There were many instances during the research period where Leon demonstrated he continually tried to make his *own* sense of the mathematics upon which the class focused. This independence of thought could explain why he (unlike many other students), did not discuss the quality of his teacher in his interview; he was not so reliant upon her for his learning. Observation during Lesson 12 suggested Leon spent the majority of his time teasing the female students seated around him. His post-lesson interview provided evidence of deep mathematical thinking about ideas that interested him.

'LOOKING-IN': THE CONTEXT

Early in Lesson 12, Mrs Milano discussed the properties of squares and rectangles in an attempt to convince class members that a square was a rectangle. Leon displayed intense interest in the properties of these figures by answering softly to himself each time Mrs Milano asked whether a certain property was present. Later in the lesson, Mrs Milano placed three large coloured triangles on the board (see Figure 1) and pairs of students were allocated a triangle and asked to find its area (without using a rule). The positioning of this exploratory activity at the beginning of a topic sequence differed from the usual lesson sequencing in this class because the exploration occurred prior to the introduction of the rule.

Leon and Pepe were meant to work together but Pepe wanted to count squares and Leon wanted to find a general method to find areas of triangles. Pepe did the writing and drawing for the pair and Leon answered Pepe's queries, and provided unsolicited advice to the female student pair (Elina and Serina) working beside him. Leon did not write or draw anything during the lesson. Pepe worked on Triangle 1 (see Figure 1) and Leon utilised the time he had manoeuvred (by letting Pepe do the written work) to think more broadly about the three triangles on the board and their respective areas. He had manoeuvred so that he was not obliged to negotiate with others about how he would proceed with his exploratory activity; he had 'cognitive autonomy' (Williams, 2005).

LEON'S EXPLORATORY ACTIVITY

Leon structured his exploration by attending simultaneously to the three images on the board and asking himself "which triangle was easiest".

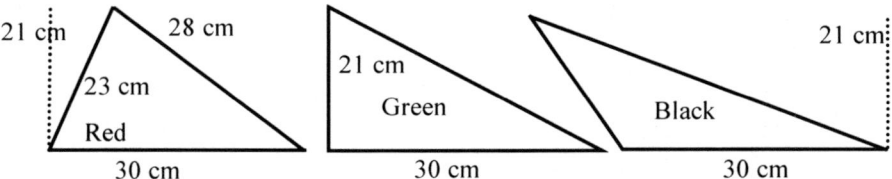

Figure 1. Triangles placed on the board as part of Task C2

This focus was evident from his interview after Lesson 12 and his interactions with the pair seated beside him: "[to Elina] You're doing two? Two's like the easiest one there. It's easy!"[iii] Leon reconstructed his thinking about how to find the area of a right-angled triangle (identified herein as Method A) in his post-lesson interview after Lesson 12:

All you have got to do is figure out what a rectangle is that has those two um (pause) lengths- length and width and ... then you can just halve it

Leon did not make explicit whether he recognized that right-angled triangles could be formed by cutting rectangles in half, or that juxtaposing two right-angled triangles would make a rectangle. He also did not discuss whether he used the properties of rectangles to justify the nature of the shape produced or whether he relied solely upon visual images. Leon's intense interest in properties of rectangles earlier in the lesson and the method he developed to find areas of acute-angled triangles (described below) suggested he juxtaposed right-angled triangles and used the properties of rectangles to justify the nature of the shape formed. Leon's initial thinking about finding areas of acute-angled triangles (Method B) involved juxtaposing two acute-angled triangles to make a parallelogram. In his interview after Lesson 13, he sketched this juxtaposition process and was not satisfied with his sketch. His accompanying comments showed he knew a parallelogram was formed and that the properties of the triangles justified that shape:

That's a parallelogram ... mmh pretend they are the same length and width and everything.

Leon's juxtaposition of two acute-angled triangles and his awareness of the properties of the figure produced, supported my interpretation that he used these ideas (juxtaposition, properties of shape formed) in Method A. He sectioned the parallelogram in Method B (see Figure 2) to find the acute-angled triangle's area.

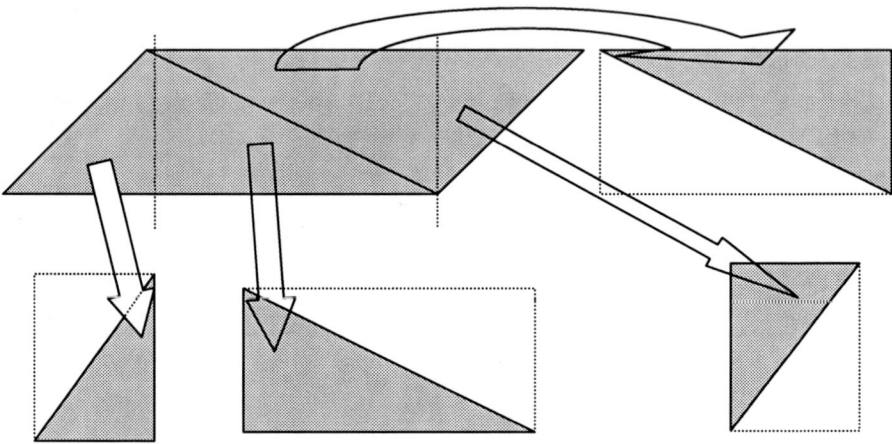

Figure 2. Interpretation of Leon's Method B

Figure out what it would be if it were a parallelogram and then halve it ... figure out what it would be if it was four because you could just trial whatever it was if it was four

Figure 2 shows the parallelogram sectioned into four right-angled triangles "figure out what it would be if it was four" and the rectangles formed by the juxtaposition of another congruent right-angled triangles with each of these four right angled triangles (multiple use of Method A as part of Method B) to form four rectangles "just trial whatever it was if it was four". The areas of these rectangles were found and totalled and the total halved in accordance with the halving of each rectangle in Method A "and then halve it". The separate parts in Method B probably made it difficult for Leon to keep all of these ideas in his head at once. The fragility of Leon's ideas was expressed in his interview as he discussed Method B:

I understood it- I didn't understand it then I understood it then I didn't understand it

Looking-In Activity Associated with Goal Change (Method C begins to Emerge)

While he was still considering Method B, Leon's Looking-In activity occurred and he began to realise there could be a more elegant way to proceed. In his interview, he described Looking-In activity that leads to goal change:

> When you look around the classroom and see how everyone else is doing it and you are doing it a completely different way- ... and you think ooh! [soft] maybe my method isn't the best and ... you think about everyone's ... and then you think about your own and they all sort of piece together and you just sort of go oh! and it pops into your head [L13-SI-Leon].

It was as Leon 'Looked-in' on what was happening around him, and considered this in conjunction with the ideas he had been developing himself that his goal changed from *finishing* the work to *understanding* the work. By Looking-In, he extracted a big idea that he had not previously been aware of: "triangles come in rectangles". He was excited by what he had found because it opened possibilities for a simpler way to proceed. He might no longer need to section up parallelograms and laboriously apply Method A to each of the four parts, total the areas, and halve the sum. There could be a more 'elegant' way. Krutetskii (1976) emphasised a quest for elegance as crucial aspect of the problem solving activity of students with high mathematical ability. Through his quest for elegance, Leon set up the conditions for flow (discovering complexity)—a spontaneous intellectual challenge (how could these enclosing rectangles help?) requiring development of mathematics beyond his present understanding. The nature of Leon's Looking-In activity is discussed in detail later. For the present, the remainder of Leon's exploratory activity is reported to emphasise the significance of Leon's Looking-In activity to the success of his exploration.

By Lesson 13, Leon knew 'triangles come in rectangles' and knew that the area of the triangle was half that of the enclosing rectangle (see Table 1, Row 3, Column 2) but he did not know *why*. This was what he was puzzling to understand. When Mrs Milano held a large pink enclosing rectangle behind the red acute-angled triangle in Lesson 13, Leon softly exclaimed "Oh!" [L13, 20:00 Mins]. He reconstructed his thinking in his interview:

> I sort of- sort of thought a little bit about why it was happening [triangle area half rectangle area] ... that it was um- if you take one part out like a triangle that's set at an angle if you take both parts out and put them together it equals the rectangle. ... I was sort of looking at them and then I just realised, like I (pause) sort of just in my head I pulled it apart and put them together so that they equalled the same.

Figure 3 includes the mental images Leon reported generating as he suddenly realised why the area of the triangle was half the area of the rectangle. He had removed two right-angled sections of the acute angled triangle (shaded parts) out of the larger rectangle. The triangles congruent to each of these shaded sections were then juxtaposed with the appropriate shaded triangle to make two smaller

227

rectangles: "if you take one part out like a triangle that's set at an angle if you take both parts out and put them together it equals the rectangle". He had simultaneously drawn upon what he knew from Method A and shifted pieces around in his head to justify that the area of the acute angled triangle was always half the area of the enclosing rectangle.

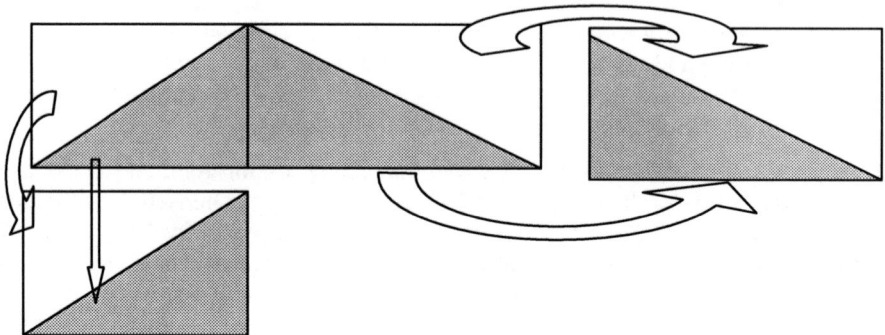

Figure 3 Leon applied Method A twice within his elegant Method

Less than a minute later, Leon clapped his hand against his cheek as he realised he could use his novel concept for another purpose. His comments in class showed what he had realised:

```
Leon    That's half of the rectangle as well
Leon    That would be three hundred and thirty centimetres squared
T       Three hundred and thirty.  Why?
Leon    It would be exactly the same as the first one [Triangle 2]ⁱᵛ
T       Why?
Leon    Because the green one is half of the rectangle too (see Figure
        1)ᵛ
```

Leon had realised that triangles enclosed in the same rectangle have the same area because each has an area of half the area of the rectangle.

The exercise set towards the end of Lesson 13 included finding areas of triangles in different orientations. Leon (unlike other members of the class) could do this exercise because he was aware of the significance of the perpendicular height of the triangle and how to recognize it in any orientation. In her interview, Mrs Milano expressed surprise at the difficulty other students had encountered:

I *assumed* … that they knew what the base and height of a triangle … is … and how to recognise it … it just sort of (pause) was made *very obvious* that … they don't understand- …ᵛⁱ

Leon had subsumed the attributes of the rectangle into equivalent attributes of triangles (length as base, width as perpendicular height) so he was able to operate with attributes of triangles to find their areas. This enabled him to calculate areas of triangles that other students could not calculate because they had not identified the critical rectangle and triangle attributes that were equivalent. They knew the rule but not why it worked (instrumental understanding, Skemp, 1976) so were not able to 'see' the perpendicular height with triangles in varying orientations. Wertheimer (1959) identified similar problems with perpendicular heights in parallelograms.

WHAT DID LEON LOOK-IN UPON?

What did Leon Look-in upon and why did this lead to a change in his goal? By analysing the activity of students seated around Leon in Lesson 12, asking Mrs Milano what she saw, and observing that other students did not volunteer general methods for finding areas of acute-angled triangles in Lesson 13, it was evident that Leon was the only student who developed methods (other than counting squares) for finding areas of acute-angled triangles. The only pair who found an area without counting squares worked with the right-angled triangle (Triangle 2). Thus, whatever Leon Looked-in upon, it was not work associated with finding the area of an acute-angled triangle by calculating the area of the enclosing rectangle.

To find what Leon Looked-in upon, an enriched transcript was developed for intervals of time over which this Looking-In activity could have occurred (see Table 1). I developed these enriched transcripts to enable data from different sources to be considered simultaneously over time intervals of interest (Williams, 2004). Table 1 shows the time interval [Column 1], Leon's interview comments related to that time interval [Column 2], the sketches he made during the interview to illustrate the internal images he was generating [Column 3], the images in the classroom that Leon had seen [Column 4], those parts of Leon's Lesson 12 talk that were relevant to this analysis [Column 5], Pepe's activity [Column 6], and other talk relevant to this analysis [Column 7]. The dotted horizontal line in Table 1 separates two intervals of time found relevant to this analysis.

Table 1 shows that at 36:14[vii] Leon knew Methods A and B because he compared the ease of two methods. Between 38:21 and 38:57, he Looked-in on Elina and Serina's diagram and Pepe's diagram. This was evident from Leon's comment to Mrs Milano that Elina and Serina could use shorter ways [see Columns 5, 4], and Leon's interview reflections about quicker ways for Pepe [see Columns 2, 4]. Leon's reconstruction of his Looking-In activity in his interview after Lesson 13 confirms what he saw, and illuminates how he attended to it idiosyncratically:

> People were drawing the actual rectangles around it- I don't know whether they knew they [triangles] were coming from rectangles ... the way they were drawing it made it look like they did ... made me think about it.

Leon Looked-in on dynamic visual displays [Table 1, Column 4] that were being generated by other students. His comment "I don't know whether they knew they [triangles] were coming from rectangles" is supported by the lesson video; students were generating displays to count squares rather than to calculate. They used rectangles to rule grids to make squares.

Table 1. Enriched transcript identifying interval in the lesson surrounding Leon's Looking-In activity

1. Time in L12	2. Leon's Interview Reconstruction	3. 'Internal' images (Leon)	4. Images in Classroom	5. Leon's Talk and Actions in Class	6. Pepe's Activity	7. Talk of Others
36:14-36:22	... I thought I could just figure out what it would be if it were a parallelogram and then halve it ... figure out what it would be if it was four because you could just trial whatever it was if it was four. ... I understood it- I didn't understand it then I understood it then I didn't understand it.	Asked to explain more, Leon prefaced his sketch with 'Um I ... think it was ... like that ... [drew 2 triangles] should have the same length and width ...'		Leon: [to T] I think I know. [Nodded] Except if I was doing Triangle 2-2's the easiest one there. [Leon did not respond to Pepe's actions]	Pepe: Bip bip [drove a pen over Leon's arm]	T: [to Leon] Do you?
38:21-38:57	... people were drawing the actual rectangles around it- I don't know whether they knew they [triangles] were coming from rectangles ... the way they were drawing it made it look like they did ... made me think about it. He's [Pepe] drawn up the grid and ... I thought ... you could do it quicker ... figure out what the area would be fully and then halve it.		Elina's and Serina's. Diagram generated by others (including Pepe)	Leon: [to T] Aren't they silly? There's so many other shorter ways. [to Pepe] [Laughs] You are doing a great job Pepe.		T: [to Elina and Serina] ... You've resorted to drawing the grid?

Key. --- Break between time intervals displayed

230

As a result of his Looking-In activity, Leon changed his focus from finishing the work to wanting to understand *why* the area of a triangle was half the area of the enclosing rectangle. Table 1, Row 3 indicates Leon was aware that the area was half the area of the rectangle at the end of Lesson 12. Leon's gasp in Lesson 13 and his explanation of this gasp in his post-lesson interview showed Leon did not know *why* the area of the acute-angled triangle was half that of the enclosing rectangle in Lesson 12.

Leon's Looking-In activity changed his focus from laborious calculation to searching for an elegant method. This led to his recognition of relationships between attributes of triangles and enclosing rectangles. Such realisation was unlikely to have occurred if he had continued with Method B. Leon's Looking-In activity compensated for his lack of appropriate cognitive artefacts; enabling him to autonomous access mathematics to progress his exploration.

THE ROLE OF LOOKING-IN IN CREATIVE MATHEMATICAL THINKING

The "Space to Think" model emerged from my analysis for students discovering complexity (Williams, 2005, pp. 365-371). It is an activity space that was manoeuvred by each of the five students including Leon (in Australia and the USA) who spontaneously undertook creative mathematical thinking (were discovering complexity). Six activities were found common to these students as they spontaneously engineered flow situations associated with mathematical explorations (Williams, 2002a):
- 'inclining to explore',
- 'spontaneously identifying a mathematical complexity',
- 'manoeuvring cognitive autonomy',
- 'autonomously accessing mathematics',
- 'spontaneously pursuing an exploration',
- 'asking questions to structure future exploration'.

These activities were found necessary (but not necessarily sufficient) for creative mathematical activity. Figure 4 summarises the cognitive activity Leon undertook. This figure is used to focus discussion of activity in the Space to Think with particular emphasis on the role of Looking-In as support to 'autonomously accessing mathematics'. Figure 4 contains: questions Leon asked himself [Column 1], Leon's solution process [Column 2], and novel ideas and mathematical structures Leon developed [Column 3]. Leon's cognitive activity was 'observable' through cognitive elements in his classroom talk (Hershkowitz, Schwarz, & Dreyfus, 2001), and his interview reconstruction of his thinking.

Enacting optimism or 'inclining to explore' is evidenced in Figure 4, Columns 1, 2. By considering that *he* would be able to adapt the mathematics he knew to find a rule for finding areas of triangles Leon enacted success as pervasive and personal. He enacted failure as temporary by searching for ways to circumvent the difficulties he encountered, and enacted failure as specific by examining 'failures' to identify what he could change (Column 2).

231

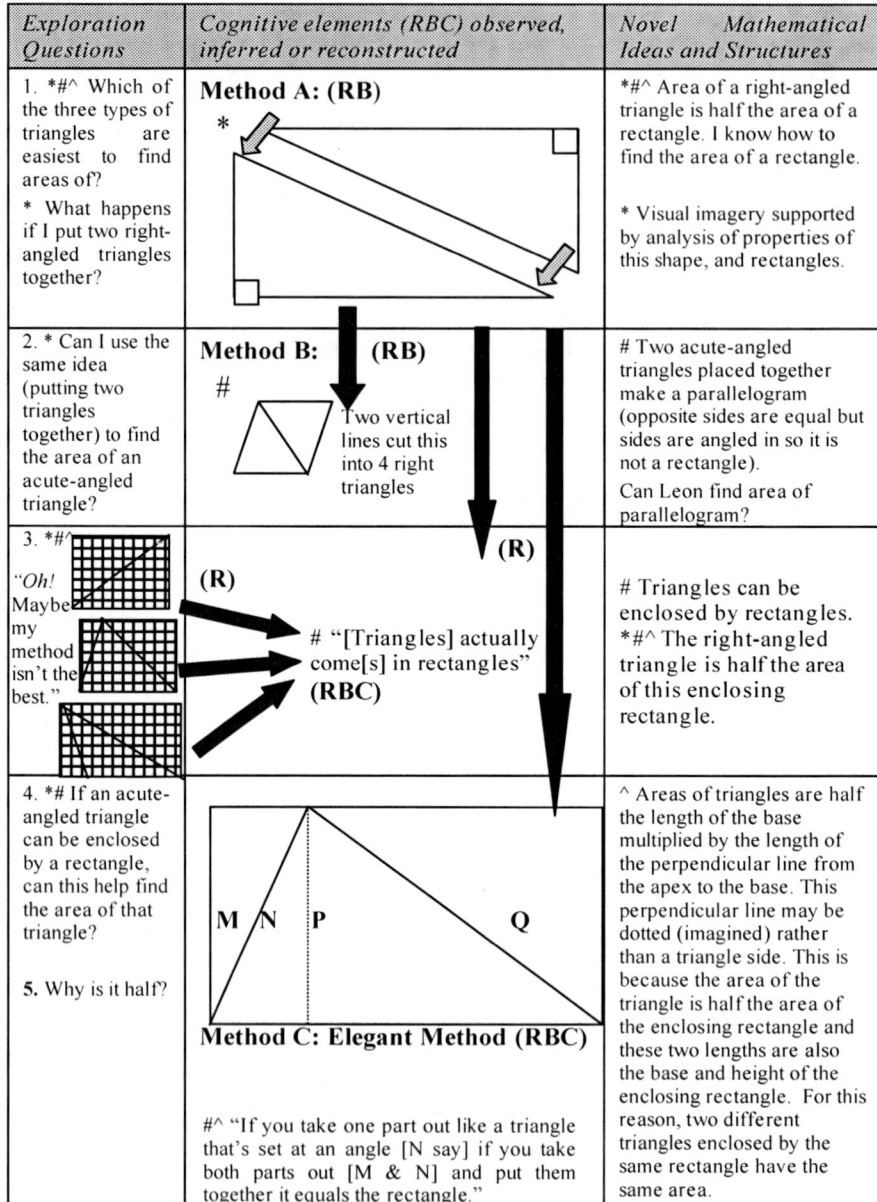

Exploration Questions	Cognitive elements (RBC) observed, inferred or reconstructed	Novel Mathematical Ideas and Structures
1. *#^ Which of the three types of triangles are easiest to find areas of? * What happens if I put two right-angled triangles together?	**Method A: (RB)**	*#^ Area of a right-angled triangle is half the area of a rectangle. I know how to find the area of a rectangle. * Visual imagery supported by analysis of properties of this shape, and rectangles.
2. * Can I use the same idea (putting two triangles together) to find the area of an acute-angled triangle?	**Method B: (RB)** # Two vertical lines cut this into 4 right triangles	# Two acute-angled triangles placed together make a parallelogram (opposite sides are equal but sides are angled in so it is not a rectangle). Can Leon find area of parallelogram?
3. *#^ "Oh! Maybe my method isn't the best."	**(R)** # "[Triangles] actually come[s] in rectangles" **(RBC)**	# Triangles can be enclosed by rectangles. *#^ The right-angled triangle is half the area of this enclosing rectangle.
4. *# If an acute-angled triangle can be enclosed by a rectangle, can this help find the area of that triangle? 5. Why is it half?	**Method C: Elegant Method (RBC)** #^ "If you take one part out like a triangle that's set at an angle [N say] if you take both parts out [M & N] and put them together it equals the rectangle."	^ Areas of triangles are half the length of the base multiplied by the length of the perpendicular line from the apex to the base. This perpendicular line may be dotted (imagined) rather than a triangle side. This is because the area of the triangle is half the area of the enclosing rectangle and these two lengths are also the base and height of the enclosing rectangle. For this reason, two different triangles enclosed by the same rectangle have the same area.

Key. Inferred from: * other evidence, video activity; # interview; ^ lesson dialogue.
Black arrows: progression of thinking (RBC; observable cognitive elements)
Cognitive Elements: C, Constructing (synthesising, evaluating); B, Building-with (novel analysing, novel synthetic analysing, novel evaluative-analysing); R, Recognizing (analysing, comprehending)

Figure 4. Cognitive activity during Leon's development of novel conceptual understanding

For example, when the juxtaposition of acute-angled triangles was not very successful, Leon focused attention on rectangles instead (because that had also previously been successful). Instead of responding to failure as though it were a pervasive attribute of himself: "I can't do this, I'm stupid", he examined the situation and identified what could be changed (Failure as Specific). The first complexity Leon spontaneously identified was that the three triangles on the board might differ in their relative ease for finding areas (see Figure 4, Column 1, Row 1). Another complexity he discovered was that triangles have enclosing rectangles. Leon manoeuvred cognitive autonomy by working alone not with Pepe who wanted to count squares. In doing so, he manoeuvred time to think without needing to take into account the way Pepe wanted to proceed. This is evident from Table 4, Column 1; Leon responded to his own questions and focused his own attention on external stimuli. Cognitive autonomy was also enabled by Leon's perception that *he* was under no obligation to undertake the task set by the teacher.

Leon autonomously accessed mathematics by assembling cognitive artefacts he possessed (e.g., properties of rectangles, area of rectangle formula, Method A as part of Method B and Method C). When he did not possess appropriate cognitive artefacts (necessary to develop an elegant solution pathway), he autonomously accessed mathematics by Looking-In on the dynamic visual displays generated by others and used these displays along with his own previous thinking to develop a new understanding—'triangles always come in rectangles' (see Figure 4, Row 3). Through this Looking-In activity Leon developed this cognitive artefact that was crucial to his construction of new knowledge.

Cifarelli (1999) identified students asking specific mathematical questions that structured their future exploration (structuring questions). My study illuminates a more general type of structuring question. As a result of Looking-In activity Leon asked: "*Oh!* Maybe my method isn't the best?" (Figure 4, Row 3) and this structured spontaneous pursuit of a new direction of exploration (see Figure 4, Row 4). This type of question is not specific to the mathematical context in which it occurred; there are many situations in which such a question could be appropriate. Spontaneous pursuit involves using increasingly complex thinking whilst trying to answer the structuring question asked, and developing novel mathematical concepts during this process. Leon's activity elaborates the nature of structuring questions.

In this study, Leon recognized (e.g., the rectangle could be used to find areas of right triangles) and built-with previously known mathematical ideas (e.g., used what he had learnt for right-triangles to calculate areas of acute-angled triangles), and constructed a novel mathematical structure thus eliminating the need to explicitly use rectangles to find areas of triangles. From least to most complex, the categories of thinking Leon used during his spontaneous pursuit were recognizing (R), building-with (analysis, synthetic-analysis, evaluative-analysis) (B), and constructing (synthesis, evaluation) (C); new conceptual understanding developed. My previous work (Williams, 2002b) integrated Krutetskii's (1976) 'mental activities' into Hershkowitz, Schwarz, and Dreyfus's (2001) observable cognitive elements of the process of abstracting to add further subcategories to assist in studying increasingly more complex thinking. Examples of each of these types of thinking are contained in Figure 4. The words italicised in the following descriptions include key aspects of each thinking category. Leon's spontaneous question at the start of Row 3 (Figure 4) involved evaluative-analysis because he *simultaneously* considered two ways to proceed (synthetic-analysis) for the purpose of making a *judgment* about the respective elegance of these solution pathways. Leon finally synthesised his previous mathematical ideas with the new cognitive artefact he had developed 'triangles come in rectangles' when he answered his question, "Why is it half?" (Figure 4, Row 4). Leon used procedures he had developed early in his exploratory activity in different ways throughout his exploration (see Figure 4, Column 2, e.g., Method A). His final constructing relied upon Method A and cognitive artefacts he developed through earlier Looking-In. Soon after, Leon *recognized the usefulness* of his new ideas in *another context* (evaluating); triangles enclosed by the same rectangle have the same area.

Without Leon's Looking-In activity, he would not have accessed the mathematics, necessary to pursue his exploration and would not have experienced the positive affect associated with flow; an optimism-building experience (Seligman, 1995). His realisation that 'angles come in rectangles' and why the area was half were accompanied by positive affect. Looking-In activity was found to involve:
- A generator of a dynamic visual display
- An extractor of mathematics implicit within this display

The generator was sometimes aware of the mathematical significance the extractor perceived and sometimes (as in this case) they were not. Looking-In activity was identified in Australian classrooms where exploratory tasks were set, opportunities to talk to other students existed, and students found time to explore idiosyncratically. This study highlights the need for further research into the process of Looking-In and situations in which it is likely to occur. Such activity could support the development of relational understanding for students with limited access to cognitive artefacts, and contribute to an increase in the frequency of creative mathematical thinking in class.

NOTES

i Key to symbols used in quotes from student interviews in this chapter:

...	Omitted text that does not affect meaning
(pause)	A pause
[L12-SI-Leon]	Lesson number, student interview, name of student
italics	Italicised text indicates emphatic speech

ii The text in square brackets is additional information provided by the transcriber.
iii [L12, 24 mins, 16 secs into the lesson]
iv The text in square brackets is additional information provided by the transcriber.
v The text in brackets is additional information provided by the author.
vi [L13-T1-Mrs Milano] Teacher interview post-Lesson 13
vii 36 minutes and fourteen seconds into Lesson 12

ACKNOWLEDGEMENTS

This analysis was undertaken at The University of Melbourne as part of my PhD studies under the supervision of Professor David Clarke whose expertise as a supervisor was much appreciated. My thanks to ICCR team for their technical assistance, and to the LPS teams for their collegiality and encouragement. I appreciate the perceptive comments made by Fritjof Sahlström on two previous drafts of this chapter, and Seah Lay Hoon's thoughtful comments, which assisted my elaboration of key constructs.

REFERENCES

Cifarelli, V. (1999). Abductive inference: Connections between problem posing and solving. In O. Zaslavsky (Ed.), *Proceedings of the 23rd conference of the International Group for the Psychology of Mathematics Education* (Vol. 2, pp. 217-224). Haifa, Israel: PME.

Csikszentmihalyi, M. (1992). Introduction. In M. Csikszentmihalyi & I. Csikszentmihalyi (Eds.), *Optimal experience: Psychological studies of flow in consciousness* (pp. 3-14). New York: Cambridge University Press.

Department of Education. (1997). *Curriculum 2005*. Pretoria, South Africa: Author.

Ericsson, K. A., & Simon, H. A. (1980). Verbal reports of data. *Psychological Review, 87*(3), 215-251.

Hershkowitz, R., Schwarz, B., & Dreyfus, T. (2001). Abstraction in context: Epistemic actions. *Journal for Research in Mathematics Education*, 32(2), 195-222.

Kilpatrick, J. (2002). Understanding mathematical literacy: The contribution of research. In L. Bazzini & C. Whybrow Inchley (Eds.), *Littéracie Mathématique à L'ere Digitale. Proceedings of Commission Internationale pour l' Étude et L'Amélioration de l'Enseignement des Mathématiques 53* (pp. 62-72). Milano: Ghisetti e Corvi Editori.

Krutetskii, V. (1976). *Psychology of mathematical abilities in schoolchildren*. (J. Kilpatrick, & I. Wirzup (Eds.), J. Teller, Trans.). Chicago: University of Chicago Press. (Original work published in 1968).

Säljö, R. (1999). Learning to cope: a discursive perspective. In E. Frydenberg (Ed.), *Learning to cope: Developing as a person in complex societies* (pp. 53-63). New York: Oxford University Press.

Seligman, M. (with Reivich, K., Jaycox, L., Gillham, J.). (1995). *The optimistic child*. Adelaide: Griffin Press.

Skemp, R. (1976). Relational understanding and instrumental understanding. *Mathematics Teaching, 77,* 20-26.

Steffe, L., & Thompson, P. (2000). Teaching experiments methodology: Underlying principles and essential elements. In A. Kelly & R. Lesh (Eds.), *Handbook of Research Design in Mathematics and Science Education* (pp. 267-306). Mahwah, NJ: Lawrence Erlbaum.

Wertheimer, M. (1945/1959). *Productive thinking*. New York: Harper.

Williams, G. (2002a). Associations between mathematically insightful collaborative behaviour and positive affect. A. Cockburn & E. Nardi (Eds.), *Proceedings of the 26th conference of the International Group for the Psychology of Mathematics Education* (Vol. 4, pp. 402-409). Norwich, UK: PME.

Williams, G. (2002b). Identifying tasks that promote creative thinking in mathematics: A tool. In B. Barton, K. Irwin, M. Pfannkuch, & M. Thomas (Eds.), *Mathematics education in the South Pacific* (Vol. 2, pp. 698-705). Auckland, New Zealand: Mathematical Education Research Group of Australasia.

Williams, G. (2003). Associations between student pursuit of novel mathematical ideas and resilience. In L. Bragg, C. Campbell, G. Herbert & J. Mousley (Eds.), *Mathematical education research:*

Innovation, networking, opportunity (Vol. 2, 752-759). Sydney: Mathematics Education Research Group of Australasia.

Williams, G. (2004). The nature of spontaneity in high quality learning situations. In M. Hoines & A. B. Fuglestad (Eds.), *Proceedings of the 28th conference of the International Group for the Psychology of Mathematics Education* (Vol. 4, pp. 433-440). Bergen, Norway: PME.

Williams, G. (2005). *Improving the intellectual and affective quality in mathematics lessons: How autonomy and spontaneity enable creative and insightful thinking.* Unpublished doctoral dissertation, University of Melbourne, Melbourne, Australia.

Gaye Williams
Faculty of Education
Deakin University
Australia

IDA AH CHEE MOK AND FRANCIS LOPEZ-REAL

CHAPTER SIXTEEN

A Tale of Two Cities:
A Comparison of Six Teachers in Hong Kong and Shanghai

INTRODUCTION

The Third International Mathematics and Science Study (TIMSS) reported by
Stigler and Hiebert (1999) identified different "scripts" used by mathematics
teachers in Japan, Germany and the USA, referring to distinctive teaching styles.
This suggestion of the existence of a national script is persuasive. However, based
on various reports as part of the Learner's Perspective Study (LPS), there appears
in general little evidence to support the existence of national teaching scripts.
Clarke and Mesiti (2003) reported that the proposed US lesson pattern in the
TIMSS video study did not correspond to any of the 25 US lessons in the data of
LPS. Jablonka (2003) showed that the German TIMSS video model is in poor
correspondence with the LPS sample of 30 German mathematics lessons. Also, as
a part of the analysis of the LPS data, Lopez-Real, Mok, Leung, and Marton (2004)
found significant variations between different lessons, taught by the same Shanghai
teacher. By analysing a sequence of five of the teacher's lessons they were able to
identify a 'pattern' of teaching that characterised his approach. But this finding
strongly suggested that analysis of a single 'snapshot' lesson is unlikely to reveal a
teacher's 'script', in the sense claimed by Stigler and Hiebert (1999). However,
despite the fact that individual lessons are unlikely to follow a definitive script, in
this chapter we discuss some interesting patterns of similarities and differences
between teachers of the same region that emerged from our analysis of data from
Hong Kong and Shanghai.

METHODS OF ANALYSIS

In the analysis, the data consisted of the videos of the lessons, and the teacher and
student interviews. There were three teachers from each of the two cities. The
number of lessons by each teacher ranged from 14 to 18 lessons. They were
consecutive lessons and all were on the topic of simultaneous equations. In this

D. J. Clarke, C. Keitel & Y. Shimizu (Eds.), Mathematics Classrooms In Twelve Countries: The
Insider's Perspective. 237–246. © 2006 Sense Publishers. All rights reserved.

chapter we focus on the analysis of the lessons, rather than the student and teacher interviews.

We used two types of coding for the videos of the lessons. The first was concerned with the class organisation and the second with the teacher's approach and activities. The categorisation of the lessons was made by using the coding facility of v-prism. At the early stage of the coding development, three researchers worked on six lessons intensively and carried out the reliability check. The coding of the remaining set of data was carried out by one of the researchers in the next stage. The explanation of the coding is as follows:

Class organisation

The activities in the lesson were coded into three types according to how the class was organised: 1) Classwork, 2) Seatwork: Individual, and 3) Seatwork: Small Group. These codes formed exclusive segments covering the complete lesson. It is quite possible that such a classification might not form mutually exclusive segments but, in our study, these segments were mutually exclusive for all the lessons by the six teachers. The codes are defined below.

1) Classwork Classwork refers to teacher talk only and teacher-led discussion carried out in a whole class setting. For example, the teacher explains a definition or gives instruction for a worksheet; the teacher asks a question which may be answered by one or several students and there may be some follow-up questions.

2) Seatwork: Individual This activity is one where students work on a task individually, without any discussion with other students.

3) Seatwork: Small Group In these activities two or more students discuss or do a task amongst themselves.

Teacher's approach and activities

With respect to the teacher's approach and activities, they were coded into five exclusive types defined below.

1) Exploratory

- The focus is on a relatively open or difficult problem which has more than one possible answer.
- The teacher gave a signal for pair or group discussion.
- A whole class discussion with the following features: inviting more than one student to give answers, inviting explanations, inviting peer comments.

2) Directive

- No comment on the student's answer, no attempt to discuss the answer with the other students, simply stating what should be done (e.g. the conventional notation).
- Emphasis is purely on following a convention
- Insistence on precise language
- Repetition of what had been learnt in an earlier lesson or in the earlier part of the lesson in a fast pace, using this as a foundation for establishing further knowledge.
- Insistence on articulation of procedures.
- Clear and directive definition of a concept or method after an illustrative example or discussion.
- Teacher plays the role of directing students to work on problems.
- Probing for 'expected' answers.
- Directive explanation by teacher.

3) Summarisation

- Teacher does summarising during the lesson, or to conclude the topics or problems discussed.

4) Exercises and Practice (sometimes includes whole class checking of exercises)

- In the situation of doing textbook exercises, there can be teacher talking about/ explaining the question, and students having seatwork.
- Teacher checks exercises with students.

5) Assigning Homework

- Teacher assigns homework or questions for students to do at home.

COMPARISON OF CLASS ORGANISATION

Using a coding system, as described in the previous section, is a useful and valuable tool for analysing and interpreting the classroom environment. However, equally important is the *way* in which that data is then presented. We have found that representing each lesson graphically, so that the progress and changes throughout the lesson are clearly illustrated, is a powerful way to compare each teacher's way of working. As an example, consider the first lesson of teacher SH2 (from Shanghai) with the twelfth lesson of teacher HK3 (from Hong Kong). In both cases the percentage of time spent under the 'Classwork' category is almost the same, that is, just over 60%. However, Figure 1 shows how the distribution of this time within the lessons is strikingly different.

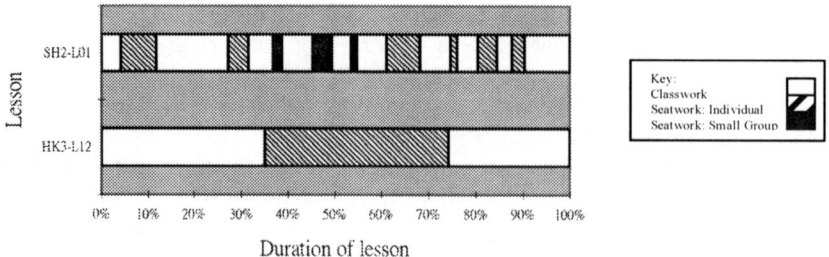

Figure 1. Comparison of Class Organisation Categories for Two Lessons

As we have reported in a previous paper (Lopez-Real et al., 2004) variations also occur between lessons of the *same* teacher, so that it is impossible to characterise a teacher's way of working from one 'snapshot' lesson. However, we also reported that over a sequence of lessons a 'pattern' of teaching does emerge that is characteristic of a particular teacher. We found evidence of the same features in this current analysis. Thus, teacher HK3 for example, taught another lesson where the ratio of 'Classwork' to 'Individual Seatwork' was almost exactly reversed to that shown in Figure 1. That is, on this occasion about 60% of the time was spent on seatwork. However, the distribution still displayed the same pattern in the sense that the seatwork was a continuous block of time within the lesson. Similarly, another of SH2's lessons also showed a reversal of the amount of classwork time, but again the distribution within the lesson exhibited a pattern of many small blocks of work as in Figure 1.

In terms of the distributions within a lesson, broadly speaking, the six teachers displayed patterns of organisation that could be described in the following way:
– Teachers SH2 and SH3: Many small blocks; frequent shifts from one category to another.
– Teachers HK2 and HK3: Large, continuous blocks; sometimes a single form of organisation for a whole lesson.
– Teachers SH1 and HK1: A 'middle road' between the above two modes.

This *may* be suggestive of a style of teaching that is typical of each city, with some teachers deviating slightly from this style, although clearly we cannot make a definitive assertion based solely on these six teachers. So, are there other features of organisation which show any clear pattern? Table 1 shows the overall percentages of the three categories.

Table 1. Overall Percentages of the Three 'Organisation' Codes for the Six Teachers

	SH1	SH2	SH3	HK1	HK2	HK3
Classwork	86.8	58.8	67.5	69.5	78.1	60.7
Seatwork (I)	11.7	38.8	21.7	30.5	21.9	39.3
Seatwork (G)	1.5	2.4	10.8	0	0	0

The first point that should be made is a note of caution. If one were comparing the six teachers based purely on this table one might be tempted to think that teachers SH2 and HK3 were quite similar. However, as we have seen, this is a quite erroneous impression, and this is also true for comparisons of the other teachers. Again, this illustrates the importance of looking at each individual lesson (as well as the totality of lessons) and also of looking at the distributions *within* a lesson.

Nevertheless, there is one striking feature that does clearly emerge from this table. This is the fact that none of the Hong Kong teachers ever used any groupwork in the organisation of their lessons. Certainly, one can hardly claim that groupwork plays a very significant role as far as the Shanghai teachers are concerned. The percentages are very small for SH1 and SH2, although rather more significant for SH3. However, analysis of the sequence of their lessons does show that groupwork is a *regular* feature of their teaching even though it is often of very short duration. The distribution shown for SH2 in Figure 1 illustrates this quite well. Typically, as in this lesson, the teacher asks the students to work in a group to find the solutions to a series of short questions, such as "What is the largest integer solution for $x < -5.4$?" Only a short time is allowed for this but the students *are* working in groups.

Figure 2 shows why SH3's percentage for groupwork is rather higher. For some of this teacher's lessons, groupwork was the dominant mode of seatwork and the students were given more time on each task because the tasks themselves were more challenging. For example, after having tackled problems where the first equation of two simultaneous equations could immediately be substituted into the second, the teacher asked the students to think about what they could do when this was not the case. After working on this open problem in groups, the students then presented their ideas to the teacher.

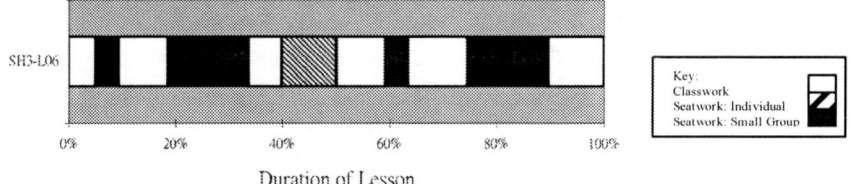

Figure 2. Distribution of Class Organisation in a Lesson by SH3

COMPARISON OF TEACHERS' APPROACHES

We now turn our attention to the teaching approaches and activities employed by the six teachers in their lessons. Here we used the five codes described earlier, namely Exploratory, Directive, Summarisation, Exercise and Practice, and Assigning Homework. Keeping in mind the reservations that one must hold when analysing aggregated data, we begin by considering Table 2 which shows the overall percentages for these categories over all the lessons.

Table 2. Overall Percentages of the Five Approach Codes for the Six Teachers

	SH1	*SH2*	*SH3*	*HK1*	*HK2*	*HK3*
Exploratory	15.3	17.7	24.7	7.4	3.7	0
Directive	57.6	10.1	30.5	46.4	57.0	36.2
Summarisation	3.0	1.3	3.6	0.3	0.2	0.1
Exercise/Practice	23.2	70.1	40.2	44.9	35.9	59.9
Assigning Homework	1.0	0.8	0.9	0.9	3.2	3.8

As before, we cannot but be struck by the individuality of each teacher's 'profile' and how this again demonstrates the futility of trying to identify a national 'script'. However, we can also identify some interesting patterns, in terms of comparing the two cities, which are suggested by the table. Perhaps the most striking is the contrast in the amounts of time spent on exploratory activities. Clearly HK3 is rather an extreme case in that this teacher never uses any exploratory activity throughout the whole sequence of 14 lessons! In fact, the profile for HK3 seems very typical of a 'traditional' mode of teaching that consists of Demonstration followed by Practice. However, the percentages for exploratory activity for HK1 and HK2 are also low, compared to the Shanghai teachers. Moreover, although the percentages for the Shanghai teachers may not be considered very high in an absolute sense, once again we find that the overall figures tend to conceal an important feature. This is the fact that the exploratory activities occurred very frequently and regularly, even though they might be short in duration. (This is rather similar to the pattern of groupwork reported in the previous section and, of course, the two are not unrelated). For example, Table 3 shows the number of lessons in which at least some exploratory activity took place.

Table 3. Frequency of Exploration Activities

	SH1	*SH2*	*SH3*	*HK1*	*HK2*	*HK3*
Total lessons in sequence	15	14	15	18	14	14
No. Lessons with some exploration	11	11	14	8	3	0
% Lessons with some exploration	73.3	78.6	93.3	44.4	21.4	0

The contrast between the Shanghai and Hong Kong teachers is now highlighted even more strongly and suggests a different underlying philosophy in the use of exploratory activities. Thus, for the Shanghai teachers it seems to be important to continually use exploration in their teaching on a very regular basis, so that almost every lesson should include an element of exploratory work if possible. The difference between this approach and that of the Hong Kong teacher who used a reasonable amount of exploratory work (i.e. HK1) can be illustrated in Figure 3. This shows the pattern of teaching in the first 4 lessons of the sequence of 18. The percentage of time for exploratory activity in each of these lessons is 50%, 25%, 14% and 0% respectively. This is followed by another four lessons, similar to lesson L04, which consist completely of Directive teaching and Exercise and Practice. Similar sequence patterns, though not quite so extreme, then occur again. This suggests that the teacher sees exploratory activity as being important at the *beginning* of a new topic but that it becomes less important, or necessary, as the topic proceeds.

The second interesting pattern that can be observed in Table 2 concerns the Summarisation category. Even though the percentages are low for all six teachers, as we would expect for such a category, however, when we compare the *relative* percentages of the Shanghai teachers compared to the Hong Kong teachers, the differences are very striking. These differences become clear when we look at the frequency of occurrence of summarising in the sequences of lessons. This is illustrated in Table 4.

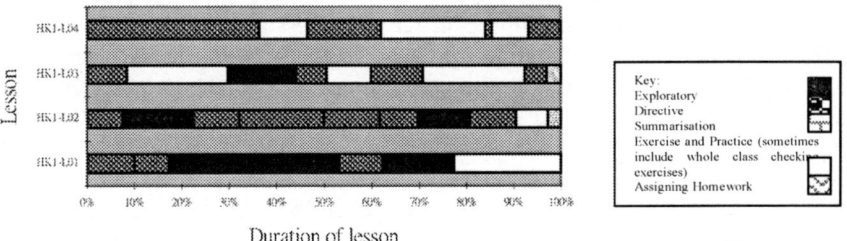

Duration of lesson

Figure 3. Teaching Approaches in HK1's First 4 Lessons

Table 4. Frequency of Summarisation

	SH1	SH2	SH3	HK1	HK2	HK3
Total lessons in sequence	15	14	15	18	14	14
No. Lessons with some summary	12	7	13	2	1	1
% Lessons with some summary	80.0	50.0	86.7	11.1	7.1	7.1

In a similar way to the Exploration category, what is significant here is that two of the Shanghai teachers are making some form of summary towards the end of almost every lesson, and the other Shanghai teacher does this in half of the lessons. However, all three of the Hong Kong teachers rarely make any kind of summary.

By contrast, when it comes to Assigning Homework, the situation appears to be almost reversed, at least as far as HK2 and HK3 are concerned. But this time, the differences in time spent on assigning homework by these two teachers, compared with the other four, are not due to the *frequency* of assigning homework. In fact, all six teachers assigned homework on a very frequent basis and, except for HK1, this happened in over half of the lessons. The difference lies in the fact that they both spent more time in explaining or discussing the assigned homework, whereas the three Shanghai teachers and HK1 spent very little in time on this aspect but simply gave the appropriate instructions.

CONCLUSION AND DISCUSSION

Our findings suggest once again that attempting to define a 'national script' for mathematics teaching, in any simplistic way, can hardly be justified when sequences of lessons by different teachers are analysed. Not only do the teachers within a given geographical or cultural environment display quite individualistic modes of teaching, but each teacher also displays significant differences from one lesson to another. This was true both for the analysis of the lesson organisation and for the analysis based on teaching approaches. Nevertheless, some interesting patterns did emerge that were suggestive of different characteristics between the group of Shanghai teachers compared to those from Hong Kong. Three distinctive elements could be identified. First, in terms of lesson organisation, there was a striking difference in the use of group work. The three Hong Kong teachers in this study never used this form of organisation in any of their lessons. By contrast, all three Shanghai teachers used groupwork in at least some lessons, albeit in very short 'doses'. For SH2 and SH3, this was a frequent and regular feature of their lessons. Some groupwork took place in 73% of SH3's lessons and in 64% of SH2's lessons. Of course, we do not wish to generalise about all Hong Kong and Shanghai teachers from this data. But the large number of lessons in each sequence does provide quite strong evidence that, at least for these particular teachers, we have a valid picture of their way of working. Even so, we need to tread carefully. All the sequences were concerned with the topic of simultaneous equations, as we mentioned earlier. It is possible that all three Hong Kong teachers felt that this topic does not lend itself easily, or profitably, to using group work, and they might use it for another topic. However, one of the reasons for studying a sequence of lessons on the same topic is *precisely* in order to see how different teachers tackle it. Hence, even if the nature of the topic might not lend itself easily to group work, we are still left with the fact that the Shanghai teachers *did* use this form of organisation, two of them on a frequent basis. This is quite strong evidence that they see it is an important part of their regular repertoire.

The second interesting pattern that emerged was in the comparative use of exploratory activities in the teaching approaches. This was even more striking than the use of group work. The Shanghai teachers' use of this approach was significantly greater than that of the Hong Kong teachers. And again, it was clear that the approach was being used in a very regular and frequent manner. As we remarked earlier, the two (group work and exploratory activity) are not unrelated, but they do not imply one another. It is certainly true that when group work is used it is very likely to be associated with an exploratory activity. But it is clearly possible to use an exploratory approach in teaching which does not require group work. In fact, in many of the lessons analysed, the exploration was taking place in a whole class setting. Nevertheless, even when whole class teaching was being used, our analysis showed that the Shanghai teachers were using an exploratory mode of teaching far more often than the Hong Kong teachers.

The third interesting pattern concerned the use of summarisation, normally, as one would expect, towards the end of a lesson. But again, it is not the amount of time spent on summarisation that is important, but rather the frequency. For the Shanghai teachers this was a feature of almost every lesson compared to the much less frequent use by the Hong Kong teachers. Taken together, the three elements we have identified do seem to constitute a clear characterisation of the Shanghai teachers that was quite different to the Hong Kong teachers.

The reader may have become aware that at no time have we commented on the *quality* of the lessons in our analysis. This has been quite deliberate in this particular analysis where we simply wanted to identify, if possible, any distinguishing characteristics between the teachers, purely in terms of organisation and teaching approaches. However, it is only fair to point out that the activities we identified as exploratory in our coding did involve the students in being *challenged* rather more than in the directive teaching. The questions that were posed tended to be more open-ended and forced the students to think more deeply about the concepts involved. Taken from this perspective, our analysis suggests that, because of the differences we identified in terms of exploratory work, the Shanghai lessons are more likely to have resulted in a richer learning experience for the students.

ACKNOWLEDGEMENTS

The work described in this paper was supported by a grant from the Research Grants Council of the Hong Kong Special Administrative Region, China (Project No. HKU 7192/02H) and a grant from the Committee on Research and Conference Grants of the University of Hong Kong.

REFERENCES

Clarke, D. J., & Mesiti, C. (2003). Addressing the challenge of legitimate international comparisons: Lesson structure in Australia and the USA. In L. Bragg, C. Campbell, G. Herbert, & J. Mousley (Eds.), *Mathematics Education Research: Innovation, Networking, Opportunity, Proceedings of the 26th Annual Conference of the Mathematics Education Research Group of Australasia* (Vol. 1, pp. 230-237). Geelong, Australia: Deakin University

Jablonka, E. (2003, April). The structure of mathematics lessons in German classrooms: Variations on a theme. In D.J. Clarke (Chair), *Mathematics lessons in Germany, Japan, the USA and Australia:*

Structure in diversity and diversity in structure. Symposium conducted at the Annual Meeting of the American Educational Research Association, Chicago.

Lopez-Real, F. J., Mok, I. A. C., Leung, F. K. S., & Marton, F. (2004). Identifying a pattern of teaching: An analysis of a Shanghai teacher's lessons. In L. Fan, N. Y. Wong, J. Cai & S. Li (Eds.), *How Chinese learn mathematics: Perspectives from insiders* (pp. 382-412). World Scientific Publishing Co.

Stigler, J., & Hiebert, J. (1999). *The teaching gap*. New York: Simon & Schuster.

Ida Ah Chee Mok
Faculty of Education
University of Hong Kong
Hong Kong SAR, China

Francis Lopez-Real
Faculty of Education
The University of Hong Kong
Hong Kong SAR, China

CHAPTER SEVENTEEN

Mathematics Lessons in Korea:
Teaching with Systematic Variation

INTRODUCTION

In the past decade or so, there has been increasing interest in the study of the mathematics classrooms in East Asian countries, or countries falling under the so-called Confucian-heritage culture (CHC), the dominant culture in East Asia. However, relatively little has been published in the international literature on classroom practices in the CHC country of Korea. In this chapter, characteristics of the Korean mathematics classroom that are deemed to be conducive to effective learning are identified through an analysis of the Korean data of the Learner's Perspective Study (LPS). The classroom characteristics identified are then interpreted in terms of the underlying cultural values that they share with other East Asian countries.

THEORIES OF VARIATION [i]

To characterise classroom features that are conducive to learning, a learning theory is needed in the identification and interpretation of the relevant data. In this study, a learning theory espoused by Marton (1999) is utilised in the analysis of the LPS Korean data. Marton hypothesised that variation, simultaneity, and discernment were critical to learning, and studies by Runesson (1999) and Mok (2000) showed that Marton's theory of variation had a demonstrated potential in revealing the salient characteristics of classroom features that are related to student learning.

This theory of variation was developed from the work of Marton and Booth (1997), which described how an "enacted space of learning" was constructed through the creation of certain dimensions of variation for the experience of the students. According to Marton, Runesson and Tsui (2003), learning is a process in which learners develop a certain capability or a certain way of seeing or experiencing. In order to see something in a certain way the learner must discern certain features of the object. Experiencing variation is essential for discernment, and is thus significant for learning, and Marton et al. argued that it is important to attend to what varies and what is invariant in a learning situation.

D. J. Clarke, C. Keitel & Y. Shimizu (Eds.), Mathematics Classrooms In Twelve Countries: The Insider's Perspective. 247–261. © *2006 Sense Publishers. All rights reserved.*

In parallel with Marton's theory of variation, a theory of mathematics teaching and learning, called *teaching with variation*, has been developed by Gu (1994). Gu's theory was based on a series of longitudinal mathematics teaching experiments in China, and was heavily influenced by theories of cognitive science and constructivism. According to Gu's theory, meaningful learning enables learners to establish a substantial and non-arbitrary connection between their new knowledge and their previous knowledge (see also Ausubel, 1968). Classroom activities can be developed to help students establish this kind of connection by experiencing certain dimensions of variation. The theory suggests that two types of variation are helpful for meaningful learning. One is called "conceptual variation", and the other is called "procedural variation" (Gu, Huang & Marton, 2004).

Conceptual variation consists of two parts. One part is composed of varying the connotation of a concept: standard variation and non-standard variation. The other part consists of highlighting the substantial features of the concept by contrasting with counterexamples or non-examples. The function of this variation is to provide learners with multiple experiences from different perspectives.

Procedural variation is concerned with the process of forming a concept logically and/or chronologically (scaffolding, transformation), arriving at solutions to problems, and forming knowledge structure (relationship among different concepts). The function of procedural variation is to help learners acquire knowledge step by step, develop learners' experience in problem solving progressively, and form well-structured knowledge.

Multi-faceted Variation and Developmental Variation

While the two kinds of variations suggested by Gu are potentially powerful conceptual tools for analysing classroom events, the terms "conceptual variation" and "procedural variation" may be misleading. The adjectives "conceptual" and "procedural" may give readers the impression that they mean something similar to the terminology of "conceptual understanding" and "procedural understanding" coined by Hiebert (1986), which are used differently from the meaning of "conceptual" and "procedural" as defined by Gu. Also, Gu's terminology may lead us to think that "conceptual variation" and "procedural variation" are mutually exclusive, but in fact according to Gu's own definition, procedural variation is also related to the formation of concept. So the terms "conceptual variation" and "procedural variation" do not reflect very well the meaning they are supposed to represent as defined by Gu.

In this chapter, the term "multi-faceted variation" will be used to denote what Gu termed "conceptual variation" because the term refers to enhancing conceptual understanding through multiple representation and varied examples of a given concept. Also, the term "developmental variation" will be used to substitute for Gu's "procedural variation", since this variation helps the learners to construct knowledge structures through progressively acquiring the knowledge.

For example, in one of the Korean lessons videotaped, the teacher familiarised students with the concept of linear equations in two unknowns through comparison with linear equations in one unknown. The teacher reminded the students that equations with one unknown and those with two unknowns are similar in the sense that a root should satisfy the equation when substituted into the unknown(s) of the equation. But the two are different because the number of roots is different. This explanation helps students to understand linear equations with one unknown and two unknowns by contrasting the similarities and differences of the two concepts, and is thus considered a "multi-faceted variation".

Another example of multi-faceted variation is found in a lesson from another school. There the teacher introduced a new concept (ratio of areas) through concrete examples in everyday life: the fact that the amount of ink needed to print a photo depends on the area of the photo. This connection between an abstract mathematical concept and a concrete example in real life can be interpreted as a multi-faceted variation as well as "mathematization" in Freudenthal's terms (Freudenthal, 1983).

An example of "developmental variation" can be found in a lesson from the first school videotaped in this study. In the lesson, the teacher marked the ordered pairs (x, y) on the coordinate plane for the case where the roots of the linear equation were limited to natural numbers. The teacher then reminded the class that if the root is not limited to natural numbers, there would be an infinite number of points between the ordered pairs, and the solution of $x + y = 5$ would be a straight line. Here the teacher progressively unfolded the mathematical concept he tended to explain.

In fact, the notions of variation suggested by Marton and Gu are similar to the "mathematical variability principle" and "dynamic principle" advocated by Dienes (1973), and are related to the notion of "zone of proximal development" proposed by Vygotsky (1978) and the "duality of mathematical concept" suggested by Sfard (1991). According to Marton's theory of variation, the "space of variation" consists of different dimensions of variation in the classroom, and they form the necessary condition for students' learning in relation to certain learning objectives. And according to Gu, it is critical for the teacher to consider how to create variations focusing on critical aspects of the learning object through appropriate activities. For the learner, it is important to experience the variations through participating in the activities.

In the data analysis in this chapter, the patterns of variation critical to learning will be described in terms of the multi-faceted and developmental variations created in the lessons. Studies by Runesson (1999) and Rovio-Johansson (1999) support the hypothesis made by Marton (2000) that variation is a key for comparing the difference in practices between the East and the West. Marton argued that the most important difference between the Chinese/Japanese classes and those in the US was the difference in the pattern of variation. Chinese and Japanese students learned to approach the same mathematics problem in different ways, whereas the American students learned to apply the same approach to different but similar problems. Hence a description of the patterns of variation is considered an appropriate way of charactering the teaching in Korean classrooms.

SAMPLING, DATA COLLECTION AND ANALYSIS

Following the methodology of LPS, three schools in the urban/metropolitan community of Seoul were sampled for study. To preserve anonymity, the three schools are referred to as School H, School K and School W in this chapter. One grade 8 mathematics teacher in each of Schools H, K and W judged to be competent by the local professional community was selected. The teacher had at least five years of experience as a qualified teacher. One of the grade 8 classes taught by the teacher was then selected for study, and a continuous sequence of at least 10 lessons were videotaped for the class.

   The videotaped lessons were then viewed carefully, and a preliminary analysis was preformed on the data. Then a lesson in each of the three schools judged to be 'typical' of lessons in the series was chosen for a more fine-grained analysis. Table 1 shows the background characteristics of the three sampled schools and the sampled teachers, as well as information about the lessons chosen for detailed analysis:

Table 1. Background characteristics of the sampled schools and lessons

	School H	*School K*	*School W*
Type of schools	Girls' school	Co-educational	Co-educational
SES of parents	Mostly middle class		
Teacher Gender, (age)	Male (47)	Female (32)	Female (33)
Teaching experience	18 years	6 years	7 years
Class size	36	34	37
Duration of lesson	45 minutes		
Topic of lesson	Linear equations with two unknowns	Area of similar geometric figures	Properties of probability

Preliminary Analysis

A cursory review of all the videotaped lessons shows that the Korean classrooms in this study, like the classrooms in other East Asian countries, were characterised by the dominance of teacher talk and reticence of students when compared with Western countries (Leung & Park, 2005). Following the practice of the Third International Mathematics and Science (TIMSS) 1999 Video Study, the number of words spoken by the teachers and the students in all the lessons in each of the three cities was counted and their ratio was then computed. The results are juxtaposed with the results of the TIMSS 1999 Video Study, as shown in Figure 1. As can be seen from Figure 1, the ratios of number of words spoken by the teacher to those spoken by the students vary between 18 and 40, with an average of 28. These ratios are higher than those obtained from the TIMSS 1999 Video Study (Hiebert et al.,

2003), especially higher than those for the Western countries in the TIMSS 1999 Video Study (Figure 2)

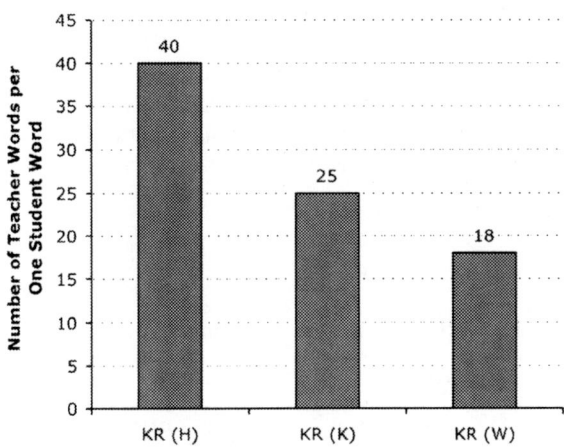

Figure 1. Ratio of number of teacher words to student words in the three Korean schools

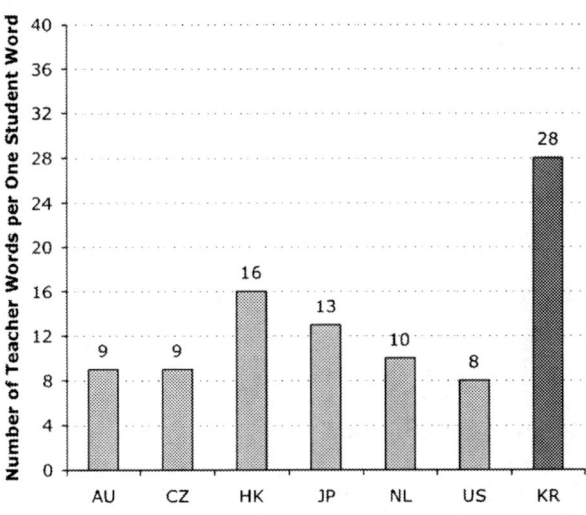

Figure 2. Ratio of number of teacher words to student words in Korean classrooms compared to those in other countries

The preliminary analysis of the video data also shows that the teaching in the three Korean schools seemed to focus more on the mathematics content to be learned rather than the process of understanding the content. Mathematics content was delivered efficiently, with mathematics concepts often stated directly. For example, in a lesson at School H on the meaning of the root of a system of linear equations, the teacher stated the relation between the root of the system of linear equations and the intersection of the two corresponding lines representing the equations directly, rather than leading students through a process in understanding the relation. It seems that what was important for the teacher is the final product (the relation between the root of the system of linear equations and the intersection of corresponding lines) rather than how the product is arrived at, as illustrated by the following scenario:

Transcript 1: School H – Lesson 3

```
Teacher:   ... We have (2, 5) as a solution of this system of linear
           equations. Here 2 means the number of stamps with a
           denomination of 150 Won, and 5 means the number of stamps
           with a denomination of 200 Won. Now, do you know what this
           pair (2, 5) represents in the graph? ... If we draw a line
           of the first equation, and then draw another with the
           second equation, the solution (2, 5) is the coordinates of
           the point of intersection of these two lines. The solution
           of the system of linear equations is the coordinates of the
           point of intersection of those two lines. You should know
           this ...
Teacher:   So, if you look at problem 4, it asks us to compare the
           solution we found and the coordinates of the point of
           intersection. We can say that the relation between the
           solution and the coordinates of the intersection is ...
           [pause] ... they are the same. Therefore, we can conclude
           that the solution of the two equations becomes the
           coordinates of intersection. This is very very important.
```

Key to symbols used in transcripts in this chapter
...   A short pause of one second or less.
[text]   Comments and annotations, often descriptions of non-verbal action.

Actually, this direct way of delivering the content seems to be a deliberate choice of the teacher. The teacher in school H made the following remarks during the interview after the lessons:

Transcript 2: Teacher School H – Interview 2

```
Teacher:   Of course, there should be lots of student activities. But
           I found that they distracted the students and made it
           difficult to proceed with the lesson. Also, the high
           achieving students seemed to get bored and would sometimes
           just sit idled. If we have activities in class, those who
           are not so good don't even know what they are for.
           Innovative lessons which try new thing in class make
           everybody tired. Just giving mathematical explanations is
           much better for both high and low achieving students.
```

> People seem to think that inquiry instruction is a good form of teaching that fits the current trend but I do my own explanation and lead the class uniformly i.e., without differentiated approaches because it [inquiry instruction] tends to loosen the lesson somewhat.

So it can be seen that the focus of the Korean lessons seemed to be on the final product rather than the process of arriving at the product.

Fine-grained Analysis

As pointed out above, one typical lesson from each of the three Korean schools was chosen for a more fine-grained analysis. In this fine-grained analysis, all the activities in the lessons were carefully scrutinised in order to determine whether there were multi-faceted variations and developmental variations in the activities. The results of the analysis show that, from the lens of Gu's theory of variation, there were indeed a lot of variations in concepts and practicing exercise covered in the three lessons. In the discussion below, we denote the variations referred to by two capital letters and a number. The first letter refers to the school (H, K or W) where the lesson took place, the second letter stands for multi-faceted variation (M) or developmental variation (D), and the number indicates the order in which the variation occurred in that particular lesson. For example, HD1 means the first developmental 1 variation that occurred in the chosen lesson of school H.

The multi-faceted and developmental variations identified in the three lessons include:
– Linkage of different concepts, introducing a new topic based on a review of the content covered in previous lessons (HD1 and WD1)
– Consolidation through summary (HM4)
– Learning concepts through comparison and contrast (HM1)
– Linkage between mathematics and concrete examples (KM1)
– Multiple representation of a concept (HM2, KM2)
– Generalisation through abstraction (WD2)

Systematic Variation

One particular kind of variation warrants highlighting for discussion. It is a kind of systematic and continuous variation that leads students to thoroughly understand the concept under discussion. In such variations, only one component of the concept under discussion is changed while other components remain constant. Then another component is changed with the rest of the components remaining unchanged, and so on. The components to be changed one at a time are chosen in a systematic manner so that the structure of the concept under discussion is made clear. Such systematic variations may be multi-faceted variations or developmental variations, although they happen more in multi-faceted variations.

It is interesting to find that such systematic variations were found in each of the lessons that we analysed: HM3, KM3, KM4 and WM1 below can all be classified as this kind of systematic variation.

HM3

Basic equation, $x + y = 5$: coefficients and domains for the variables are natural numbers

- Domain being natural numbers unchanged, coefficients changed systematically, first to another set of natural numbers ($3x + y = 15$), and then to include a negative whole numbers ($-3x + y = 15$), i.e., the coefficients are now integers
- Next, domains for the variables extended to include negative whole numbers
- Then the domains are further extended to real numbers
- Finally, the general form of a linear equation with two unknowns is introduced ($ax + by + c = 0$) where the domains for the unknowns (x and y) are real numbers.

In the analysis above, we can see that there are systematic variations starting with the basic equation $x + y = 5$ and moving step by step to the general form of $ax + by + c = 0$. In each variation, all but one of the components of the equation concerned are kept constant, so that the effect of the varied component is elucidated.

KM3 and KM4

Basic diagram: pair of rectangles with ratio of similarity 1 : 2

- Then ratio of similarity kept constant (1 : 2), and basic diagram changes to a pair of triangles
- Next, geometric figures (pair of rectangles) unchanged, and ratio of similarity changes to 1 : 3
- Then both figures (from rectangles to triangles) and ratio of similarity (from 1 : 2 to 2 : 3) change
- Ratio of similarity generalised to $m : n$
- Finally, keeping ratio of similarity (2 : 3) constant, figure further changed to a pair of pentagons

With these systematic variations, students are guided to understand the concept that for a pair of *any* similar polygons, if the ratio of similarity is $m : n$, then the ratio of areas is $m^2 : n^2$

WM1

Basic situation: pouch with three red stones

- Situation varies systematically: each time a red stone being replaced by a blue one until eventually there are three blue stones
- These systematic variations bring out the facts that:
  a) probability of an impossible event is 0
  b) probability of a certain event is 1
  c) probability is a value lying between 0 and 1

In the three examples above, the teaching all started with a certain simple basic situation. Then only one of the different aspects of the basic situation was varied at a time, and the variations followed a systematic pattern until the situation reached a target form. It is argued that such systematic variations constitute a kind of *exploration* of the concept concerned on the part of the students. Exploration is in essence a mental activity that examines different aspects of an idea. Many educators argued that some forms of external activities (such as manipulation of physical materials) foster such an internal activity, but from the reactions of the students in the Korean lessons observed, it is found that a systematic set of stimuli presented by the teacher will also provide the necessary conditions for the different aspects of the mathematical idea concerned to be examined. And it seems that the systematic variations were carefully designed by the teacher, leading students to discern attributes of the object of learning or the concepts involving the final situation.

In addition, in most of the lessons analysed above, the exercise given to students also varied in such a systematic manner. So students after being exposed to systemic variations in the presentation of the concepts would now have an opportunity to practice the application of the concepts systematically in class and/or at home. According to Marton's theory of variation, these combined experiences of the students on the systematic variations of the concepts will help establish their understanding.

DISCUSSION

It was mentioned above that the systematic variations identified constitute a kind of *exploration* on the part of the students. Exploration in the Western context often means students were given open-ended tasks and engaged in free exploration activities, usually conducted in a small group or individualised setting. This is in sharp contrast to the teacher directed Korean classroom reported above. However, the fine-grained analysis of the data shows that in the seemingly teacher-directed Korean classroom, students still had the opportunity of exploring mathematics ideas under the close guidance of the teacher (such kind of exploration may be referred to as "teaching directed exploration" or simply "directed exploration"). According to Marton's theory of variation, such variations where different components of a concept change systematically one by one will create the necessary condition for different features or critical attributes of the object of learning to be experienced by the students (Marton & Booth, 1997).

The descriptions above fit well with the findings of another study on classroom practices in Hong Kong and Shanghai. Huang and Leung (2004) reported that the Hong Kong and Shanghai mathematics classrooms in their study were characterised by teacher dominance and student active engagement, with much emphasis on exploration of mathematics and exercises with systematic variation.

The East Asian Culture

How do we account for the classroom practices in Korea as identified in this study? To what extent can these classroom characteristics be attributed to the underlying East Asian culture?

In the literature, various scholars have tried to attribute differences in classroom practices and achievements to cultural factors (Watkins & Biggs, 1996; Wong, 1998). In particular, Leung (1999) discussed the traditional Chinese views of mathematics and education which might have an impact on the classroom practices in the current Chinese classroom. Leung (2001) extended the argument from the Chinese classroom to the East Asian classroom and identified features of East Asian mathematics education in contrast to features in the West, and presented the differences in terms of six dichotomies. Leung argued that the different practices between East Asian classrooms and those in the West are based on different deep-rooted cultural values and paradigms, whether explicit or implicit, that have been built up over centuries. In the next section, we will try to account for some of the classroom practices identified in this study through referring to the underlying cultural values that Korea shares with the other East Asian countries.

Teacher dominance and whole class teaching

Teacher dominance and whole-class teaching accord well with the traditional East Asian philosophy which emphasises integration and harmony (Sun, 1983), in contrast to the Western culture which stresses independence and individualism (Taylor, 1987). Related to this tendency in the East Asian culture, which Yang (1981) labelled as 'social orientation' (as opposed to 'individual orientation'), are characteristics such as compliance, obedience, respect for superiors and filial piety (Lin, 1988; Liu, 1986). East Asians are known to have a tendency of complying with rules or orders more than Westerners, giving rise to a strong tendency for uniformity (Sun, 1983) and conformity (Bond & Huang, 1986). In such a cultural environment, it is not surprising that classrooms are found to be teacher dominated, with whole-class teaching being commonplace.

Teacher dominance may also be related to the high regard given to teachers in the East Asian culture. In the East Asian culture, the image of the teacher is that of a scholar held with high respect. So it is just natural that in the classroom setting, teaching and learning activities should be directed by the scholar-teacher. Teacher dominance and whole-class teaching however do not necessarily mean that students are not actively engaged in the lesson. As can be seen from the results of this study presented above, active student engagement is still possible in a classroom where the class size is large and the activities are dominated by the teacher.

Content versus process

It has been reported in the literature that "Chinese teachers held the more rigid view of mathematics being more a product than a process, (and) the more important thing for them in mathematics teaching was to have the mathematics content expounded clearly." (Leung, 1995, p. 315). How do we account for the classroom practices in Korea as identified in this study? To what extent can these classroom characteristics be attributed to the underlying East Asian culture?

The emphasis in the East Asian mathematics classroom was on the mathematics content and the procedures or skills in dealing with the content rather than the process of handling mathematics. There is an underlying belief that

> the critical attribute of mathematics is its distinctive knowledge structure, and it is this distinctive structure which distinguishes mathematics from other forms of knowledge. So the most important goal of mathematics learning is to understand and get hold of this distinctive knowledge structure, and the foremost task of the mathematics teacher is to help students acquire the mathematics content. The process of doing mathematics is part of the process of learning the content, but the process needs the content as its foundation. Without content, there is nothing for the process to be applied to (Leung, 2001, p. 39).

The findings of this study agree well with the reported views above. This stronger stress on the content rather than the process of mathematics reflects how the nature of mathematics is perceived in the East Asian culture.

The emphasis on directed exploration and practice

The finding in this study on the emphasis of the Korean teaching on directed exploration may seem to contradict the stereotype of the East Asian classroom. The learning styles in East Asia are often portrayed in the literature as "learning by rote" or "passive learning" (Biggs & Watkins, 1996), and the teaching strategies characterised as "procedural" (Zhang, S. Li & J. Li, 2003). But results of this study show that behind the seemingly procedural teaching and passive learning, the Korean students are actually heavily involved in exploration when following the prescribed classroom activities designed by the teacher.

On the other hand, the finding that there are a lot of practicing exercises in the Korean lessons is consistent with the stereotype many held for the East Asian classroom. However, the results of this study also suggest that the exercises that Korean students worked on were not simply repetitive drills, but were carefully designed problems with systematic variations.

257

In the East Asian culture, practice has always played an extremely important role in the learning process. Actually, the word or term in Chinese for "learning" consists of two characters (学习), and the second character (习) conveys the meaning of practice. So in the CHC tradition, practice is an inherent part of the learning process. The idea of learning without practicing is absurd in the CHC. The well-known saying (熟能生巧) which is often translated as "practice makes perfect", reflects this philosophy of learning well. As Confucius put it, "Is it not a pleasure, having learned something, to try it out (i.e., practice) at due intervals?" (学而时习之, 不亦说乎?) (*Analects*, I. 1).

Underlying this stress on practice are the traditional East Asian cultural values which lay a strong emphasis on the importance of education and which attribute achievement more to effort than to innate ability. Under the influence of such values, education or study is considered a serious endeavour, and there is a high expectation for students to put in hard work and perseverance in their study and to achieve. This is reinforced by a long and strong tradition of public examination, which acts as a further source of motivation for learning. All these add up to form an important source of motivation for students to learn well and to excel.

CONCLUSION

The analysis of the Korean LPS data from the perspectives of theories of variation has yielded some interesting results which helped reveal the kind of teaching in Korea. From the discussions above, it can now be seen how the seemingly traditional teaching in the teacher dominated Korean classrooms may still have contributed to their students' superior performance in international studies of mathematics achievement. A focus on mathematics content in the teaching is not in itself good or bad for the learning. It depends on how well the content is organised. As can be seen from the analysis in this chapter, if the content varies systematically when presented to students, a lot of exploration may still take place on the part of the students in a teacher-directed classroom. And ample practice of mathematics skills does not necessarily imply rote learning or learning without understanding. With well-designed practising exercises that vary systematically, effective learning may still take place. According to Marton's theory of variation, such experience of systematic variations in classroom learning and practicing exercise on the part of the student will lead to discernment and understanding. As Leung (2001) pointed out, understanding is "not a yes or no matter, but a continuous process or a continuum". The process of learning often starts with gaining competence in the procedure, and then through "continuous practice with increasing variation" (Marton, 1997), students gradually gain understanding.

Amidst the global tide of educational reform, there is a pressure on governments of East Asian countries, Korea included, to change the educational practices in their countries as well, and a common strategy taken is to send a team of policy makers to a number of 'more developed' countries and shop around for new ideas and practices. But too often, those new ideas and practices have not been well tested even in those 'developed' countries, and the cultural differences between the East Asian countries and the 'developed' countries being visited have not been attended to in the adoption of the reforms. What is needed in Korea and other East Asian countries for policy decision are systematic collection and analysis of relevant data, and reflection on the strengths and weaknesses of the existing system and the interaction between existing educational practices and the underlying culture. And what is reported in this chapter represents exactly one such endeavour.

Of course the rich dataset of the Korean component of LPS provides scope for a lot more analysis than what have been presented in this chapter. Further research may build on what has been presented above, and should focus on the important characteristics of the dataset. In particular, further analysis of the characteristics of the mathematics classroom based on the sequence of lessons, supplemented by analysis of the teacher and student interview data to incorporate the views of the participants in the classroom should prove to be fruitful.

NOTES

i. A theory of variation based on phenomenography was developed by a Swedish research group in the early 1970s. The word 'phenomenography', coined by Marton in 1979, was derived from the Greek words 'phainemenon' and 'graphein', which mean appearance and description respectively. Thus 'phenomenography' concerns about the description of things as they appear to us. According to phenomenography, a way of experiencing something is defined in terms of the critical aspects of the phenomenon as discerned and focused upon by the experiencer at the same time. Nobody can discern an aspect of a phenomenon without experiencing variation in a dimension which corresponds to that aspect (Marton & Booth, 1997; Pang, 2003). This provides the basis for a theory of variation.

ACKNOWLEDGEMENTS

This research was supported in part by a grant from AP-EPRI/KEDI (Korean Educational Development Institute) under a program of the Global Development Network. All opinions expressed are those of the authors and do not necessarily reflect the views of AP-EPRI/KEDI or World Bank/GDN.

REFERENCES

Ausubel, D. P. (1968). *Educational psychology: A cognitive view*. New York: Holt, Rinehart and Winston.

Biggs, J. B., & Watkins, D. A. (1996). *The Chinese learner: Cultural, psychological, and contextual influences*. Hong Kong: CERC and ACER.

Bond, M. H., & Huang, K. K. (1986). The social psychology of Chinese people. In M. H. Bond, (Ed.), *The Psychology of the Chinese People*. Hong Kong: Oxford University Press.

Dienes, Z. P. (1973). A theory of mathematics learning. In F. J. Crosswhite, J. L. Highins, A. R. Osborne, & R. J. Shunway (Eds.), *Teaching mathematics: Psychological foundation* (pp. 137-148). Ohio: Charles A. Jones Publishing Company.

Freudenthal, H. (1983). *Didactical phenomenology of mathematical structures*. New York: Kluwer Academic Publishing Company.

Gu, L. Y. (1994). *Theory of teaching experiment: The methodology and teaching principle of Qinpu* [青浦实验的方法与教学原理研究]. Beijing: Educational Science Press.

Gu, L., Huang, R., & Marton, F. (2004). Teaching with variation: An effective way of mathematics teaching in China. In L. Fan, N. Y. Wong, J. Cai, & S. Li (Eds.), *How Chinese learn mathematics: Perspectives from insiders* (pp. 309-345). Singapore: World Scientific.

Hiebert, J. (1986). *Conceptual and procedural knowledge: The case of mathematics*. Hillsdale, NJ: Erlbaum.

Hiebert, J., Gallimore, R., Garnier, H., Givvin, K. B., Hollingsworth, H., Jacobs, J., Chui, A. M. Y., Wearne, D., Smith, M., Kersting, N., Manaster, A., Tseng, E., Etterbeek, W., Manaster, C., Gonzales, P., & Stigler, J. (2003). *Teaching mathematics in seven countries. Results from the TIMSS 1999 Video Study*. Washington, D.C.: National Center for Education Statistics.

Huang, R., & Leung, F.K.S. (2004). Cracking the paradox of the Chinese learners — Looking into the mathematics classrooms in Hong Kong and Shanghai. In L. Fan, N. Y. Wong, J. Cai, & S. Li. (Eds.), *How Chinese learn mathematics: Perspectives from insiders*. New Jersey: World Scientific.

Leung, F. K. S. (1995). The mathematics classroom in Bejing, Hong Kong and London, *Educational Studies in Mathematics, 29*, 297-325.

Leung, F. K. S. (1999). The traditional Chinese views of mathematics and education: Implications for mathematics education in the new millennium. In C. Hoyles, C. Morgan & G. Woodhouse (Eds.), *Rethinking the mathematics curriculum*. London: Falmer Press Ltd.

Leung, F. K. S. (2001). In search of an East Asian identity in mathematics education. *Educational Studies in Mathematics, 47*, 35-51.

Leung, F. K. S., & Park, K. (2005). *Behind the high achievement of students: A study of mathematics lessons in Korea*, Report of a project funded by the World Bank–GDN Regional Research Competition, KEDI.

Lin, F. L. (1988). Social differences and their influences on Children's mathematics understanding. *Educational Studies in Education, 19*, 471-497.

Liu, I.M. (1986). Chinese cognition. In M. H. Bond (Ed.), *The psychology of the Chinese people*. Hong Kong: Oxford University Press.

Marton, F. (1997, March). *Student learning: East and West*. Public lecture delivered at the Chinese University of Hong Kong.

Marton, F. (1999, August). *Variatio est mater studiorum*. Opening address in 8th European Conference for Research on Learning and Instruction (Earli 99), Göteborg, Sweden, 24-28 August.

Marton, F. (2000, March). *Some critical features of learning environment*. Invited Keynote Address, The Bank of Sweden Tercentenary Symposium on Cognition, Education, and Communication Technology, Stockholm, Sweden, March 30-April 1.

Marton, F., & Booth, S. (1997). *Learning and awareness*. Mahwah, NJ: Lawrence Erlbaum Associates.

Marton, F., Runesson, U., & Tsui, A. B. M. (2003). The space of learning. In F. Marton, A. B. M. Tsui, P. P. M. Chik, P. Y. Ko, M. L. Lo, & I. A. C. Mok (Eds.), *Classroom discourse and the space of learning* (pp. 3-40). N.J.: Lawrence Erlbaum.

Mok, I.A.C. (2000). The anatomy of an "open" mathematics lesson. In, Z. P. Dai (Ed.), *Open-ended questions: New models in mathematics pedagogy*. [In Chinese]

Pang, M. F. (2003). Two faces of variation – On continuity in the phenomenographic movement. *Scandinavian Journal of Educational Research, 47*(2), 145-156.

Rovio-Johnsson, A. (1999). *Being good at teaching. Exploring different ways of handling the same subject in higher education*. Göteborg: Acta Universitatis Gothoburgensis.

Runesson, U. (1999). *The pedagogy of variation: Different ways of handling a mathematical topic*, [Variationens pedagogik: Skilda sätt att behandla ett matematiskt innehåll]. Göteborg: Acta Universitatis Gothoburgensis.

Sfard, A. (1991). On the dual nature of mathematics conception: Reflections on processes and objects as different sides of the same coin. *Educational Studies in Mathematics, 22*(1), 1-36.

Sun, L. K. (1983). *The "deep structure" of Chinese culture*. Hong Kong: Chap Yin Co. [in Chinese]

Taylor, M. J. (1987). *Chinese pupils in Britain*, Windsor, Berkshire: NFER-Nelson.

Vygotsky, L. (1978). *Mind in society*. Cambridge, MA: Harvard University Press.

Watkins D. A., & Biggs, J. B. (Eds.) (1996). *The Chinese learner*. Hong Kong: Comparative Education Research Centre.

Wong, N. Y. (1998). In search of the 'CHC' learner: Smarter, works harder or something more?, Plenary lecture. In H. S. Park, Y. H. Choe, H. Shin, & S. H. Kim (Eds.), *Proceedings of the ICMI-East Asia Regional Conference on Mathematical Education*, 1, 85-98.

Yang, K. S. (1981). Social orientation and individual modernity among Chinese students in Taiwan. *Journal of Social Psychology*, 113, 159-70.

Zhang, D., Li, S., & Li, J. (2003). *An introduction to mathematics education* [In Chinese: 数学教育导论]. Beijing: Higher Education Press.

Kyungmee Park
Department of Mathematics Education
Hongik University
Korea

Frederick Koon Shing Leung
Faculty of Education,
The University of Hong Kong
Hong Kong SAR, China

261

RONGJIN HUANG, IDA AH CHEE MOK AND
FREDERICK KOON SHING LEUNG

CHAPTER EIGHTEEN

Repetition or Variation:
Practising in the Mathematics Classrooms in China

INTRODUCTION

There have been many studies about the teaching in Confucian-Heritage Culture (CHC) classrooms and the psychological and pedagogical perspectives about Chinese teaching and learning (Fan, Wong, Cai, & Li, 2004; Leung, 2001; Stigler & Hiebert, 1999; Watkins & Biggs, 2001). The results of Third International Mathematics and Science Study (TIMSS) 1999 Video Study showed that 80% of the duration of the lesson in its data set was devoted to solving problems. Moreover, it was found that more of the lesson time in Hong Kong SAR was spent on practising new content than in other countries (Hiebert et al., 2003). These results imply that a deep investigation of practice in Chinese mathematics classroom may shed new light on the understanding of mathematics classroom teaching in CHC. Moreover, some educators argue that routine practice, if properly carried out, is a prerequisite for concept formation and problem solving (Li, 1999; Wong, 2006). Recently, by examining the teaching of Pythagoras' theorem in Hong Kong and Shanghai, Huang and Leung (2004) used a distinction of explicit variation and implicit variation to feature the nature of practice in the classrooms of Hong Kong and Shanghai. It was found that in both cities, exercises of both types of variation were used. However, the Hong Kong teachers tended to use explicit variation while the Shanghai teachers seemed to prefer providing implicit variation. All these findings and arguments lead to the research agenda in our chapter: to study in depth the meaning of 'practice' in mathematics lessons.

Our chapter will delineate the nature of the practice problems used in the mathematics lessons taken from the three cities in China, that is, Hong Kong, Macau and Shanghai. The lessons were similar in topic, namely, the method of elimination for solving simultaneous equations. In the analysis we apply the framework of variation based on the works of Marton and Booth (1997), Huang (2002) and Gu, Huang and Marton (2004). This chapter includes three parts. Firstly, the theoretical framework and the source of the data will be briefly

D. J. Clarke, C. Keitel & Y. Shimizu (Eds.), Mathematics Classrooms In Twelve Countries: The Insider's Perspective. 263–274.

elaborated. Then, the characteristics of the practice problems in Hong Kong, Macau and Shanghai will be examined and compared. Finally, we will discuss the pedagogical implications in mathematics classrooms in CHC.

THEORETICAL FRAMEWORK

A Theory of Variation

According to Marton et al. (2004), learning is a process in which learners develop a certain capability or a certain way of seeing or experiencing. In order to see something in a certain way learners must discern certain features of the object. Experiencing variation is essential for such discernment, and thus is significant for learning. They further argue that it is important to pay attention to what varies and what is invariant in a learning situation. Furthermore, learning always involves an object of learning. What is more important is how a teacher structures the lessons so that it is possible for the object of learning to come to the fore of the students' awareness, which is called the enacted object of learning.

Independently, based on a series of longitudinal mathematics teaching experiments in China, and heavily influenced by cognitive science and constructivism, a theory of mathematics teaching/learning, called "teaching with variation", has been developed by Gu (Gu et al., 2004). According to this theory, meaningful learning enables learners to establish a substantial and non-arbitrary connection between new knowledge and their previous knowledge. Classroom activities are developed to help students establish this kind of connection by experiencing certain dimensions of variation. This theory suggests that two types of variation are helpful for meaningful learning. One is called "conceptual variation", which provides students with multiple experiences from different perspectives. The other is called "procedural variation", which is concerned with the process of forming a concept logically or chronologically, arriving at solutions to problems and forming knowledge structures (relationships among different concepts).

Although the studies led by Marton et al. and Gu et al. developed independently, their theories agree on one important fact, that variation is essential for enhancing learning.

Practice

In this chapter, applying the just learnt knowledge to solve problems (routine or non-routine) as well as extending the learnt knowledge to generate new knowledge via problem solving is referred to as 'practice'.

264

Distinctions Between Explicit Variation and Implicit Variation

Identified by Huang (2002), and further illustrated by Huang and Leung (2004), there are two different types of varying problems in mathematics classroom: explicit variation and implicit variation. If the changes from the prototype of problems in which the learnt knowledge can be applied directly by learners to their variations are identified visually and concretely such as variations in numbers, positions of figures etc., and the conditions for applying the relevant knowledge are still explicit and direct, then this kind of variation is regarded as explicit. On the other hand, if the changes from the origins to their variations have to be discerned by abstract and logical analysis by learners such as variations in parameters, subtle changes or omissions certain conditions, or changes of contexts, or reckoning on certain strategies etc., so that the conditions or strategies for applying relevant knowledge are implicit and not obvious, then this kind of variation is characterised as implicit.

In this study, perspective of variation will be used as a theoretical orientation in general, and distinctions between two variations will be used as a zoom in particular.

DATA ANALYSIS

Data Source

The data discussed in this chapter were taken from the Learner's Perspective Study (LPS) in Hong Kong, Macau and Shanghai. The set of the data from the three cities consisted of 135 lessons from three schools in each of the three cities. In order to focus our analysis on lessons with compatible topics, we selected all the lessons in which the method of elimination was taught. That made up a total of 21 lessons including six lessons from Hong Kong, seven lessons from Macau and eight lessons from Shanghai.

Data Analysis

First of all, we sorted out all the practice problems from the selected lessons of each school, and then started to examine all the problems with an open mind to look for implicit variation and explicit variation. Finally, we carried out a fine-grained categorisation which derived from our initial analysis. More details will be reported in the next section.

In order to understand how the teachers use the practice problems from one level of variation to another, we watched the videos and read the relevant transcripts several times. Then, one event, referred to as summarising, comes to our focus. It seems that it can catalyse the transformation of practising between different levels and enhance students' learning through practice. Finally, we capture the characteristics of summarising as described in next session.

RESULTS

All the lessons in the three Chinese cities included the following three phases: suggesting the method intuitively, introducing the new method and practising the new method. In our analysis, we focus on two aspects: the variations in the practice problems and the ways of using the problems.

Varying Problems Systematically

In the analysis, the distinction between explicit and implicit variation is still applicable but with more refined specifications. There are three levels of variation within explicit variation and there are two levels within implicit variation. The prototype of explicit variation has the form "$ax + by = c$" whereas the equations with the implicit variation need a transformation which requires a deeper understanding of the meaning of unknowns in equations. The variation will be illustrated in the following examples.

Explicit variation level 1 (EV 1). The signs of the coefficients of the same unknown are different. These equations consist of at least one pair of coefficients with the same values or values with opposite sign such that the unknown can be eliminated directly by addition or subtraction. For example:

$$(I) \begin{cases} 8x - 5y = 4 \\ 3x + 5y = 18 \end{cases}; (II) \begin{cases} 5x - 4y = 12 \\ 5x + y = 7 \end{cases}; (III) \begin{cases} 3x + 2y = 19 \\ 3x - 2y = 11 \end{cases}$$

Explicit variation level 2 (EV 2). The values of the coefficients of the same unknown are different. The coefficients are all different and a transformation (either by multiplication or division with a whole number) to the prototype of level 1 is needed before applying the method.

$$(I) \begin{cases} 5x - 2y = 11 \\ x + y = -2 \end{cases}; (II) \begin{cases} 2x + 3y = 8 \\ 3x + 4y = 11 \end{cases}$$

Explicit variation level 3 (EV 3). The representations of coefficients of the unknowns are multiple such as fraction, decimal, percentage and so on. The method can only be applied after transforming the equations into level 2.

$$(I) \begin{cases} 0.3x + 0.6y = 2.4 \\ \dfrac{x}{4} + \dfrac{y}{3} = 1 \end{cases}; (II) \begin{cases} 25\%x + 5\%y = 8 \\ x + 2y = 140 \end{cases}$$

Implicit variation level 1 (IV 1). The forms of linear equations are different. The method can only be applied after transforming the equations into the prototype demonstrating the explicit variation mentioned above. In working with these examples, students may come to awareness that the linear equations will eventually be transformed into the same prototype $ax + by = c$.

$$(I) \begin{cases} 4(x+2) = 1-5y \\ \dfrac{y+3}{2} = 1 - \dfrac{x}{3} \end{cases} \quad ; (II) \ 5x - 4y = 3x + 2y = 33$$

Implicit variation level 2 (IV 2). New unknowns need to be defined. Students need to see an abstract expression containing unknowns as an object and define it with new unknowns in order to transform the given into the prototype. For example, the students need to treat $\frac{1}{a}$ as an unknown and $\frac{1}{b}$ as another unknown in (I), which is more advanced.

$$(I) \begin{cases} \dfrac{1}{a} + \dfrac{2}{b} = -2 \\ \dfrac{1}{a} - \dfrac{3}{b} = 8 \end{cases} \quad ; (II) \begin{cases} |x| + y = 3 \\ 2y + |x| = 4 \end{cases}$$

By practising with explicit variation, students could be familiarised with the procedural knowledge. However, practising with implicit variation should be helpful in associating the new method with learnt knowledge such as operations of fraction, decimal, and percentage, the transformation of algebraic expression. Compared with the teachers in Hong Kong and Macau, the Shanghai teachers paid more attention to using implicit variation (see Table 1). The numbers in this Table refer to the numbers of teachers who adopted relevant levels of varying problems.

Table 1. The number of teachers using different levels of variations

City	EV 1	EV 2	EV 3	IV 1	IV 2
Hong Kong	3	3	1	1	1
Macau	3	3	2	2	1
Shanghai	3	3	3	3	2

Summarising Key Points Frequently

Summarising in this chapter refers to summarising the invariant underpinning principles of the particular method, the procedure of solving problems by using the particular method, and cautious reminders of solving problems. It was found that the teachers in the three cities often adopted the technique of summarising to move practice from one level of variation to another.

Let us take a Hong Kong lesson (HK3-L10) for an example. After solving the following simultaneous equations by the method of elimination: $\begin{cases} x - y = 8 \\ x + y = 22 \end{cases}$, the teacher discussed the steps and the method with the whole class: (1) labelling two equations as equation one and two; (2) subtracting one equation by the other in order to get a simple equation; (3) the answer being got right away. Furthermore, the teacher invited students to solve a simultaneous linear equation $\begin{cases} 3m + 2n = 4 \\ \dfrac{m}{2} - \dfrac{n}{6} = 1 \end{cases}$, and then the following summary was made: (1) finding out the L.C.M. of the denominators of the second equation; (2) multiplying the entire equation by six; (3) doing addition or subtraction; (4) the final answer being found out.

In the lesson (MC3-L06), starting with a review of the following points: (1) the mathematical thinking of solving simultaneous linear equations is to eliminate unknowns; (2) the methods of eliminating unknowns are substitution and addition/subtraction, the teacher presented the following simultaneous linear equations in two unknowns (see the following (I)). First, the teacher induced students to transform the equations into the standard form of linear equation (see the following (II)), then, the students found out the solution by subtraction directly.

$$\text{(I)} \begin{cases} \dfrac{x+2}{3} + 1 = \dfrac{y+5}{4} \\ 2(x+1) - 3(y+1) = 4 \end{cases} ; \text{(II)} \begin{cases} 4x - 3y = -5 \\ 2x - 3y = 5 \end{cases}$$

With the guidance of the teacher, the following methods were discussed: (1) multiplying the second equation by 2 and subtracting it from the first equation in order to illuminate unknown x; (2) substituting $3y$ in the first equation with $2x - 5$ ($3y = 2x - 5$ according to the second equation) in order to illuminate unknown x; (3) replacing $2x - 3y$ in the first equation with 5 ($2x - 3y = 5$ according to the second equation) in order to eliminate unknown y. After that, the teacher made a summary as follows:

An effective way of solving this equation is the substitution of whole expression. Why can we find so many different solving methods? Of course, there are some unique features in this equation, aren't there? However, the most important thing is the coefficients of the two unknowns x and y. If the coefficients of the x or y in the two equations are the same, then, we need to use subtraction elimination. The method is determined by the characteristics of the coefficients. Thus, we can flexibly use the methods to eliminate unknowns. In addition, if the equations are not in standard forms, how can we deal with them? It is necessary to transform the non-standard form of equation into the standard form of equation.

In Shanghai, some similar cases were found. For example, in one Shanghai lesson (SH1-L07), after completing practising at Level 1, the teacher received successful answers to the following questions: (1) Under what circumstance can the method of elimination by direct addition be applied to a system of linear equations in two unknowns? (2) Under what circumstance can the method of elimination by direct subtraction be applied to a system of linear equations in two unknowns?

In the following lesson (SH1-L08), echoing the key points of the previous lesson, the teacher reviewed the conditions under which we can use addition elimination or subtraction elimination through questioning. Then practice of the problems at explicit variation level 3 and implicit variation level 1 was conducted. Once the exercises were completed, the teacher encouraged students to summarise the procedures of solving linear equations in two unknowns by themselves and one student was invited to present his/her summary. After that, the students were asked to read in chorus the following steps from their textbook: (1) using the equality properties to transform the equations to make the absolute values of the coefficients of one of the unknowns the same; (2) after transforming the equations, adding them together or subtracting one from the other; (3) solving the linear equation in one unknown obtained by eliminating an unknown in order to find the value of the other unknown; (4) substituting the value of the unknown into any of the equations in the original system of equations; (5) writing down the values of the unknowns and enclosing them in bracket ,which means they are the solutions to the system of equations.

Furthermore, following practising at implicit variation level 2, the teacher summarised what they had learnt in this lesson: (1) making the absolute values of the coefficients of the same unknowns to be equal; (2) eliminating by the method of addition or subtraction.

We found that it is a common scenario in the three cities: practising how to solve simultaneous equations with varying problems, then summarising critical points to help students consolidate and reorganise knowledge (see Table 2)

Table 2. Summarising crucial points of solving problems

City	Lesson	Number of summarising	Summarising in per lesson
Hong Kong	6	6	1
Macau	7	9	1.3
Shanghai	8	16	2

By doing so, certain underpinning invariant principles and mathematical thinking are highlighted. All the teachers in the three cities adopted summarising at least one time. Table 2 also shows the differences between the three cities. The Shanghai teachers used summarising more often (twice per lesson) than the others did, and the Hong Kong teachers were least often in making use of the strategy (once each lesson).

Summary

Regarding the classroom practising when the particular topic was taught, it was found that the teachers in Hong Kong, Macau and Shanghai all paid much attention to practising with variation, which was definitely not repetitive practising. Again, the Shanghai teachers preferred to practise more with implicit variation than Hong Kong and Macau teachers. Approaches in Macau seemed to lie between Hong Kong and Shanghai. With regard to the method of summarising, the Shanghai and Macau sample teachers more preferred to highlight the underpinning principles and mathematical thinking than Hong Kong teachers did. In a word, by adopting practising with varying problems (sometimes multiple approaches to solving problems) and summarising key points in due course, the students seemed to be offered many more opportunities to understand and apply the particular method and recognise the underpinning principles such as properties of equation and mathematical thinking (e.g. transformation, elimination).

DISCUSSION

Based on the analysis of the teaching of the particular procedural method of elimination in Hong Kong, Macau and Shanghai, the following observations could be made: (1) The teachers in the three cities emphasised classroom practising with both explicit variation and implicit variation. (2) Compared with Hong Kong and Macau teachers, Shanghai teachers tended to practise more with implicit variation. However, the approaches in Macau seem to lie between Hong Kong and Shanghai. (3) The teachers in the three cities all summarised critical points of solving equations in due course to highlight the underpinning principles and mathematical thinking. The first two findings fully support and extend the observations by Huang and Leung (2004), and further indicate that the distinction of implicit variation and explicit variation may be a useful tool to examine classroom practice.

Regarding the first finding, it was argued that according to the Chinese notion of learning (學習, Xuexi), learning and practice are inseparable. Moreover, the widely accepted axiom, "practice makes perfect"(熟能生巧), suggests that practice should go beyond routine repetition. In the categorisation of the different kinds of variations in our study, we differentiate problems with different demands for different mathematical abilities. However, they are basically applying the same method representing a phenomenon well described by an old Chinese saying, "remaining essentially the same despite all apparent changes" (萬變不離其宗). That is to say, it will enable students to solve the problems with the same content regardless of the contexts in which the problems are embedded, if the students master the knowledge and skills by practising with extensive variations of exercises focusing on the same content. It may be the reason why the teachers in the three cities pay attention to exercises with variations.

With regard to different preferences in practising with variation: Teachers in Shanghai attempt to vary the problems implicitly, while the teachers in Hong Kong prefer to vary the problems visually and explicitly. This difference may reflect the differences in teachers' beliefs in teaching mathematics. Western teachers believe that it can make learning meaningful and visible to connect teaching mathematics to concrete models, while Mainland Chinese teachers believe it can develop students' critical thinking and logical reasoning abilities to adopt abstract representation (An, Gerald, Wu, Ma, & Wang, 2006; Cai, 2006). It may be reasonable to assume the teachers in Hong Kong tend to accept the Western beliefs in this aspect. However, we need to be cautious in interpreting the similarities and substantial differences between the three cities. Is it related to the cultural differences or topic specified or comparative levels concerned (Huang & Leung, 2004)? To sum up, it will be dangerous to generalise the teaching pattern for one city or country from the selections of a "snapshot" of a lesson or a serial of lessons (Lopez-Real, Mok, Leung, & Marton, 2004).

IMPLICATIONS OF VARIATIONS

What are the implications of these findings to mathematics teaching? As argued by Gu et al. (2004), by adopting teaching with variation, even with large classes, students could still actively involve themselves into the process of learning and achieve a meaningful learning.

Regarding the distinction between explicit variation and implicit variation, the former serves for mastering basic skills of using particular methods and procedures, and the latter serves for enhancing the integration of knowledge and developing advanced thinking (in this case, such as elimination thinking, substitution thinking, transformation etc.). The different patterns of variation in classroom practices may provide different learning opportunities for students (Häggström, 2006). As described by Ma (1999), many ways to solve one problem (一題多解) in China contributed to students' profound understanding of knowledge. This study further suggests that practising with extensive problems (both implicit variation and explicit variation) (一題多變) may be meaningful for students to build up an interrelated knowledge structure and to develop flexible problem solving abilities.

With respect to the alternation of practising with variation and summarising critical points, it may be helpful for organising a coherent lesson which is also a feature of Chinese mathematics classroom (Huang & Leung, 2005; Wang & Murphy, 2004). Due to historical and cultural traditions, mathematics teaching in Mainland China is "two basics" (i.e., basic knowledge and basic skills) oriented which demonstrates some characteristics such as emphasising effectiveness of classroom teaching, refined talk with plentiful practice, and logical deductive thinking. Practising with variation plays a positive role in Chinese mathematics classroom (Zhang, Li, & Tang, 2004). Consequently, it is possible for students to master basic procedural knowledge and to have a deep understanding of these principles and methods. This observation may help to explain why the Hong Kong students performed well in international comparative surveys (Mullis, Martin, Gonzalez, & Chrostowski, 2004; OECD, 2004).

However, if practising with variation is not used appropriately, such as undue varying in technical aspects for specific skills, students may be limited in a narrow space of learning and will not be challenged enough. Therefore, practising with variation seems to have its positive and negative impacts. How to use it in a positive way has practical implications. Principally, it is important to vary problems in terms of its openness of context and process. Furthermore, an important objective to provide students with more experiences of variation of problems is to enable them to meet uncertain situations and challenging problems in future with confidence.

ACKNOWLEDGEMENTS

The study is a part of the project "Chinese pedagogy of mathematics: An exploration from an international comparative study perspective", funded by the Academic Committee of the University of Macau, Macau SAR, China (Project no. 2050). Any opinions expressed herein are those of the authors and do not necessarily represent the views of the University of Macau.

REFERENCES

An, S., Gerald, K. Wu, Z. Ma, F., & Wang, L. (2006). The impact of cultural differences on middle school mathematics teachers' beliefs in the U.S. and China. In F.K.S. Leung, K. D. Graf, & Lopez-Real, F. J. (Eds.), *Mathematics education in different cultural traditions- A comparative study of East Asia and the West* (pp. 449-465). Dordrecht/Boston/London: Springer.

Cai, J. (2006). U.S. and Chinese teachers' cultural values of representations in mathematics education. In F.K.S. Leung, K. D. Graf, & Lopez-Real, F. J. (Eds.), *Mathematics education in different cultural traditions- A comparative study of East Asia and the West* (pp. 465-482). Dordrecht/Boston/London: Springer.

Fan, L., Wong, N. Y., Cai, J. & Li, S. (Eds.). (2004). *How Chinese learn mathematics: Perspectives from insiders*. Singapore: World Scientific.

Gu, L., Huang, R., & Marton, F. (2004). Teaching with variation: An effective way of mathematics teaching in China. In L. Fan, N. Y. Wong, J. Cai, & S. Li (Eds.), *How Chinese learn mathematics: Perspectives from insiders* (pp. 309-345). Singapore: World Scientific.

Häggström, J. (2006). The introduction of new content: What is possible to learn? In Chapter 9 of D. Clarke, J. Emanuelsson, E. Jablonka & I. A. C. Mok (Eds.), *Making connections: Comparing mathematics classrooms around the world*. Rotterdam: Sense Publishers.

Hiebert, J., Gallimore, R., Garnier, H., Givvin, K. B., Hollingsworth, H., Jacobs, J., Chiu, A.M.Y., Wearne, D., Smith, M., Kersting, N., Manaster, A., Tseng, E., Etterbeek, W., Manaster, C., Gonzales, P., & Stigler, J. (2003). *Teaching mathematics in seven countries: Results from the TIMSS 1999 Video Study*. U.S. Department of Education. Washington, DC: National Center for Education Statistics.

Huang, R. (2002). *Mathematics teaching in Hong Kong and Shanghai: A classroom analysis from the perspective of variation*. Unpublished doctoral dissertation, The University of Hong Kong, Hong Kong.

Huang, R., & Leung, F. K. S. (2004). Cracking the paradox of the Chinese learners: Looking into the mathematics classrooms in Hong Kong and Shanghai. In L. Fan, N. Y. Wong, J. Cai, & S. Li (Eds.), *How Chinese learn mathematics: Perspectives from insiders* (pp. 348-381). Singapore: World Scientific.

Huang, R., & Leung, F.K.S. (2005). Building a coherent and progressive lesson by adopting teaching with variation: A case study of a Shanghai lesson. In N. Y. Wong (Eds.), *Revisiting mathematics education in Hong Kong for the new millennium* (pp. 282-305). Hong Kong: Hong Kong Association for Mathematics Education.

Leung, F. K. S. (2001). In search of an East Asian identify in mathematics education. *Educational Studies in Mathematics, 47*, 35-51.

Li, S. (1999). Does practice make perfect? *For Learning of Mathematics, 19*(3), 33-35.

Lopez-Real, F., Mok, A. C. I., Leung, K. S. F., & Marton, F. (2004). Identifying a pattern of teaching: An analysis of a Shanghai' teacher's lessons. In L. Fan, N. Y. Wong, J. Cai, & S. Li (Eds.), *How Chinese learn mathematics: Perspectives from insiders* (pp. 382-412). Singapore: World Scientific.

Ma, L. (1999). *Knowing and teaching elementary mathematics: Teachers' understanding of fundamental mathematics in China and the United States.* Mahwah, NJ: Lawrence Erlbaum Associates.

Marton, F., Tsui, A. B. M., Chik, P. P. M., Ko, P.Y., Lo, M. L., Mok, I. A. C., Ng, D. F. P., Pang, M. F., Pong, W. Y., & Runesson, U. (2004). *Classroom discourse and the space of learning.* N.J.: Lawrence Erlbaum.

Marton, F., & Booth, S. (1997). *Learning and awareness.* Mahwah, NJ: Lawrence Erlbaum Associates.

Mullis, I. V. S., Martin, M. O., Gonzalez, E. J.,& Chrostowski, S. J. (2004). *TIMSS 2003 international mathematics report: Findings from IEA's trends in international mathematics and science study at the fourth and eighth Grades.* TIMSS & PIRLS International Study Center, Lynch School of Education, Boston College.

OECD (2004). *Learning for tomorrow's world: First results from PISA 2003.* OECD, Organisation for economic co-operation and development.

Stigler, J. W., & Hiebert, J. (1999). *The teaching gap: The best ideas from world's teachers for improving education in classroom.* New York: The Free Press.

Wang T., & Murphy, J. (2004). An examination of coherence in a Chinese mathematics classroom. In L. Fan, N. Y. Wong, J. Cai, & S. Li (Eds.), *How Chinese learn mathematics: Perspectives from insiders* (pp. 107-123). Singapore: World Scientific.

Watkins, D. A., & Biggs, J. B. (2001). The paradox of the Chinese learner and beyond. In D. A. Watkins & J. B. Biggs (Eds.), *The Chinese learner: Cultural, psychological, and contextual influences* (pp. 3-26). Hong Kong: Comparative Education Research Centre, The University of Hong Kong; Melbourne, Australia: Australian Council for Education Research.

Wong, N. Y. (2006). From "entering the way" to "exiting the way": In search of a bridge to span "basic skills" and "process abilities". In F.K.S. Leung, K. D. Graf, & Lopez-Real, F. J. (Eds.), *Mathematics education in different cultural traditions – A comparative study of East Asia and the West* (pp. 111-128). Dordrect/Boston/London: Springer.

Zhang, D., Li, S., & Tang, R.(2004). The "Two Basics": Mathematics teaching and learning in Mainland China. In L. Fan, N. Y. Wong, J. Cai, & S. Li (Eds.), *How Chinese learn mathematics: Perspectives from insiders* (pp. 189-207). Singapore: World Scientific.

Rongjin Huang
Faculty of Education,
University of Macau
China

Ida Ah Chee Mok
Faculty of Education,
The University of Hong Kong
Hong Kong SAR, China

Frederick Koon Shing Leung
Faculty of Education,
The University of Hong Kong
Hong Kong SAR, China

HELENA BINTEROVÁ, ALENA HOŠPESOVÁ
AND JARMILA NOVOTNÁ

CHAPTER NINETEEN

Constitution of the Classroom Environment: A Case Study

INTRODUCTION

One of the central issues currently discussed is that of analysing the culture of the mathematics classroom (e.g. Seeger, Voigt, & Waschescio, 1998). The importance of the social dimension of learning is discussed in relation to an individual's ways of acquiring and using knowledge of mathematics. Laborde and Perrin-Glorian (2005, p. 2) stated

> ... [the classroom] is the place of social interrelations between the teacher and students shaped by the difference of position of the two kinds of actors with respect to knowledge and giving rise to *sociomathematical norms* (Yackel & Cobb, 1996) or to a *didactical contract* (Brousseau, 1989, 1997).

Our analysis of the set of videotaped lessons in this chapter is based on the theory of didactical contract. The implicit nature of Brousseau's concept of 'didactical contract' is fundamental when explaining environment effects on learning mathematics (Sarrazy & Novotná, 2005).

BACKGROUND

Theory of Didactical Situations

In Brousseau's Theory of didactical situations (TDS) (Brousseau, 1989, 1997), teaching is seen as *devolution* of a learning situation from the teacher to the student. Brousseau (1975, in Warfield, 2005) described the learning process as follows:

> ... a learning process can be characterized in a very general way (or even determined) by a sequence of identifiable situations (natural or didactical), reproducible and leading regularly to the modification of a set of behaviours of the students, modifications which are characteristic of the acquisition of a particular collection of knowledge.

D. J. Clarke, C. Keitel & Y. Shimizu (Eds.), Mathematics Classrooms In Twelve Countries: The Insider's Perspective. 275–288. © 2006 Sense Publishers. All rights reserved.

Acquired knowledge can appear in many different forms. Knowing mathematics is not only knowing definitions and theorems and recognising where to apply them. Knowing also means doing mathematics, which includes solving problems, production and construction of models, formulation and justification of proofs and proving and so on.

Brousseau considers the conditions of a particular use of a piece of mathematical knowledge to form a system, which he calls a *'didactical situation'*. In *non-didactical situations*, the evolution of the learner is not submitted to any didactical intervention whatever. *Didactical situations* are situations in which an actor, for instance a teacher, organizes a plan for an action which is intended to modify or create new knowledge in another actor, the learner. Models of effective teaching combine the two approaches: didactical situations that are partially liberated from direct interventions are called *a-didactical situations.* In TDS, situations are classified according to their structure (action, formulation, validation, institutionalisation, etc.)[i] which determines different types of knowledge (implicit models, languages, theorems, etc.).

The process by which the teacher manages a didactical situation in putting the learner in the position of a simple actor in an a-didactical situation is called *devolution.* Devolution does not only propose a situation to learners which should provoke them to an activity not previously agreed, but also makes them feel responsible for obtaining a proposed result, and that they accept the idea that the solution depends only on the use of knowledge which they already have.

Environmental effects on learning mathematics are explained using Brousseau's concept of *didactical contract* presented in the 1980s, i.e. the set of the teacher's behaviours (specific to the taught knowledge) expected by the student and the set of the student's behaviour expected by the teacher. It equally concerns subjects of all didactical situations (students and teachers). This contract is not a real contract; in fact it has never been 'contracted' either explicitly or implicitly between the teacher and students and its regulation and criteria of satisfaction can never be really expressed precisely by either of them.

The interplay of relationships and constraints between the teacher and students may also produce certain unwanted effects and developments that can be observed (e.g. the Topaze effect, the Jourdain effect, metacognitive shift, the improper use of analogy). They are inappropriate for the learning (especially from the metacognitive point of view) but often inevitable. It is more their systematic use that is detrimental. In our analysis of videotaped lessons we will focus on the occurrence of two of these, the Topaze effect and the Jourdain effect (Brousseau, 1997). The occurrence of these effects in a teaching unit influences significantly the quality of the envisaged learning process.

The *Topaze effect*[ii] can be described as follows: When the teacher wants the students to be active (find themselves an answer) and they cannot, then the teacher disguises the expected answer or performance by different behaviours or attitudes without providing it directly. In order to help the student give the expected answer, the teacher 'suggests' the answer, hiding it behind progressively more transparent didactical coding. In this way, the teacher usually tries to achieve the optimum meaning to the maximum number of students. During this process, the knowledge, necessary to produce the answer changes.

The *Jourdain effect*[iii] is a form of Topaze effect. The teacher who, intentionally or unintentionally, does not want to admit student's lack of knowledge of an issue claims to recognize indications of scholarly knowledge in the behaviour or responses of a student, even though they are in fact motivated by trivial causes (such as analogy with a different problem, using lower mathematical knowledge, coincidence etc.).

DATA

In the Czech Republic data for this study were gathered in the eighth grade (students aged 14-15) of a junior secondary grammar school, the alternative to more academic education. The framework was based on the method used in Learner's Perspective Study (Clarke, 2001). The school is located in the county town České Budějovice (with approximately 100 000 inhabitants). The teacher was chosen on the recommendation of and by the agreement with the headmaster. This fact is of crucial importance for further analysis of her lessons. The observed teaching is rated as 'outstanding' in the school. Parents, teachers and professionals respect the teacher as one of the best mathematics teachers in the town. This is largely because her students are successful when passing entrance exams to institutions offering further education and parents do not have to help their children cope with the given homework and tasks. She is an experienced teacher approaching the end of her professional career. The fact that she agreed to being recorded reveals that she is confident in her professional skills.

THE CLASSROOM FROM A SOCIAL PERSPECTIVE

Let us now ask how the work of the students and their teacher can be characterized with regard to the theoretical framework described above. To put it simply, what kind of teacher's work is applauded by professionals, parents and students?

When evaluating the teacher's approach to students, the following questions were raised. Can we trace hidden didactical contract established in this particular classroom and illustrate it by suitable teaching episodes? What influence of the didactical contract on the pupils' mathematical knowledge can be presupposed? How does it support or constrain learning? How does the teacher create a secure, confident work environment for the students in the classroom?

TEACHING EPISODES

Episode 1: Didactical Contract and its Breach

The recognition of didactical contract in the classroom discourse is not easy. We can say that it can be best recognized at the moment when it is breached.

Let us illustrate this idea by the following episode. The main characteristic of the teacher's work is that she keeps returning to reasoning about rules that were taught and validated a long time ago. In our opinion, this is the source of students' confidence. The teacher even expressed that explicitly:

CZ1-L04, 00:37:57[iv]

T: It is important to have some system, step by step, and not, Vojta, to discover America again on your own! That's how you must cope with it. I also haven't come up with the procedure. It's not my invention. It has been tried and tested.

Students' expectations that the teacher would refer to previous knowledge that would be useful when solving the assigned problem sometimes led to mistakes. In Lesson 3, in assigning homework with equations containing mixed numbers (e.g. $5\frac{2}{3}$), the teacher did not refer to the students' former knowledge about mixed numbers. As a consequence of this, several students worked with them in an incorrect way. As a reaction to their mistakes the teacher explained again, in great detail, how to work with mixed numbers. We can observe that she took responsibility for the explanation herself, letting the pupils respond only to simple questions. She did this although the subject matter should have been well known:

CZ1-L04, 00:00:39

T 11:   Let's start with the homework. You will have noticed that the equation written on the whiteboard is the equation from your homework. And some of you have made there a cardinal mistake! Not in the solution of the equation, I mean when using equivalent adjustments or anything like that, but a numerical mistake. And because I don't want you to make any similar mistake again, let's have a look at it again. It might be a good idea to open the exercise book and to check that the mistake isn't yours! [On the whiteboard, there is the equation from their homework $5\frac{2}{3}x - \frac{3}{2} = 4\frac{1}{6}x + \frac{1}{2}$.] That you haven't made this mistake. Watch carefully [Points at the whiteboard.], do you know? Where the mistake has been made? Well? Those who have made the mistake have it marked in the exercise book. So they should know! Well here [Circles the fraction $4\frac{1}{6}x$.] and here [Circles the fraction $5\frac{2}{3}x$.]! What mistake could you have made?

Class 11:   [Humming.]
T 12:   Well! In that equivalent adjustment that you have written down here [Points at the beginning of the equation.], which could be done at once mentally, but I must take care doing it, you found the lowest common multiple, the common denominator, which is?
Class 12:   Six.
T 13:   Six. And you raised the whole equation by six. That was OK but! Five and two thirds doesn't equal five times two thirds and another cardinal mistake! That those of you who have made the mistake multiplied by six the wholes and also

	the fraction! If you remember when I first showed you equivalent adjustments I asked what the expression on the left side consisted of, Pavel? And you told me: it is a binomial, it is a trinomial. And how do terms differ from each other? Of the expressions? What is the difference? Hanka! How do I know what polynomial is on the left side?
Hanka 1:	It depends on whether there is a plus.
T 14:	Or minus, yes! So even if there was multiplication, it is still a monomial. But there is no multiplication, is there? It is neither five point two thirds, nor five times two thirds. What should you have done beforehand?
Class 13:	[Humming.]
T 15:	Say it out loud!
Jirka:	Transform it into fraction!
T 16:	Yes! To fraction! You needn't have written this down. You could have done it mentally, couldn't you? So how many thirds is this? [She points at the fraction $5\frac{2}{3}x$.]

Key to symbols used in transcripts in this chapter:

T12   Episode 1, the teacher's second utterance.

[text]   Comments and annotations, descriptions of non-verbal action.

...   A pause of 3 seconds or less.

Episode 2: Local Topaze Effect

The class were solving parametric equations. For their homework, the students had been assigned to solve the equation $dx + 1 = 2(4x + 1) - 5x$ by substituting the day of their birth for d. Some of them substituted 3 and then the equation had no solution. Together they tried to find a universal solution of such an equation. In the teaching episode, they adjusted the equation to the form $x(d - 3) = 1$.

CZ1-L05, 00:16:42

T 21:	What will I do with the equation now?
Class 21:	divide ...
T 22:	Yes! I will divide it by that $d - 3$. And because there is a common number and I don't know its value, I must write down the condition because if the result by any chance was that I would have to divide ...
Class 22:	by zero
T 23:	... by zero, I mustn't divide by zero, yes, you are already giving me the result [Hanka said 3 before], that wouldn't make sense, this adjustment would be nonsense, so, Hanka! What did you say?
Hanka:	d mustn't equal 3.
T 24	[Records this condition on the whiteboard.] That's why you got the result that you got [she means at home]. And I will finish it! So x must equal 1 over $d - 3$ [records on the whiteboard]. And now children! I will not only carry out the verification, but we will have to do one more step, which is called discussion. Because it is more complicated and because we are working with a parameter, the common number, so we have to carry out the discussion now. Which means we will have a look at how these two numbers work. Let's try to substitute d by that 3, you already know it, you two have tested it already at home, but let's do it

	again, let's substitute it in the equation and see what will happen. So I will get here [points at the left side of the equation].
Class 23:	$3x$
T 25:	$3x + 1$. I will write that down; $3x + 1$ should equal [points at the right side of the equation] here is no d, so I can do the calculation at once mentally as we have done before, so $8x + 2 - 5x$, yes? We will solve it with what result, Michal?
Michal 21:	… the result is …
T 26:	Take your time with the calculations.
Michal 22:	… $3x$, the result is that zero doesn't equal one.
T 27:	Yes, zero should equal one which isn't true, … so, we can write down the conclusion later and now! The other proof. So, the only number we eliminated was that 3. You tried some other numbers at home. You know what? Let's try whether in case we don't substitute only dates of birth but also for example decimal numbers or fractions, whether they could also be the result. Let's substitute d by for example what?
Class 24:	1.25
T 28:	1.25, why not, $d = 1.25$. So how will it work?

In the course of explanation the teacher provided the students with opportunities to fill in single words of her explanation, which they accepted, speaking in chorus (Class 21-24). The occurrence of the didactical contract mentioned earlier can be seen again in that the teacher, explaining the new procedure, kept referring to students' earlier knowledge and relations to other topics (e.g. T 22, T 25).

In their homework, most students had forgotten to discuss the conditions for division by the expression $d - 3$. When checking the homework, the teacher tried to help them discover their mistake. In turns T 21 to T 23 the teacher used a local Topaze effect. For a certain period (locally), she replaced division by an algebraic expression by division by numbers which she discussed with her students. She drew their attention to the condition of a non-zero divisor. Then she returned to the original problem and related it to other content.

We can question whether the pupils were able to grasp the explanation albeit the discussion was influenced by the Topaze effect. In the post-lesson interviews with the pupils, the experimenter asked about that:

CZ1-L05 (post-lesson interview with Michal 00:01:03)

Exp. 1:	Would you like to tell me anything about the homework?
Michal 1:	… First of all, I did not know why the teacher assigned it to us, but when we went through it together, I knew the result $\dfrac{1}{3-d}$. We tried it before the lesson and one boy — Adam — showed it to me.
Exp.2:	You knew why you were given that homework?
Michal 2:	I knew.
Exp 3:	And you knew the explanation?
Michal 3:	He [Adam] didn't tell us. We knew the solution, but we couldn't explain it.

The fact that a contract was at stake is confirmed in the post-lesson interviews with the teacher. Clearly, she feels responsible for her students' grasping of the subject matter:

CZ1-L05 (post-lesson interview with the teacher, 00:11:21)

T 1:	Well, it's certainly not easy. My feeling is that they didn't quite cope with it. Every equation will be an exception. I will have to go over it again.
Exp 2:	Those who substituted three got the result. They will have understood.
T 2:	It seems they discussed before the lesson.

Episode 3: Creation of the Didactical Contract

In her lessons the teacher shows the importance of appropriate handling of the problem, its mathematisation and logical argumentation. She creates a belief that in mathematics all solution steps must be reasoned. To these ends she guides the students step by step giving clues to correct answers. This was manifested for example during the solution of the following problem from the textbook "Find two numbers whose sum is twenty and the difference of their squares is 120."

CZ1-L07, 00:36:43

T 31:	So … do they expect you to produce the result by heart? Not likely, is it? You will have to calculate that! So what? What shall we do with it?
Class	[Humming.]
T 32:	Luboš! I just don't want to hear "I don't know"! I won't take an answer like that! So what should be done?
Luboš:	I would try it out! [Class laughs.]
T 33:	You would try it out! Oh my god! If you were, let's say … six years younger, I would accept it! Such an answer. [Vojta puts up his hand.] Vojta!
Vojta 1:	A set of equations?
T 34:	Very good! That's more mature. I will simply write it down and I will solve an equation! Even more, Vojta proposes a set of equations! How comes? Why?
Vojta 2:	There are two variables.
T 35:	Well, they could be there and still it needn't be a set of equations! Denisa! Am I distracting you from something important? It needn't be a set of equations! What is the value of the first number? I don't know, x. So what is the other number?
Jirka:	$20 - x$
T 36:	Well done, Jirka! Twenty minus x. But you proposed a set of equations [points at Vojta], so let's use a set of equations! Well, let's write down the first number [writes on the whiteboard], now the other number. What is the value of the first number then? x, the other number therefore y. Lenka, dictate the first equation!
Lenka:	$x + y = 20$
T 37:	Read the rest of the assignment and dictate the other equation, Marek!
Marek:	$x^2 - y^2 = 120$

281

One student, Luboš, proposed to solve the problem by trial and error[v]. It seemed as if the teacher was giving the students a free choice how to solve the problem, but actually what she expected was mathematisation via an algebraic expression, even with square numbers. She strongly refused the trial and error strategy (T 33). Students noticed it and did not propose it any more (even in the following lessons, this strategy was not used). If similar teacher behaviour occurs whenever a trial and error type of strategy is proposed or used by students, the unsuitability of the trial and error strategy becomes a part of didactical contract. Students will, as in this episode, use those strategies which are valued as 'mathematically higher level'.

From the post-lesson interviews it could be seen that some pupils searched for their own ways of solution but respected the authority of the teacher:

CZ1-L07 (interview with Luboš)

```
Exp 1:      You solved the problem with a system of equations. Is it
            convenient for you? Do you solve it in a different way?
Luboš 1:    I tried it with the calculator. I knew the results within
            several seconds.
Exp. 2:     Aha, you found it without an equation?
Luboš 2:    Yes. Then I did it in the manner required by the teacher.
            It is simple.
Exp. 3:     Would it be possible to solve it always with a calculator?
Luboš 3:    Definitely not, an equation gives the whole result.
```

Other pupils did not hesitate about the correctness of the teacher's statements:

CZ1-L07 (interview with Roman, 00:09:38)

```
Exp 1:      You solved the problem with a system of equations, in the
            same way as you did it, or differently?
Roman 1:    With a system.
Exp. 2:     Did you solve it by yourself or with the teacher?
Roman 2:    By myself.
Exp. 3:     Were you right?
Roman 3:    Yes.
```

Episode 4: Topaze Effect

The Topaze effect can occur with different purposes:
(a) In order to avoid mistakes the teacher simplifies the tasks by
– recalling needed previous knowledge, or previous activities (e.g. T 13)
– breaking up the procedure into simpler steps (e.g. T 32-T 36)
– giving partial answers to the questions posed (e.g. T 25)
– posing 'warning' questions (e.g. T 401)
without always making explicit the relationship between the original task and the simplified one.
(b) In order to draw students' attention to an occurrence of a mistake, the teacher 'fails to notice' the mistake when students perform it and continues with the procedure in the form (a) until they reach a point where they recognise the false result. Then she draws their attention to the moment where the mistake was born.

282

CZ1-L03, 00:01:24

T401:	So write down! Square root 4*x* + 6 equals minus 4. No objections? … That's fine! You won't have any objections because you've had no experience with it so far. Otherwise some of you would object. And now you see, the variable, the *x* is under the radical. How can I get it from under there, how can I make it stand-alone? It would be easiest for us if the radical disappeared. If I could somehow remove it form the equation! And for those ends I could use the reverse mathematical …
Class 41:	operation
T 402:	… operation, which is? …
Class 42:	exponentiation …
T 403:	Exponentiation, and what do we do with the whole equation? Well? We say we raise it to the second power. The left side to the second power, the right side to the second power. So, what's the outcome on the left?
Michal:	The radical is cancelled.
T 404:	Well, the radical there is alone so if I raise it to the second power, I cancel it. And what remains, Michal?
Michal:	4*x* + 6
T 405:	Yes, equals the right side, also raised to the second power, Jirka!
Jirka:	16
T 406:	Speak up!
Jirka 2:	16
T 407:	16. Yes, that would be nifty, let's continue. … When we are doing these adjustments, what equivalent adjustment am I doing now? As an exception, let's write this adjustment down. Hanka?
Hanka:	We subtract 6.
T 408:	Yes, only if you could speak up, we subtract six, from both the left and the right sides of the equation and so we get … Dominika!
Dominika:	4*x* = 10
T409:	Yes 10, that's it. And Dominika, finish it off …
Dominika:	… one *x* equals 10 quarters, which means …
T410:	five halves, yes? I will record that in the following manner, five halves, two and one half. So now you will think. All right, we have the root …
Pavel:	But it doesn't work!
T 411:	Excuse me?
Pavel:	But it doesn't work …
T 412:	What doesn't work? Speak up!
Pavel:	The verification doesn't work!
T 413:	You've already carried out the verification?
Pavel:	No, I haven't had time to do it but a radical can never have a negative result!
T 414:	Yes!!! This is what I expected you to object at the very beginning. But now, yes? Well, let's pretend nobody has noticed and let's carry out the verification, shall we? The left side of the equation equals, we substitute the result into the equation! Four times five halves, how much is that?

283

This unusual reaction to a mistake was confusing for the pupils as is again revealed in post-lesson interviews:

CZ1-L03 (post-lesson interview with Pavel 00:01:18):

```
Pavel 1    I wanted to tell, that I objected to something, but I told
           it softly.
Exp 1:     What do you want to tell?
Pavel 2:   That the square, we said before, could not be … negative
           number … radix can not be minus …
```

The teacher commented on the situation with surprise:

CZ1-L03 (post-lesson interview with the teacher 00:22:40)

```
Exp 1:     One of the boys in the first bench said that he knew that …
T 1:       And why didn't he tell it?
Exp. 2:    He said he was afraid that it is not correct.
T 2:       You see. They should get it? I will return to it again.
```

DISCUSSION AND CONCLUSIONS

Sarrazy and Novotná (2005) present three teaching styles (devolving, institutionalising and intermediary) which are in strong contrast with one another. The characterisation is based on the modes of teacher's actions described by the following three dimensions (variables v_1 to v_6):

- *Didactical structure of the lesson* (v_1. type of didactical dependence; v_2. place of institutionalisation; v_3. types of validation.
- *Forms of social organisation* (v_4. interaction modes; v_5. management with regard to the students' groupings)
- *Variability of the problem assignment* (v_6. teacher's 'capacity' to consider diverse modalities for the same didactical variable in editing the problem assignment)

Teachers with a *devolving style* use a strong variability in class management:

These teachers regularly use group work without inevitably restricting to this form of students' grouping; generally speaking, the problems are complex; classroom work is very interactive (students interact spontaneously, 'choral' answers are not rare, …); in the lesson, institutionalisation is diverse. (Sarrazy & Novotná, 2005, p. 39).

Teachers with the *institutionalising style* mostly use the scheme 'show-remember-apply'.

These teachers institutionalise very quickly one solving model and then present students with exercises of growing complexity. First, the exercises are corrected locally – the teacher passes through the rows and corrects them individually. Then the teacher gives the complete correction on the blackboard; here he gives details of the solution and, depending on the time he has, occasionally invites some students to the board either to make sure that they are paying attention, or to remind of certain knowledge. (Sarrazy & Novotná, 2005, p. 39)

The *intermediary style* combines features from both previous cases. "The students have more chances than those of 'institutionalising' teachers to encounter research situations, and debate, but markedly less than those exposed to the devolving style." (Sarrazy & Novotná, 2005, p. 40)

The teaching style of our teacher is intermediary. Referring to variables v_1 to v_6 we can say that:

v_1: In most cases, the teacher proceeds from simple to more complex tasks. If there are didactical reasons, she does not hesitate to insert a more difficult task in between.

v_2: She does not have a fixed place for institutionalisation in the meaning mentioned above. She uses her pedagogical and didactical experience to decide when institutionalisation is appropriate.

v_2: She tries to keep students informed about the validity of their results by using a stable means of evaluation. In the teaching sequence on solving linear equations the usual way of evaluating the results is by verification. Occasionally, she applies the validation by Topaze effect (see e.g. Episode 4) or by exchange (especially of written works) in pairs.

v_4: The prevailing interaction is between the teacher and one or more students. In this organisation, she offers students a wide space for participation in lessons.

v_5: The prevailing class organisation is whole class teaching. Individual work is included for practising of taught algorithms. Group work did not occur in any of the ten videotaped lessons. In order to avoid difficulties and mistakes, she offers a limited space for students' independent investigations.

v_6: She has sequences of problems where the mathematical formulation of the problem as well as the solution procedure needed for its solution are of increasing difficulty.

What can be said about the teacher's style and behaviour from this perspective?

– The selection of tasks and problems is the result of the teacher's rich, long term experience. Her ability to predict possible obstacles leads her to pose only questions that she expects students will be able to answer. She does not insist on individual justification. We can say that she concentrates on creating 'routine' skills. She leaves only a limited space for students' independent investigations.

– She creates a work atmosphere in the classroom. Most of the time, she works with the whole class at once. She asks questions and reacts to all student contributions. She also explicitly urges the students to be diligent and responsible in the dialogue, for example:

CZ1-L05, 00:10:45

```
T:        The most important is that you want to. It won't just
          happen, Michael! Every success is hard work! Well, if you
          are not willing to invest anything! To say 'I can't cope or
          I don't get it' is just an excuse. There must be some
          effort, some input!
```

- If a student finds a mistake in the teacher's calculation, reasoning or method, she is glad. She regards this correction as a proof of their attention; it shows that students are following her instruction, for example:

CZ1-L04, 00:41:17:

```
Vojta:     There is a y missing.
T:         Where? I see! Here! Yes, thanks.
```

- By asking certain questions, the teacher tries to eliminate students' fear of the unknown world of 'new', non-standard tasks. In her lessons, she shows the importance of appropriate handling of the problem, its mathematisation and logical argumentation, for example:

CZ1-L04, 00:40:42:

```
T:         Word problems. Let's return to word problems again. We will
           again return to the word problems. How shall we solve using
           equation? How shall we begin? Denisa.
Denisa:    First we have to record it and set the unknown.
T:         And you will record it with the …
Student:   unknown …
T:         with the unknown. You express the relations with … if you
           solve it using an equation. Alternatively you might solve
           it using logical thinking. If you use an equation, it is
           this way.
```

- From the interviews it was apparent that the students appreciated her approach:

CZ1-L10 (post-lesson interview 00:03:36):

```
David:     I am quite good at those problems about cooperation, it is
           drill, they're always the same.
I:         Even here where the assignment was not so unambiguously set
           as before? Have you solved this one also at once?
David:     About as fast as on the whiteboard, but I haven't been
           copying it!
I:         I wonder if you needed help of the teacher.
David:     No. I checked from time to time with the whiteboard that I
           was proceeding well and I continued solving. … So, I
           checked this first part of the problem with the whiteboard,
           it was trouble. I did the second part on my own.
```

- The teacher maintains students' attention mainly by calling on any student in the class. When observing them, we found out that in reaction to this situation some students try to be a little in advance with their work, to ensure that when called out they will be able to respond without difficulty. However, it cannot be said that they do it in fear that their answer would be incorrect. The students called on do not seem to be afraid that their answer will be wrong or that they will not be able to solve the problem or individual work, because the atmosphere in the classroom is calm and the teacher's reaction to a mistake is not rejecting or negative. The students accept positively even her gentle irony. For example:

CZ1-L07, 36:32:

```
T:         Read it out, Hanka!
           [Hanka reads something else]
```

286

```
T:        Page 89, which means an 8 followed by a 9 [Other students
          laugh.]
```

Sarrazy and Novotná (2005) raise the question of whether there is one 'best' teaching style. In this paper we have tried to demonstrate that it would be wrong to say that either the devolving or the institutionalising style produces the best results in developing students' knowledge of mathematics and their personalities. The recognition accorded to our teacher from among professional, parental and students communities confirms this conclusion.

The question of which of the six variables v_1 to v_6 has the biggest impact on the success of the teaching remains unsettled.

ACKNOWLEDGEMENT

This research was partially supported by the project Czech Science Foundation 406/05/2444.

NOTES

i   *Situation of action*: the actor decides and acts on the milieu, it is of no importance whether the actor can or cannot identify, make explicit or explain the necessary knowledge
     Situation of formulation: at least two actors are put into relationship with the milieu; their common success requires the formulation of the knowledge in question
     Situation of validation: a situation whose solution requires that the actors establish together the validity of the characteristic knowledge of this situation; its effective realization thus depends on the capacity of the protagonists to establish this validity explicitly together
     Situation of institutionalisation: a situation which reveals itself by the passage of a piece of knowledge from its role as a means of resolving a situation of action, formulation or validation to the new role of reference for future personal or collective uses. (Brousseau and Sarrazy, 2002, pp. 4-5)
ii   The name comes from the first scene of Marcel Pagnol's play *Topaze* where Topaze is dictating to a weak student and suggests the spelling by concealing it in more and more transparent ways.
iii It is called so by reference to the scene in *Le Bourgeois Gentilhomme* by Molière, where the philosophy tutor reveals to Jourdain what prose and vowels are. The whole humour of the scene is based on the absurdity of repeatedly giving familiar activities the status of learned, scholarly discourse.
iv   The transcripts from the classroom are labelled as follows: CZ1 (Czech school 1), L05 (Lesson 5), time of the start of the episode.
v   By a trial and error strategy, the solution of the problem takes only a few steps, e. g. 10 + 10 (difference is unsuitable), 11 + 9 (difference unsuitable), 12 + 8 (difference unsuitable), 13 + 7 (difference 120).

REFERENCES

Brousseau, G. (1975). *Etude de l'influence des conditions de validation sur l'apprentissage d'un algorithme* [Study of the influence of the validation conditions on the learning of an algorithm]. Bordeaux: IREM de Bordeaux.
Brousseau, G. (1989). Le contrat didactique: le milieu [The didactic contract: The milieu]. *Recherches en Didactique des mathématiques*, 9(3), 309-336.

Brousseau, G. (1997). *Theory of Didactical Situations in Mathematics*. (N. Balacheff, M. Cooper, R. Sutherland, V. Warfield, Eds. & Trans). Dordrecht: Kluwer.
Brousseau, G. & Sarrazy, B. (2002). *Glossaire de quelques concepts de la théorie des situations didactiques en mathématiques* [Glossary of some concepts of the theory of didactical situations in mathematics.]. Bordeaux: DAEST. Unpublished manuscript, (pp. 4-5).
Clarke, D. (2001). Complementary accounts methodology. In D. Clarke (Ed.), *Perspectives on practice and meaning in mathematics and science classrooms*. Dordrecht: Kluwer.
Laborde, C. & Perrin-Glorian, C. (Eds.). (2005). Teaching situations as object of research. Empirical studies within theoretical perspectives [Special Issue]. *Educational Studies in Mathematics*, 59, 1-3.
Sarrazy, B. & Novotná, J. (2005). Didactical contract: Theoretical frame for the analysis of phenomena of teaching mathematics. In J. Novotná (Ed.) *Symposium of elementary mathematics teaching 05* (pp. 33-45). Prague: Univerzita Karlova v Praze, Pedagogická fakulta.
Seeger, F., Voigt, J. & Waschescio, U. (Eds.). (1998). *The culture of the mathematics classroom*. Cambridge: Cambridge University Press.
Warfield, V. (2005). Invitation to didactics. Unpublished Manuscript.

Helena Binterová
Pedagogical Faculty
University of South Bohemia České Budějovice
Czech Republic

Alena Hošpesová
Pedagogical Faculty
University of South Bohemia České Budějovice
Czech Republic

Jarmila Novotná
Faculty of Education
Charles University Prague
Czech Republic

YASUHIRO SEKIGUCHI

CHAPTER TWENTY

Mathematical Norms in Japanese Mathematics Lessons

INTRODUCTION

This chapter reports an analysis of lessons taught by three eighth-grade Japanese teachers who participated in the Learner's Perspective Study (LPS). The analysis focuses on classroom mathematical norms introduced by the teacher.

As Clarke (2004) points out, one of the goals of the LPS is to complement the TIMSS 1999 Video Study. In Hiebert et al. (2003), the analysis of mathematics teaching focused on the mathematical knowledge, procedures, and reasoning involved in the problems presented in the lessons. The teaching of mathematical norms was beyond their analysis. Although mathematical norms are often neither taught explicitly by teachers nor written in textbooks, they are crucial when the process of learning mathematics is conceived as mathematical activities.

Mathematical norms are knowledge 'about' doing mathematics, so they belong to the domain of metaknowledge in mathematics. It is hypothesized that beginning teachers are often preoccupied with covering curriculum content, paying attention to mathematical knowledge and skills; competent teachers, as selected by the *LPS* design, would invest more time and effort in teaching metaknowledge. The major questions that guided this analysis of Japanese data are: What mathematical norms would surface in the lessons? How would the teacher introduce, negotiate or utilize those norms during the lessons?

THEORETICAL FRAMEWORK

Norms

Scientists search for patterns, regularities, rules, or laws in the real world, and try to build causal theories, so as to be able to explain the phenomena in which they are interested. An understanding of causal relationships is useful for prediction and control. Social sciences, likewise, study those patterns, norms, regularities, rules, or laws appearing in human activities (cultures), so that they can explain and understand human activities. Positivist sociologies are known to have assumed a

D. J. Clarke, C. Keitel & Y. Shimizu (Eds.), *Mathematics Classrooms In Twelve Countries: The Insider's Perspective. 289–306. © 2006 Sense Publishers. All rights reserved.*

normative conception of human action: this conception has three main components, *actors*, *rules* and *situations* and presumes that "actors know and follow rules in social situations" (Mehan & Wood, 1975, p. 74). This closely parallels that of natural phenomena: "Physical objects follow natural laws in the physical world."

Ethnomethodologists have also studied people's rule use in social situations, but have made strong attacks on the normative conception described above. They claim that actors, rules, and situations are mutually shaped in practice, in their terminology, 'reflexively' related to each other:

> Actors, rules, and situations ceaselessly inform one another. The situation is not independent of the actors who are within it. And because of actors' ever-shifting corpus of social knowledge and practical interests ... a situation is never judged once and for all. Every judgment is situationally absolute, based on the realization that some later determinations may change the certainty of the here and now.
>
> The very invocation of a rule alters the situation. Rules, like actors and situations, do not appear except in a web of practical circumstances. Intertwined, the actor, rules, and the present definition of the situation *constitute* the situation. No single one of these can be abstracted out and treated as either cause or effect. Actors cannot be seen as outside of the situations judging them, for they are an integral and reflexive constituent of those situations. (Mehan & Wood, 1975, pp. 75-76)

Thus, even if a norm is taken as shared among people, it cannot prescribe their actions. Norms are cultural knowledge or "tools" (Waschescio, 1998, p. 235) that may help people to accomplish something in some specific situations:

> Members of society do not simply follow internalized norms or rules in the manner of 'cultural dopes' ... but rather practically analyse situations in terms of the relevance of such norms and rules. (Francis & Hester, 2004, p. 206)

Mathematical Norms

Cobb and his colleagues have been developing the most sophisticated arguments on studying norms in the mathematics classroom (for a comprehensive review of research on norms, see Herbel-Eisenmann, 2003). They began their research with a constructivist framework that conceived learning mathematics as an active construction process by individuals. Adopting the frameworks of symbolic interactionism and ethnomethodology, they have come to incorporate into it a sociological framework. They investigated how students developed beliefs and values about mathematics. Their focus of analysis was on classroom processes of the 'inquiry mathematics' tradition, where children actively participated in exploring, explaining, justifying, and arguing mathematics. For the analysis, they introduced the notion of 'norms' of classroom process as a device to interpret classroom processes and clarify how children's beliefs and values developed. They identified several classroom social norms working in their project classroom:

- Students were obliged to explain and justify their reasoning.
- Students were obliged to listen to and to attempt to understand others' explanations.
- Students were obliged to indicate non-understanding and, if possible, to ask the explainer clarifying questions.
- Students were obliged to indicate when they considered solutions invalid, and to explain the reasons for their judgment. (Cobb, 1996, p. 88)

As their research progressed, they proposed norms specific to mathematical learning, 'sociomathematical' norms, distinguished from the above social norms:

[W]e extend our previous work on general classroom norms by focusing on normative aspects of mathematics discussions specific to students' mathematical activity. To clarify this distinction, we will speak of *sociomathematical* norms rather than social norms. For example, normative understandings of what counts as mathematically different, mathematically sophisticated, mathematically efficient, and mathematically elegant in a classroom are sociomathematical norms. Similarly, what counts as an acceptable mathematical explanation and justification is a sociomathematical norm. (Yackel & Cobb, 1996, p. 461)

They contend that mathematical activity has norms as constituent, and that norms are reflexively related to beliefs and values about mathematical activities. Sociomathematical norms are interactively constructed in each mathematics classroom, and may be different from one classroom to another (Yackel & Cobb, 1996, p. 474)

In this chapter I propose to use the different phrase 'classroom mathematical norm' to refer to a norm specific to classroom mathematical activity, rather than 'sociomathematical' norm, because I consider that mathematics is intrinsically a sociocultural activity, as claimed by current philosophies of mathematics and sociocultural theories (e.g., Bloor, 1976; Ernest, 1994). If we accept this understanding, the prefix 'socio-' of sociomathematical norm is redundant. Since if we say a 'mathematical norm', it may be taken to belong to the community of the professional mathematicians, I put the word 'classroom' before it to indicate that it is inherent in classroom situations. When no confusion between the norms of a classroom and those of mathematicians is anticipated, the word 'classroom' would be omitted. (This usage is consistent with the use of the term 'social norm' by Yackel and Cobb (1996). They often use the term 'classroom social norms' to refer to social norms specific to a particular classroom.)

From the methodological point of view, the distinction between social norms and sociomathematical norms ('mathematical norms' in the current chapter) requires some clarification. Yackel and Cobb (1996) illustrated it with examples:

The understanding that students are expected to explain their solutions and their ways of thinking is a social norm, whereas the understanding of what counts as an acceptable mathematical explanation is a sociomathematical norm. Likewise, the understanding that when discussing a problem students should offer solutions different from those already contributed is a social norm, whereas the understanding of what constitutes mathematical difference is a sociomathematical norm. (Yackel & Cobb, 1996, p. 461)

This contrast between social and sociomathematical norms is incomplete, however. Each social norm was formulated in a sentence that described an expectation or obligation. For instance,

SOCIAL NORM A: Students are expected to explain their solutions and their ways of thinking.
SOCIAL NORM B: When discussing a problem, students should offer solutions different from those already contributed

Sociomathematical norms were not formulated in that way. If we tried to do so, we would obtain the following formulations:

SOCIOMATHEMATICAL NORM Am: Students are expected to explain *mathematically* their solutions and their ways of thinking.
SOCIOMATHEMATICAL NORM Bm: When discussing a problem, students should offer solutions *mathematically* different from those already contributed.

Once sociomathematical norms are formulated this way, first it is clear that the classroom participants should be able to recognize the difference between sociomathematical norms and social norms. As far as we say that the above sociomathematical norms were interactively constituted among the classroom participants, the participants need to understand that explanations presented in their classroom must be 'mathematical', and that the solutions presented in their classroom must be 'mathematically' different from those already contributed. Otherwise, the sociomathematical norms would just be constructs of observers (cf. Herbst, 1997). Second, it is assumed that when social norm A or B is invoked in mathematics lessons, the participants can take it as sociomathematical norm Am or Bm, respectively, without difficulty.

Thus, when we identify a sociomathematical norm in the data, we need to be certain that the classroom participants understand that something 'mathematical' is required in their activity, and what 'mathematical' means there. Also, we need to be certain that the classroom participants can understand social norms invoked in mathematics lessons in terms of sociomathematical norms. These points may not be so problematic in researching secondary mathematics classrooms like *LPS*, however, because students seem to be deeply enculturated into mathematics lessons.

Another issue to be clarified concerning norms from the methodological point of view is about the situation where a norm is considered relevant and utilized. As mentioned earlier, ethnomethodologists consider that norms emerge in a web of practical circumstances. Norms are reflexively related to the situations where they are generated, or taken to be shared. As the norms identified by Yackel and Cobb (1996) are those of an 'inquiry' tradition classroom, when we analyse mathematical norms in other classrooms, we need to take into consideration the traditions in which each classroom is located. We may need to further look at the types of situations a classroom process contains. Mathematics lessons in regular classrooms consist of different types of activities or situations: Teacher's explanation of new content, problem solving, students' presentations to the class, exercises, checking homework problems, tests, and so on. It is unlikely that the same norm is considered relevant or utilized through all the situations even in the same lesson. In fact, symbolic interactionism and ethnomethodology warn about overuse of norms in explaining social conduct:

> Rather than the major criterion people employ to regulate their own and other's conduct, social norms are one of several forms of knowledge that people employ in their everyday conduct. ... It should not be thought, however, that norms are constantly implicated in acts, nor that people behave by finding the appropriate norms that govern each and every social situation. (Hewitt, 1994, p.160)

Mathematical Norms As Objects of Learning

Classroom mathematical norms are important constituents of mathematical activities in the classroom. Though they are not about individual learning, they facilitate the development of personal mathematical orientations consistent with mathematical norms. This is the internalisation of mathematical norms within individual students. Therefore, from the viewpoint of curriculum, mathematical norms can be considered to be objects of learning. In the constructivist teaching experiments of Cobb and their colleagues, classroom mathematical norms are described as emerging through interaction, and their emergence as 'unanticipated' (McClain & Cobb, 2001). Experienced mathematics teachers in regular classrooms like the teachers of the *LPS*, however, would often be able to anticipate the emergence of a norm, and pay explicit attention to its development and internalisation. When analysing norms in their lessons, we should be aware of their understandings of the goals of the unit from the viewpoint of curriculum.

For instance, one of the goals of the unit on simultaneous linear equations in eighth-grade mathematics in Japanese schools is for students to be able to solve simultaneous linear equations by algebraic operations. This is explained in the official guidebooks of the national course of study, teachers' guides to textbooks, and so on. Through the unit, students are expected to solve simultaneous equations and explain their solutions by using algebraic operations, and only those solutions and explanations would become acceptable. This is a mathematical norm, and teachers have the responsibility of achieving the internalisation of this norm by the end of the unit.

DATA ANALYSIS

The following is an analysis of norms from the three Japanese data sites, J1, J2, and J3. In each site a teacher taught ten consecutive lessons. The videotapes of the lessons, their transcripts, and the interview data of the teachers and students were analysed qualitatively. To let mathematical norms emerge from the data, any piece of the data that appeared to indicate patterns on how to work on mathematics was coded, and the normative aspects behind those patterns were repeatedly analysed. The lessons at Sites J1, J2, and J3 were about linear functions, geometry, and simultaneous linear equations, respectively.

It was evident in the data that every lesson utilizes some norms introduced in previous learning. All the norms found in the data seem to be more or less related to previous norms. Therefore, the analysis first looks into this common feature. Next, the analysis turns to the problem-solving activity because it is common in all three sites, and forms an important part of the Japanese lessons.

In the following, lesson data is referred to by indicating site name, lesson number, and the time from the start of the lesson. For example, (J2-L3, 16:35) means that the data was taken from the third lesson of Site J2, and occurred at 16 minutes and 35 seconds from the start of the lesson.

Utilizing Previously-Introduced Norms

Most of the lessons in the eighth grade are considered to be related to lessons of previous grades or units. Teachers often reminded students about previous learning:

J1-L1, 26:50

```
T:      OK, well, now, how should we examine this? For example, I
        weigh myself on the scale everyday, and when I record it, I
        write it like this. Then well I can write it and examine the
        changes. How did you arrange these in seventh grade? Do you
        remember?
S:      Graphs.
T:      Graphs. Oh, sure. We use graphs. What else?
```

Key to symbols used in transcripts in this chapter:
...   Pause of three seconds or less
Ss   Simultaneous talk by two or more students
[text]   Descriptions of non-verbal actions, or other additional explanation provided by the transcriber.

In the seventh grade, students had studied proportional and inversely proportional relations by using tables, graphs, and algebraic expressions. By referring to "the seventh grade," the teacher tried to remind students about the use of tables, graphs, and algebraic expressions for exploring changes. Thus, the teacher was proposing a classroom mathematical norm for this unit: *Changes should be explored by using tables, graphs, and algebraic expressions.* This norm had been developed through the unit of proportional and inversely proportional relations at the previous grade.

At the eighth grade mathematics, once the norm was recalled at the beginning, it was then assumed as a tool for investigation.

At Site J2, in the unit on similarity, problems that involved proving the similarity of geometric figures were discussed. The unit was situated immediately after the unit on congruence. The teacher told students to write proofs of similarity just as they did proofs of congruence:

J2-L8, 22:34

> You can write in the style of proof, which we have been using until today. You can use the same writing structure as the case of congruence. Okay? In the case of congruent triangles, these were three lines. So, you can just start writing as the same. Its content, of course the condition as well, is a little different though. Write it as the same. Okay, so, firstly, for triangle something and triangle something, …

Here, the teacher was expecting students to follow a classroom mathematical norm: *Proofs should be written the same way as in proofs of congruence of triangles.*

At Site J3, the teacher began to discuss how to check the solution of simultaneous equations. Then, he asked students to remember how they had checked solutions of linear equations in the seventh grade:

J3-L2, 48:52

T: Before that, in the next class which is tomorrow, we would like to go over how to check the two that we studied here. I think you've studied how to do check when you were in the seventh grade. I just want to quickly review something we did in the seventh grade. There is another minute left so I need your attention up here. Um, for instance, [writes on the blackboard] 5X plus 3 equals, let me see, um, um, six. Yes six. If there was something like this, what would the answer be? Can you do it mentally? Maybe it's difficult. Should I do it on the board? [writing on the blackboard] 5x equals three, so x is?

Ss: Three fifths.

T: … So the answer is x equals three fifths. … Now you have to check this. I think you've studied how to do it in the seventh grade. … Does anyone remember how to do this?

The teacher was here expecting students to extend the classroom mathematical norm of checking linear equations to checking of simultaneous equations.

By referring to previously-learned norms, the teachers seem to have expected that students would be able to connect the current topic to the previous ones. Also, the use of already learned norms seems to be easier than creating a totally new one. Applying previously-learned norms to new topics was not always trivial, however. Students needed to reflect on their previous learning and norms, and generalize them beyond the contexts of the previous learning.

Norms in the Problem-Solving Process

In Japanese mathematics lessons, a whole-class problem-solving process is very common. Basically, it consists of four phases (cf. Becker, Silver, Kantowski, Travers & Wilson, 1990; Stigler & Hiebert, 1999):

- Phase 1: The teacher presents one problem.
- Phase 2: Students first try to solve it individually. Then, they may work with neighbours, or in small groups.
- Phase 3: Some students present their solutions on the blackboard, and the class discuss the presented solutions.
- Phase 4: The teacher summarizes important points.

This pattern of problem solving process was often found at all three data sites. The whole process was not necessarily completed within one lesson, however. For example, a lesson may end at Phase 2. Then the problem was assigned as homework, and the next lesson began from Phase 3. Also, there were some variations of the process. For example, at Phase 3 the teacher, instead of students, may write solutions on the board. Phase 4 may be very quick, or may be very extensive.

Classroom mathematical norms situated in this Japanese tradition of problem solving process are described in the following. As noted earlier, classroom mathematical norms are related to curriculum objectives. Therefore, some norms were very visible in some units, but not in the other units.

Diverse ideas. In the history of mathematics, difficult mathematical problems were tackled by many mathematicians with diverse ideas. Mathematical creativity develops when diverse ideas are generated. Perceptions of the diversity here are based on the participants' understanding of "mathematical difference" (Yackel & Cobb, 1996).

In Phases 2 and 3 of the problem solving process, finding more than one solution method is usually encouraged. In Lesson 1 of Site J1, the teacher presented figures of "staircases" (Figure 1), and asked students to find as many changing aspects as possible in the staircases. They came up with many different aspects: size, area, height, perimeter, the number of right angles, the number of edges, and so on, and she commented:

J1-L1, 23:50

```
T:     OK, well, it takes a little time but if you really put your
       mind to it, I'm sure there are lots more. Um, those of you who
       come up with new ones as we go along, please come up and tell
       me, later.
```

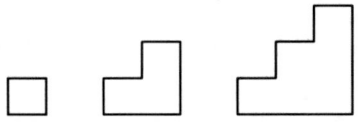

Figure 1. Staircase problem.

Also, when the class investigated the perimeter of staircases, the teacher asked how to get the perimeter of six-step staircase:

J1-L1, 37:09

```
T:     How many centimetres is it when there are six steps?
Ss:    Twenty-four.
T:     Twenty-four. Is that right? I see. Twenty-four. Now, why is it
       twenty-four, then? You can just share your idea. It doesn't
       have to be the same as everyone else.
```

Students provided three different methods. The teacher commented, "Hmm, there are lots of different methods, aren't there?" In the interview the teacher expressed the importance of diverse ideas: "[About my ideal lesson] Though I suppose it depends on the materials, in the materials of function like this it is important that students come up with various views." Thus, the classroom mathematical norm that *students should explore mathematically-diverse ideas* was clearly supported by the class.

Also, at Site J2, in Lesson 2 the teacher was encouraging the students to find two different constructions for a construction problem. All the students except two found the same construction, however. Only two students found a different construction. In Lesson 3 the teacher chose the latter construction first and discussed it extensively. The teacher's expectation of diverse ideas was evident. He referred to his expectation of the students in the interview: "Uh, not like this is the only way of doing, but pursuing various ways of approach, I hope students could acquire such an attitude."

Efficiency. The value of pursuing efficient ways of solving problems is generally shared among mathematicians. Many theories, theorems, and formulae in mathematics have been produced to improve efficiency.

At Site J2, in Lesson 8 the class was reviewing a homework problem. In the problem seven figures of triangles were given: The information about side lengths and degrees of angles of them was only partially given. The task was to choose similar triangles, and justify the choice referring to appropriate conditions for similar triangles. Students were able to find some pairs of similar triangles quickly, but they expressed difficulties in finding other pairs. Then, the teacher began to discuss an efficient way of using similarity conditions, comparing them with congruence conditions, and advised students to look at angles first:

J2-L8, 16:07

> For this one [congruent condition] you always need to use
> side. Therefore, we should start it by looking at the sides
> first. But there's one exception for the similar condition,
> you see? Here, we have one without using sides. Then, in other
> words, you can start looking at angles for this time. If you
> found a figure with the data of only angles, then you have
> only one condition to use. Right? You can't use these two
> conditions if you don't have any side. Therefore, we look at
> the angles first. And then we should know if it is possible to
> use this condition with two angles. Also, you need some
> calculations for other two conditions with the ratio of sides.
> But if you could use 'Two angles equality' condition, you
> don't have to do any calculation. Isn't that the easiest one
> of all after all?

Then the teacher presented a proof problem on similar triangles. When discussing the solution, the teacher repeated that we should look at angles first. Here, the teacher seemed to emphasize a classroom mathematical norm: *Students should pursue as efficient solutions as they can.*

At Site J3, the teacher often encouraged the students to pursue efficient ways of solving simultaneous equations. In Lesson 1 of Site J3, the class discussed a pair of simultaneous equations:

$$5x + 2y = 9 \quad \text{...} \quad (1)$$

$$-5x + 3y = 1 \quad \text{...} \quad (2)$$

The teacher asked a student Kori to write his solution on the board. He subtracted (2) from (1), obtaining $10x - y = 8$. Solving it for x, he put the result into (1), obtaining the value of y. Finally, he put the value of y into (1), and got the value of x. After Kori explained his solution to the class, the teacher asked the class:

J3-L1, 10:54

> OK, any question? Can you understand? Well, do you have any
> thoughts as you worked out this problem? Any impressions of
> this explanation?

A student Suzu responded to it: "I think there is much simpler one." Suzu wrote his solution on the board: He added (1) and (2), and got an equation without variable x. And he solved it for y, and got the value of y. He then put it into one of the given equations, and got the value of x.

The students were then seeing two different solutions on the board. The teacher explained the reason why he asked Kori to write his solution on the board. The teacher intentionally chose Kori because he had observed at the previous lesson that Kori had solved the problem differently from the other students: "Almost, actually almost all students had this opinion as far as I saw yesterday. And in fact, the way which Kori did was different so that I wanted them to write both ways on the blackboard" (15:27).

The teacher thought that by comparing solutions with different degrees of difficulty, students would be better able to appreciate an easier one: "I think you can know which point was difficult as you compare the difficult way and the easier one" (15:48). Finally, the teacher concluded that Suzu's solution was easier and better than Kori's:

J3-L1, 16:24

> Now, actually that way is much better than this way, when we compare the calculations so far. As a result, it is better to notice that this way, which Suzu wrote, is better, you know?

He asked the students where they thought Kori's solution was more complicated than Suzu. This question tried to elaborate inefficiency of Kori's solution.

Up to this point, the teacher seems to be putting more value on efficient solutions, encouraging the classroom mathematical norm of efficiency. Kori's solution seems to be devalued. This does not mean that inefficient solutions are useless, however. First, the teacher later pointed out that Kori's method gave the same result as Suzu. Second, he suggested that Kori's method contained an important idea: "There are some important ideas in this [Kori's] process, I think" (19:21) which I discuss next.

Important ideas. Efficiency is not the only value in pursuing mathematics. New ideas for developing new ways of solving problems are equally important in mathematics. Those could be discovered through numerous inefficient, or failed attempts as the history of mathematics shows.

At Site J3, in Lesson 1 after the inefficiency of Kori's solution was discussed as described above, the teacher pursued "Kori's idea", and went into the idea of the substitution method, which was formally introduced at Lesson 7. This pursuit continued well into the next lesson. The teacher gave an opportunity for the whole class to appreciate an important idea found in the 'inefficient' solution. Thus, he seems to indicate a classroom mathematical norm: *We should consider if any mathematically important idea is found in unaccepted attempts.*

In addition to this normative action, the teacher paid respect and care to both solutions. Devaluing one's idea may hurt his or her feelings. When Kori received negative opinions to his solution, the teacher encouraged Kori:

J3-L1 13:42

> It's OK. Don't be depressed as it didn't go well. It is better to get some comments, right? Don't worry.

By pursuing Kori's idea with the whole class, the teacher showed further care towards the student whose idea had been devalued.

At Site J1, from a rather long discussion an important mathematical idea emerged. In Lesson 4 the class began to work on the 'origami' (paper folding) problem. Each student was given a rectangular paper of length 12 and width 15 centimetres. The right side of the paper was coloured, and the reverse side was white. The teacher first asked the students what changes would occur if they continued to fold the paper by changing the place of crease (Figure 2). Then the class investigated some of the changes they found.

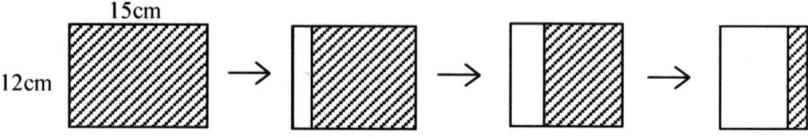

Figure 2. Origami problem.

In Lesson 5 a student Taka investigated a relation between the width (x) and the perimeter (y) of the unhidden coloured area [the shaded parts in Figure 2] using a table and graph. When she made the table, she put $y = 0$ when $x = 0$. In Lesson 6 photocopies of her table and graph were handed out to the class, and a student Shima suggested that $y = 24$ when $x = 0$ because the length of the unhidden coloured area did not change. The teacher temporarily accepted both ideas about the value of y for $x = 0$.

J1-L6, 25:43

> There are people who think there is nothing and people who
> think go twelve and come back twelve, and the total would be
> twenty-four. There are two ways to think. I want to choose
> one. But let me write twenty-four, too. Are they reasonable?
> Let's write twenty-four, here.

In Lesson 7 the teacher again brought up the issue of the value of y for $x = 0$ in the problem of Taka. Students' opinions were divided. Some even suggested another idea that $y = 12$ because the area looked like a single segment when $x = 0$. The teacher confessed to the students that she had struggled with it:

J1-L7, 32:29

> Huh, so, there has been many opinions, and cannot decide which
> is right. It is difficult. But you have decided your own
> opinion, haven't you? I have thought about this all night, and
> didn't sleep at all.

Finally, the teacher proposed that the value zero should be excluded from the domain. Thus, the ideas of the three students were not fully accepted, but the teacher capitalized on this discussion as an opportunity to emphasize the importance of determining the domain of a function, and to introduce the symbol for marking the endpoints on graphs that are out of the domain.

Valid writing. Mathematics is traditionally written in the deductive way: It must begin with axioms, definitions, or already proved theorems, and proceed logically. Therefore, you cannot write what you have not yet shown to be true.

At Site J3 in Lesson 3, the teacher reviewed the solution of a pair of simultaneous equations: $3x + 2y = 23$, $5x + 2y = 29$. As homework, he had asked the students to check the solution. First, he asked Uchi to put up his work on the board (see Figure 3).

By putting $x = 3$, $y = 7$ into $3x + 2y = 23$ and $5x + 2y = 29$

$$3\times3 + 2\times7 = 23 \qquad\qquad 5\times3 + 2\times7 = 29$$
$$9 + 14 = 23 \qquad\qquad 15 + 14 = 29$$
$$23 = 23 \qquad\qquad 29 = 29$$

Figure 3. Uchi's writing on the board.

As a "different way," the teacher then asked Kizu to put up his work on the board (see Figure 4).

By putting $x = 3$, $y = 7$ into $3x + 2y$
$3\times3 + 2\times7$
$= 9 + 14$
$= 23$
By putting $x = 3$, $y = 7$ into $5x + 2y$
$5\times3 + 2\times7$
$= 15 + 14$
$= 29$

Figure 4. Kizu's writing on the board.

The teacher posed to the class the question: What differences did they notice between the two methods? The students discussed the question with nearby students. After that, Uchi and Kizu explained their work at the front. The teacher mentioned that most of the students did it the same way as Uchi did. Reviewing the checking of the solution of linear equations studied in the previous year, the teacher pointed out that Uchi's writing used an unconfirmed fact:

J3-L3, 36:28

```
This is just substituting x as three, and y as seven into the
equation, right? It's just substitution, right? It's just
substitution but this is already an equality, so the right
side and the left side have to be equivalent, doesn't it? But
```

> you can't confirm that yet, can you? Right? Which means, if
> you write it this way, actually, [writes on the blackboard]
> you've already shown that the right side and the left side are
> equivalent at this point. But you haven't confirmed that yet.

Here the teacher was trying to let the students become aware of the classroom mathematical norm of valid writing that if you write an equation in your solution, it means that you have already shown the equality, or that *in a mathematical explanation you cannot write what you have not yet shown to be true.*

Based on this norm, the teacher accepted Kizu's way of checking, and devalued Uchi's way. Again, the teacher did not forget to remind the students that most of the students had done it in Uchi's way: You were not the only one who did wrong.

The lessons of Site J2 were about geometry. The idea of mathematical proof had been introduced to the class in the unit on congruence of geometric figures before the data collection of Site J2 started, and the students seemed to be accustomed to it. Basically, mathematical proofs were considered necessary when any proposition was accepted. This was the case even in construction problems. When students made constructions as a solution of a problem, the teacher often responded to them by asking if they could prove that it is correct: "No, that's not really right. When I ask you to prove that, how would you do it?" (J2-L2, 24:18)

Doing proofs for all the propositions accepted in the lesson would be unrealistic, however. Writing and explaining a proof takes time, but the lesson needs to accomplish other various objectives. For instance, introducing the conditions for similar triangles, the teacher did not discuss their proofs because he wanted to put more emphasis on the use of those conditions to investigate other properties:

J2-L2, 03:34

> Here, I won't talk about why they are conditions for
> similarity, but I will put more importance on how to use them
> instead. All right, let me write them down on the blackboard
> quickly.

This does not mean that the norm of valid writing was ignored. Rather, the norm was well-accepted; otherwise, the teacher would not have to give an excuse for skipping the proofs. Acceptance of the norm does not mean that whenever you want to use a new proposition, you actually have to produce its proof beforehand. Proving the similarity conditions was simply not given priority over other plans of the teacher.

Awareness of meanings. Discussion of solution methods sometimes leads to awareness of meaning, indicating a classroom mathematical norm: We should reflect on the meanings of what we are dealing with when doing mathematics. This is because meanings are an important base for generating and justifying mathematics.

In Site J1, after a number pattern "when the number of steps increases by one, the perimeter of the staircase increases by four" was found from the table, the teacher asked students to go back to the staircase, and explain the meaning of the pattern in terms of the staircase:

J1-L1, 41:38

> Those of you that found that it was increasing by four, can you actually use this figure [of the staircase] and see that it's increasing by four? It's sure. For example, from the second step to the third step, it's increasing by four, increasing by four centimetres. Right? It's also increasing here.[writes on the paper on the blackboard] OK, what do you think is increasing by four centimetres? Now, in the process of going from here to here, what in the world is increasing by four centimetres?

At Site J3, in Lesson 3 when the class was comparing two ways of checking simultaneous equations, the teacher asked the students to go back to the checking of linear equations at the seventh grade, and reflect on the meaning of checking:

J3-03, 21:46

> T: Well, OK, I want to ask you something. What is the meaning of checking when you get the solution of an equation? How should we think here?

And this question led to discussion of the meaning of solution of an equation, and the question "What does it mean when it says it can make up any equation?"

PATTERNS IN NORM DEVELOPMENT

From the data discussed above, there seem to be at least four strategies the teachers used to develop classroom mathematical norms.

Use of Previously-Developed Norms

The teachers re-introduced previously-developed norms into new leaning contexts. A norm is introduced in a particular context, and the students' understanding of it is situated in that context. What the previously-learned norm means in a new learning context needs to be considered, and negotiated. Which previously-learnt norms are to be used in this new context is usually indicated by the teachers. The use of previously-introduced norms seems to be an important opportunity to connect the learning of the previous and present topics, to form a coherent experience of mathematics.

Using Students' Work

The teachers explained a norm by using students' work to exemplify what it means to follow the norm. Since any norm has generality, it can be communicated only by using general terms like *in a mathematical explanation you cannot write what you have not yet shown to be true*. But the teachers in this study almost always talked about norms by using students' work. In addition, the teachers did not use any artificial examples but actual student work.

Making a Comparison

Important classroom mathematical norms emerged during Phases 3 and 4 of the problem solving process. Comparing solutions and opinions seems to be a very important way to develop classroom mathematical norms (cf. McClain & Cobb, 2001).

This comparative discussion among the classroom participants seems to correspond to 'neriage,' which is an instructional strategy common in Japanese elementary schools. Japanese elementary teachers often ask children to present their own ideas or solutions on the blackboard. Then comparing their writing the children discuss what they notice about them. This process of comparative discussion is called 'neriage.' (kneading). Since 'neriage' accompanies comparative discussion, it is not just 'sharing ideas.' This creates an important opportunity to highlight, negotiate, and share classroom mathematical norms with the students.

Being Considerate of Those Students Who Did Not Follow a Norm

The teachers often discussed the fact that a student's work did not follow a norm. When doing so, they took careful measures to reduce psychological and social damage to the student.

DISCUSSION

The present paper has identified four patterns in developing classroom mathematical norms. Developing a norm that functions fully in various contexts cannot be done in just one lesson; it requires a long patient effort. The use of previously-developed norms, therefore, is a necessary part of regular lessons. The use of students' work seems very important. Since a norm is about how to work on mathematics, the use of mathematical work is a natural way to communicate a norm. Also, since students are familiar with their own work, the use of students' work facilitates students' understanding of the norm. Comparison of students' work would also be very helpful for students in developing a clear understanding of the norm as well as their metacognition about their own work. Since pointing out students' violation of a norm may hurt their feelings, being considerate of those students who did not follow the norm seems to be a hallmark of 'competent' teachers.

Studying norms requires an understanding of the relationships between various norms. A classroom in Japanese schools constitutes a community where a teacher and students stay together, negotiate meanings, share common goals, and shape their identities. It forms a "community of practice" (Wenger, 1998). A community generates, maintains, modifies, or eliminates various kinds of patterns called norms, standards, obligations, rules, routines, and the like. Consider a mathematical norm that I identified above, *in mathematics you cannot write what you have not yet shown to be true*. This is consistent with a general social norm: *You cannot write what you have not yet shown to be true*, and more general moral *You should not tell a lie to people*. Mathematical norms seem to be backed or authorized by social norms. This makes mathematical norms appear to be reasonable and not arbitrary rules.

Relations between mathematical norms can be problematic. Norms may cause a dilemma. In fact, the mathematical norms of efficiency and important ideas appear contradictory. On the other hand, in the geometry unit, the mathematical norm of valid writing comes into prominence among the mathematical norms. Which norms to utilize depends on the context in which the participants are situated.

Behind these mathematical norms of diverse ideas, efficiency, important ideas, and valid writing, there seems to be a value system on mathematical activity, which has at least three components, validity value, efficiency value, and creativity value (Figure 5). To keep mathematical activities productive, the delicate balance between these values needs to be considered. If we pursue validity too far, it may spoil creativity, and vice versa. If we pursue efficiency too far, validity might be sacrificed, and so on. Mathematical norms reflect such a delicate balance in the value system. In each practical circumstance, we need to decide which norms would be appropriate to utilize for the meaningful and fruitful development of mathematical activity.

validity

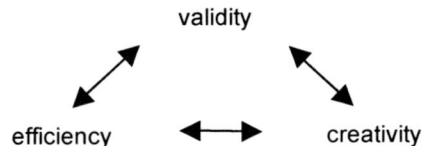

efficiency   creativity

Figure 5. A value system of mathematical activity.

ACKNOWLEDGEMENTS

This research was in part funded by 2004-2006 Grant-in-Aid for Scientific Research (B) of Ministry of Education, Science, Sports and Culture (No. 16300249). I would like to thank Alena Hošpesová and Mary Barnes for their helpful comments on the previous version of this paper. Especially, Mary's detailed suggestions on the organization of the data analysis section helped me improve the clarity of the paper.

REFERENCES

Becker, J. P., Silver, E. A., Kantowski, M. G., Travers, K. J., & Wilson, J. W. (1990). Some observations of mathematics teaching in Japanese elementary and junior high schools. *Arithmetic Teacher, 38*, 12–21.

Bloor, D. (1976). *Knowledge and social imagery*. London: Routledge & Kegan Paul.

Ernest, P. (1994). *Mathematics, education and philosophy: An international perspective*. London: Falmer Press.

Clarke, D. (2004). Learner's perspective study: Developing meaning from complementary accounts of practice. In M. J. Høines & A. B. Fuglestad (Eds.), *Proceedings of 28th Conference of the International Group for the Psychology of Mathematics Education* (Vol. 1, pp. 212-216). Bergen, Norway: Bergen University College.

Cobb, P. (1996). Accounting for mathematical learning in the social context of the classroom. In C. Alsina, J. M. Alvarez, B. Hodgson, C. Laborde, & A. Pérez (Eds.), *8th International Congress on Mathematics Education: Selected lectures* (pp. 85-99). Sevilla: S.A.E.M.'THALES.'

Francis, D., & Hester, S. (2004). *An invitation to ethnomethodology*. London: Sage.

Herbel-Eisenmann, B. (2003). Examining "norms" in mathematics education literature: Refining the lens. Paper presented at *NCTM Research Pre-Session, Symposium: The role of beliefs, values and norms in mathematics classrooms: A conceptualisation of theoretical lenses*. San Antonio, TX.

Herbst, P. G. (1997). Effectiveness of a strategy as a sociomathematical norm. In E. Pehkonen (Ed.), *Proceedings of 21st Conference of the International Group for the Psychology of Mathematics* (Vol 3, pp. 57-64). Helsinki, Finland: University of Helsinki Lahti Research and Training Centre.

Hewitt, J. (1994). *Self and society: A symbolic interactionist social psychology* (6th ed.). Boston: Allyn and Bacon.

Hiebert, J., Gallimore, R., Garnier, H., Givvin, K., Hollingsworth, H., Jacobs, J., Chui, A., Wearne, D., Smith, M., Kersting, N., Manaster, A., Tseng, E., Etterbeek, W., Manaster, C., Gonzales, P., & Stigler, J. (2003). *Teaching mathematics in seven countries: Results from the TIMSS 1999 video study*. Washington, DC: U.S. Department of Education, National Center for Education Statistics.

McClain, K., & Cobb, P. (2001). An analysis of development of sociomathematical norms in one first-grade classroom. *Journal for Research in Mathematics Education, 32*(3), 236-266.

Mehan, H., & Wood, H. (1975). *The reality of ethnomethodology*. Malabar, FL: Robert E. Krieger Publishing.

Stigler, J. W., & Hiebert, J. (1999). *The teaching gap*. New York: Free Press.

Waschescio, U. (1998). The missing link: Social and cultural aspects in social constructivist theories. In F. Seeger, J. Voigt & U. Waschescio (Eds.), *The culture of the mathematics classroom* (pp. 221-241). Cambridge, UK: Cambridge University Press.

Wenger, E. (1998). *Communities of practice*. Cambridge, UK: Cambridge University Press.

Yackel, E., & Cobb, P. (1996). Sociomathematical norms, argumentation, and autonomy in mathematics. *Journal for Research in Mathematics Education, 27*(4), 458-477.

Yasuhiro Sekiguchi
Faculty of Education
Yamaguchi University
Japan

JONAS EMANUELSSON AND FRITJOF SAHLSTRÖM

CHAPTER TWENTY-ONE

Same from the Outside, Different on the Inside:
Swedish Mathematics Classrooms from Students' Points of View

INTRODUCTION

This chapter aims at understanding how students and teachers in two Swedish classrooms constitute learning in interaction, with a particular focus on mathematical and citizenship contents.

'En skola för alla,' 'One school for all,' summarises the ideology behind the Swedish comprehensive education. The Swedish school is characterised by little or no organisational differentiation, but by a curricular expectation of great pedagogical differentiation realised through individualisation. In recent years, the restructuring of education governance has challenged the 'One school for all' ideology, with increased differences between schools (cf. Gustafsson, 2006).

In the Swedish discourse, the comprehensive school, both in terms of content and organisation, has been discussed using the term 'differentiation.' On this concept, there were and are, different opinions (cf. Callewaert & Nilsson, 1974; Dahllöf, 1967, pp. 37-59; Dahllöf, 1999; Husén, 1962; Lindblad, 1994). For our purposes here, it is sufficient, as Lindblad (1994) and Callewaert and Nilsson (1974) do, to distinguish between organisational differentiation, used to represent all kinds of arrangements which systematically separate different students from each other (such as streaming, tracking, ability grouping), and pedagogical differentiation, used to represent the ways actions in the classroom are carried out in different ways for and by different students (such as working at one's own pace, working on different tasks, etc.). In short, the Swedish comprehensive school is characterised by small central regulations on organisational differentiation but by a curricular expectation of great pedagogical differentiation. This pedagogical differentiation is above all discussed in terms of, and expected to be realised through, individualisation.

Individualisation can be characterised, using Husén's (1962, p. 4) description, as "all the more formal or informal procedures taken within the class in order to provide for individual differences." Both the goals of the Swedish comprehensive school (democracy, equity, the development of individuals) and the ways chosen to

D. J. Clarke, C. Keitel & Y. Shimizu (Eds.), Mathematics Classrooms In Twelve Countries: The Insider's Perspective. 307–322.

reach them (student-centred teaching, inductive inquiry, the increase of student work) were bold and ambitious. These reforms have spawned a substantial amount of research, which in more or less explicit terms evaluates the comprehensive reform. Not surprisingly, the general result found in these evaluations, done in relation to the different curricula, is that neither the goals nor the prescribed work methods have been achieved in full.

In one review, Urban Dahllöf (1999, pp. 205-212) argues that the Swedish reforms have been successful in terms of enrolment and system design, but less successful in terms of curriculum structure and teaching. According to Dahllöf, the price of comprehensivisation has been "quite heavy" with respect to a loss of effectiveness due to difficulties in individualising instruction in heterogeneous classes. As Dahllöf concludes, much research has showed that in practice, the pedagogical differentiation from a teaching point of view has been much less than expected (cf. Lindblad, 1994, for a survey in Swedish).

Despite recent changes, there is still a strong orientation to the notion of one school for all in Sweden. The following two citations from the current national curriculum of the Swedish compulsory school exemplify this idea:

> All children and young persons shall have equal access to education and within each type of school receive an education of equivalent value. The school shall provide pupils with knowledge and skills and in partnership with the home, support their harmonious development into responsible persons and members of society. The education shall take into account those pupils with special needs. (Swedish ministry of education and science, 1996, p 9-10).

> However, equivalent education does not mean that the education shall be the same everywhere or that the resources of the school shall be allocated equally. Account shall also be taken of the varying circumstances and needs of pupils as well as the fact that there are a variety of ways of attaining these goals (ibid, p. 11).

Furthermore the national curriculum describes different competencies that the students are supposed to develop. "The main task of the school is to impart knowledge, and together with the home, help pupils develop into responsible persons and members of society" (ibid, p. 14).

For our purposes here, the competences described in the above citation, can be described as being of two different types. One of the types is of course related to the what the students are supposed to learn in the various school subjects (e.g. mathematics, mother tongue, English, handicraft etc.). The other type is of a more general character and concerns competences needed to function as a citizen in society. These more general competencies are supposed to be developed in all school subjects. Among these more general aims we find formulations about learning to work both independently and together with others, and to use knowledge as a tool to reflect over experience and critically examine propositions and value statements and relationships (ibid, p. 17).

In practice, the Swedish education system has not been able to live up to its own expectations: there have always been injustices with respect to social class and ethnicity (cf. Eriksson & Jonsson, 1993), and in the last ten to fifteen years these injustices have grown (Gustafsson, 2006). These injustices have been based entirely on measuring student success in achieving subject-specific goals, thus focusing on only one of the two curricular competencies emphasised in this chapter. In practice, this means that reported findings of unequal distribution of school success in relation to social class and ethnic background rely on subject-specific grades in Year 9 in subjects such as mathematics and Swedish, whereas competencies related to taking part as a citizen in society are almost completely ignored in measuring school outcomes. In fact, these competencies are scarcely measured or evaluated in any systematic way, in either Swedish or international education statistics.

PERSPECTIVES

In this chapter, we will attempt to address both issues of mathematics and issues of citizenship, and to treat 'citizenship' and 'mathematics' as comparable, interactionally constituted contents. Our notion of content is argued within social interaction perspective of learning and socialisation (cf. Nuthall, 1997; Säljö 2000 for comprehensive reviews). In this field, learning is viewed as taking place in-between people, in interaction, rather than in the individual mind. The view on development within this perspective can be exemplified by Rogoff (2003, pp. 52) describing it as "people's changing participation in socio-cultural activities of their communities", and Lave (1993) proposing that learning is a matter of "changing participation in the culturally designed settings of everyday life".

In recent educational research a growing number of researchers have successfully attempted to find ways of working empirically with learning and socialisation within a conversation analysis (CA) framework, arguing that conversation analysis research has an understanding of 'participation' that substantially adds precision and understanding to educational research on learning and socialisation (e.g. Evaldsson, 2005; Liljestrand, 2002; Macbeth, 2004; Martin, 2004; Sahlström, 1999; 2002; Tholander, 2002). Of particular value to educational research on learning and socialisation is the conversation analysis interest in, and methodology for studying, social action from the perspectives of the participants in the interaction.

Conversation analysis has been characterised as a type of discourse analysis that tends to focus on the form rather than the content of interaction, and justified criticism has been directed at this aspect of the perspective. This criticism is correct and motivated with respect to the interests reflected in most of the empirical work within the tradition. However, we believe that there are theoretical openings for analyses of content orientation within the discursive traditions. Thus, we take a slightly different position than the one taken, for example, by Pimm (1994, p 138).

In addition to the analysis of the organisation of interaction, we have a focus on the content and the meanings constituted by the participants. In doing so, we take

309

the participants' points of view as our points of departure (cf. Emanuelsson, 2001; Marton & Tsui, 2004). For a more elaborated argument for this dual focus, and a more elaborated exemplification of its potentials and pitfalls, see Emanuelsson and Sahlström (in press).

Taking the participants' points of view is here a matter of analysing what they orient themselves to and how they possibly understand aspects of the interaction, not how they report their positioning or understandings in the post-lesson interview. These interviews are not analysed here. Hence, learning is in this chapter conceptualised and analysed as changes in patterns of interaction, including the use of intellectual artefacts, that are produced 'in situ' in the classroom, rather than as the outcome of teaching as measured by test or as probed in interviews.

The results reported here, on the differences found within seemingly similar Swedish classes, are products of our on-going work at attempting to expand the somewhat narrow views on content found in many current learning theories (see also Melander and Sahlström (2006). The expansion of content beyond strict school subjects also makes it possible to address issues of justice and equity, as will be made visible in the discussion of our results.

MATERIALS AND METHODS

The chapter compares and contrasts the practices of two different classrooms (below referred to as SW1 and SW2, cf. Lindblad & Marton, 2004 for a discussion of comparisons). We have chosen two classrooms where the mathematical topic taught is similar. In both classroom examples of mathematical representations are worked with and the pupils are expected to interpret and formulate such representations. This similarity makes it possible to do comparisons on a detailed level with respect to differences and similarities in the interactional constitution of content. Further, the analysed episodes are chosen as representative examples of patterns of interaction in the respective classroom, in the sense that they are typical with respect to the two ten-lesson sequences taken as a whole. The selected episodes also represent common teaching formats, plenary teaching and teacher-student desk work.

However, the classes are different with respect to how they score on an international benchmark test, where the first class, called Swedish class one, or SW1, has a considerably lower average score in mathematics than the second studied class, called SW2. As expected, these differences are related to considerable differences with respect to parents' occupations. In SW1, there is a larger number of unemployed and working class parents than in SW2, which is dominated by middle- or upper middle class professions. With respect to ethnic backgrounds, the classes are comparable. The differences found between the classes does not reflect a difference at school level, where in fact both schools are quite similar, but is a result of the SW2 mathematics class being a class chosen by and reserved for the best performing students in their year-group within the school.

PATTERNS OF INTERACTION

Swedish mathematics lessons analysed in other studies, seem to have a fairly stable pattern of interaction over time, and across materials and schools. When analysed at the level of collective activities, Swedish lessons generally turn out to be organised according to three patterns (cf. Sahlström, 1999): Lessons which were given from the teacher's desk from start to finish, lessons which contained both teacher lectures and work in small groups, and lessons where students spent almost all their time working in small groups.

The most common lesson is the mixed pattern. The most frequent activity in classrooms observed is work at the desks, usually in small groups of two or three, but occasionally alone, based on the analysis presented in Sahlström (1999). This crude categorisation is supported by a survey study of Swedish secondary school mathematics teachers. Bentley (2003) finds ten different teaching approaches in teachers' self reports. Five of them relate to whole-class instruction, two to within class grouping and three to different types of individualised instruction.

Lindblad and Sahlström (1999) conducted a study in which they compared recordings from a Swedish eighth-grade class in 1972 with recordings made in the same eighth-grade classroom subjects in 1993. The results of this small scale study indicate that there has been an important shift in the organisation of classroom interaction at this level. The major difference between the older and newer materials was that the dominance of small-group work in the material from the nineties was absent in the material from the seventies, when there were no lessons of the kind we call desk work lessons at all. The mixed lessons of the seventies devoted less time to desk work than the nineties lessons. Lindblad and Sahlström observed a much higher frequency of plenary lessons in the tapes from the 1970s material than in the tapes from the 1990s and the 2000s.

Interestingly, Lindblad and Sahlström (1999) note that mathematics is an exception in their analysis, with no lessons being plenary only, and with a higher frequency of lessons where students spend almost the entire time of the lesson working at their desks. The lessons in mathematics retained their similarity in the 1972-73 and 1993-1995 materials. This is also the case in the 2002 materials, where we find no plenary-only lessons, and a substantial number of lessons dominated by student work.

According to the Swedish National Agency for Education, one model for mathematics teaching dominates in the later years of compulsory school. This model consists of a pattern where the teachers presentations are followed by individual tutoring in the pupils desks. Planned student cooperation and whole class discussions about mathematical problems and solving strategies are considered scarce in Sweden (Skolverket 2003, pp. 17-18).

The episodes analysed below have been chosen, as described above, to be representative for each classroom. Hence we can make only modest claims in relation to representativeness and generalisation. With Eisner (1999) we argue that generalisations of the type that can be made here has a soft character and is related to how our claims as authors are in resonance with the reader's experiences.

'CITIZENSHIP' AND MATHEMATICS CONSTITUTED IN TEACHER AND STUDENT PLENARY INTERACTION

The contrast found between the kinds of content found in mathematics classrooms can be illustrated using the following two excerpts, from two different mathematics classrooms (referred to below as SW1 and SW2). In the first excerpt, taken from the beginning of a lesson on functions and graphs, the teacher has spent several minutes attempting to get the lesson started. At the time the transcript begins, the students and teacher have spent some time taking off hats, turning down earphone music, sharpening pencils, and arguing about what and what not to do. Many of the students argue that they do not want to spend time on mathematics. This discussion comes to a turning point in the transcript below, where teacher and students seem to reach agreement about spending the lesson on mathematics.

Transcript 1. Negotiating relevance (SW1)[i]

```
 1. Mattias:    we will never use this ever
 3. Mattias:    John do you think we [will use this ever
 4. Dino:                           [I   can't see a thing
 5. Teacher:    YES: ((voice breaking))
 6. Mattias:    for?
 7. Teacher:    [for sure
 8. student:    [OOO: ((mimics T))
 9. Johan:      AAH [mimics T loudly))
10. Mattias:    for wha[t
11. Viktoria:          [I know John I know
12. Teacher:    ((tries to pull the blinds, does not succeed)) yes
               tha- that is the point is that you in the end should
               learn that you know how to read curves that you
               [see in newspapers and stuff ((angles the blinds))
13. Ottilia:    [but [that we know
14. Beata:           [but we know that now
15. Teacher:    no that well tha- what I have seen so far can I say
               that you do not know not at all now ((angles blinds))
16. Viktoria:   ok give me a task then
17. Teacher:    bu- but I can to the contrary eh say that you
               hopefully after a while here will know that (1.0)
               that can I definitely say (1.0)
```

Key to symbols used in transcripts in this chapter:

[ ]   Left and right brackets mark the beginning and end of temporal overlap among utterances.
(.)   A pause of less than 0.2 seconds.
(1.0)   A pause of 1.0 seconds.
°text°   Speech produced more softly than surrounding talk.
>text<   Speech uttered significantly more quickly than surrounding talk.
( )   Empty single parentheses represent untranscribed talk. The talk may be untranscribed because the transcriber cannot hear what was said.
(text)   Plausible speech but difficult to hear.
((text)) Comments and annotations, often non-verbal action.
x   Underscoring indicates stress on a word, syllable, or sound.

At the beginning of the transcript, in line 1, Magnus initiates an argument about the relevance of the topic chosen by the teacher. In this classroom, discussions of relevance in relation to mathematics is a quite common topic. He contests the usefulness of the mathematical content. The teacher John asserts that it is useful in relation to reading curves in newspapers. Beata then argues that they already are able to do this. The teacher denies this, and Viktoria demands to be given the opportunity to show her ability by solving a task. In the excerpt, students display intricate knowledge not only to coordinate themselves in interaction, but also to collectively produce an argument for relevance. In this case, the argument is also in accordance with the curriculum. Thus, they act competently in relation to one of the two types of contents that is emphasised on the national and intended level.

In this school and classroom, the students are not streamed into different groups of ability. Hence, the class is a heterogeneous group in relation to mathematical ability. The class has a mixed ethnical background, with several students having arrived to Sweden only recently. The social background is also mixed. On the International Benchmark Test (IBT), used in all classes in the Learner's Perspective Study (cf. Chapter Two, this volume), the mean score for the class is 26, with a standard deviation of 7.7 (max: 50). In short: a classroom for all, in line with 'One school for all' (and with the problems of individualisation following from that).

At the time of recording, the class is working with mathematical relationships described as linear functions in one variable. The mathematical focus of the class is on arriving at the correct solution and an understanding of the rationale for this solution. The citizenship focus of the class is on argumentation, participation and involvement. During the fifteen documented lessons in this class, all instructional formats described above are present. The most typical in this classroom is the variation in instructional formats both within and between lessons. Instead of streaming, individualisation is handled by allowing students to work at different paces through the textbook. Hence, a wide range of tasks is worked on at the same time in the classroom.

MATHEMATICS ONLY IN TEACHER AND STUDENT INTERACTION

As a contrast to the above interaction on the relevance of the mathematical topic, in the first classroom, we offer a transcript from the other school. This excerpt illustrates how the teacher invites the students to discuss the issue of what to use equations for. In a funnel-like manner (cf. Wood, 1994; Bauersfeld, 1988) the initial, and highly relevant and interesting, problem of utility is collapsed to a matter of the lexical meaning of the term equation.

Transcript 2. Mathematics only (SW2).

1. Teacher: I see on the roll here that all were not here on
              Saturday one has other- (.) other things to do so
              that I think that it is maybe needed to repeat this
              with equation not from the beginning but what it is
              about and what are to do with them and stuff .hh yes
              we can take that question first (.) what are we to do
              with this equation solving really in maths (4.0) have

313

```
                    you thought of that is it just to (.) get an equation
                    and then I should solve it and then (.) that is it
                    (3.0) .hh yes that is one part out of the whole but
                    the most important is to know to use these as one
                    part of of maths to solve (1.0) problem (.) and we
                    have a few like that (.) that I think are quite
                    tricky stuff right at the beginning of the chapter
                    (1.0) ptt. on page 102 but if we start from the
                    beginning then ((writes "equation" on board)) quite
                    short I will be (1.0) what does the expressions stand
                    for (5.0) Ken
2. Student:         Similarity
3. Teacher:         yes.
```

In this school, the students are streamed into different groups of ability. The current group is the top group out of four. Their mean score on the IBT-test was 39, with a standard deviation of 6.0. The class is a relatively homogenous group in relation to mathematical ability. The mathematical focus of the studied lesson is on problem solving with equations, and the formulation of linear equations in one unknown that represents word problems.

The general pattern of instruction in this classroom can be described as a cyclic process. While cruising the classroom the teacher monitors the students' work seemingly seeking for, from the students' point of view, critical problems where many seem to have got stuck or having other problems. When such problems are identified exemplary solution are made public either by the teacher or by a student who is invited to present his or her solution in the front of the class. After a discussion on the solution the teacher continues his walk around the classroom now checking how the solution is understood. After a while the monitoring turns to focusing on new problematic problems.

Students are being kept together in the sense that they are scattered over very few problems in the textbook. Students who have missed a class are instructed to catch up and are given a note on what problems they worked in previous lesson(s). In this way plenary presentations are relevant with respect to the particular concrete problems most students currently are working on. The general pattern of instruction fits well with the functioning of the textbook. The used way of making solutions to problems public requires that the students work within a narrow span of problems. Furthermore the excerpt illustrates a prototypical Initiation-Response-Evaluation (IRE) pattern (cf. Bellack, Kliebard, Hyman, & Smith, 1966; Mehan, 1979; Wells, 1994), where the teacher initiates interaction, one pupil responds and the response is followed by a positive evaluative statement from the teacher. Also notable are the relative lengths of the turns. The first initiating turn is very long compared to the one word response and evaluation.

In this second school, the topical relevance is initiated by the teacher, and eventually collapsed to lexical meaning in a funnel-pattern of interaction. The relevance of mathematics does not seem to become an issue in this classroom. Not for the students, since they do not pick up the proposed discussion, nor for the teacher, since he does not pursue the problem further. The topic of mathematical relevance is hence not constituted as relevant. This is in contrast to the first classroom discussed, where the discussion of topical relevance is pupil-initiated and treated 'as taken for granted relevance' that will be obvious for the pupils later 'when they are more competent'. The demand for a discussion on relevance is taken seriously by the teacher, and other students pick up the student Mattias's initiation. Hence relevance is constituted as something that can be, and indeed is, argued about.

DESK WORK INTERACTION AND THE CONSTITUTION OF MATHEMATICS AS CLASSROOM ISSUE ONLY

The following excerpt is taken from a lesson on algebra in school one, where the students are working on tasks from their text books. Prior to the excerpt, the student Faro has tried several times to get the attention of the teacher, and has succeeded in doing so by means of shouting out remarks about immigrants (of which he is one himself), and by eating candy in class. This was followed by the teacher leading Faro to the rubbish bin to throw away the candy. As they reach Faro's desk, the teacher comments to Faro that he is a Turk, provoking the response from Faro that he is Kurdish. Having settled the preliminaries (having spent a long time in doing so) Faro initiates interaction by beginning to read out loud the problematic task, interrupted by more joking from the teacher, followed by Faro's reading (lines 1-12). Faro's displayed understanding of the relationship of functions and graphs is problematic: he points at one graph for the "hundred and twenty five" and at another for "three x" from the function $125 + 3x$, as shown in the transcript below.

Transcript 3. Turk (SW1).
```
1. Faro:      (damn) Irishman (1.5) thirtyseven
2. Teacher:   turk. ((nods at Faro)) that was [a good one
3. Faro:                                       [I   am Kurdish
4. Teacher:   th- yes I   know
5. Faro:      hundredandtwentyfive that is this one isn't it
6. Teacher:   mm (.) °turk°
7. Faro:      but three x (1.5) is that [this bastard
8. Teacher:                             [wh- what
9.            (2.0)
```

Faro's way for dealing with this problem is to commence reading the task aloud from the book, while at the same time pointing to the function $125 + 3x$ with his pencil.

Transcript 4. Lines showing relationship (SW1).
```
10. Faro:     check: (.) A (.) [wh-] (2.0) which of the lines show
              relation
11. Teacher:                  [aa:]
12. Teacher:  yes ri[ght
```

Having finished reading, he holds the teacher accountable for having argued that one can see the line only by looking at the slope coefficient; they point out the varying slopes of the three curves. In line 5 the teacher introduces "125" in a way that aligns with a presumed understanding of Faro's, arguing that "they" (i.e.

315

somebody else, not me, not you) have "thrown in a thing in front" (using colloquial terms instead of mathematics).

Transcript 5. A thing in front (SW1).

```
12. Teacher:    yes ri[ght
13. Faro:            [yo- you said you said that one could see eh
                the lin-
15. Faro:       cur[ve ] of only [that there
14. Teacher:       [yes]        [yes this here yes that's it this
                there slopes the most (.) that slopes a little less
                (.) and this slopes the least
17. Faro:       mm=
18. Teacher:    =bu there they have thrown in a thing in front too
                (.) what does this mean (.) if X is zero (1.0) then
                this will be zero right (.) but then it begins on
                hundredandseventyfive
```

The representations (the graph and the function) are here discussed in terms of their properties. The focus is on how the line slopes, and where it starts. As displayed above, the focus is on issues such as what happens with y when x equals zero. What is made relevant refers to the graphs and the functional expressions in themselves, and not to the outside world that they are supposed to depict. However, the mathematical expressions are not talked about as mathematics. On the contrary, both Faro and the teacher implicitly construct the mathematical vocabulary as problematic, and rely on highly vernacular expressions to 'translate' mathematical content into something else. It is also interesting to note in this transcript the time and effort spent by both the teacher and the student in order to establish a common ground, and how this construction is done in relation to what is perceived as the reasonable amount of time a student can get teacher help.

DESK WORK INTERACTION AND THE CONSTITUTION OF MATHEMATICS AS TOOL FOR REPRESENTING REALITY

In the second school class, we have analysed a similar situation, where the teacher helps a student with a task at the student's desk. The task in question is the following: (519) "A father is four times his son's age. 5 years ago the father was 7 times as old. How old are the father and the son today?"

In contrast to the first example, the representations (the equation and its formulation) are discussed in terms of the phenomenon it represents. Here, the focus is on ages at different points in time, and how 'now' and 'then' ought to be formulated to represent the actual differences in age. What is made relevant is how the phenomenon is described by the equation. The solution and the exact steps to solve it are not made a theme in the interaction.

Transcript 6. Finding a similarity (SW2).

```
1. Teacher:     then you have to find a similarity here (2.5) and
                that is well I mean (4.0) that yes (4.0) those are
                not worth the same it can't be that because he is of
                different age then and now (1.5) and the father the
                same way
```

```
 2. Kajsa:    yes (1.0) that is what messes up everything
 3. Teacher:  >one more time<
 4. Kajsa:    that is what messes up ever[ything
 5. Teacher:                              [yes but-
 6. Kajsa:    cause really this is with y or z because this is a
              entirely different age (.) but the you cannot put it
              into an equation
 7. Teacher:  how can you compens- yees how can you compensate .hh
              eh: well. how old were you (1.0) five years ago?
 8. Kajsa:    yees?
 9. Teacher:  you were (.) as you a- as old as you are today (.)
              minus five (1.5) and that goes for all people
              (3.0)
10. Kajsa:    so if I am x now [then]I was x minus five
11. Teacher:                   [mm- ]
12. Teacher:  mmm? exactly like (2.0) the son (1.0) your father
              (2.5) if it is him this is about
              (10.0)
13. Kajsa:    so if he would be four times older than I
14. Teacher:  now yes
15. Kajsa:    now
16. Teacher:  that he was not then
17. Kajsa:    then he was older than me ((laughter))
18. Teacher:  nowellyes ((laughter)) that he was (.) that's right
              (2.5) seven times (3.0) as old (8.0) this age (.)
              four times his son's age (1.0) except for five if you
              take off five (1.0) then you get sa- equal to what he
              was five years ago (3.0) nineteenhundredninetyeight
              (1.5) and twentythree we can say (3.0) then the
              father was (3.0)
19. Kajsa:    noe
20. Teacher:  no wait now we have to think here so that I (2.5)
              opposite (T erases, 3.0) four x (1.0) and (1.0) I put
              the seven in front (2.0) I can not put equal signs in
              between there ((continues))
```

In the first of these two examples (SW1) no arguments are made by referring to the phenomenon which the representations (graph and function) are supposed to depict. The teacher and Faro have a within-mathematics discussion on how to interpret the two forms of describing a mathematical relationship and how these two forms relate to each other. In the second example (SW2) the representations are rationalised/motivated by referring to real persons and real points in time. Further the first transcript is an example of interaction where it is clear that the teacher takes a didactic position in the way of arguing for a particular and preferred way of seeing the graph. In the second example the interaction is more uncertain in relation to how the equation can be formulated. Both the teacher and Kajsa contribute to the interaction by suggesting and trying different possibilities. Hence the pattern of interaction is an example of negotiation of meaning rather than the teacher arguing for a particular way of formulating the equation. This example is, in some respects close to how people who use mathematics professionally use it as a tool to understand and model reality. Both the ways of approaching problems and the solutions are typically unknown here.

A SUMMARY OF THE MAIN ARGUMENTS

The above results can be summarised in the following table:

Table 1. Summary of results.

| Content aspect | Patterns of interaction and meanings | | | |
| | SW1 | | SW2 | |
	Plenary	*Desk*	*Plenary*	*Desk*
Citizenship	Participating in discussions, co-ordinate speech in groups,	Taking initiative, relying on authority for answers	Limited to being a receiving silent student	Participating in discussion, negotiation of meaning
Topical relevance	The relevance of the chosen topic is something to argue about	The relevance of the chosen topic is taken for granted	The relevance of the chosen topic is taken for granted	The relevance of the chosen topic is taken for granted
Mathematics	Math as a tool to understand math, everyday language	Math as a tool to understand math everyday language	Math as a tool to understand the world, technical language	Math as a tool to understand the world, technical language

In this chapter, we have reported findings that show that interaction in Swedish classrooms, when analysed in detail, show substantial differences between classrooms in different schools, despite the ideology of 'One school for all', despite the similarity of content expectations and despite the similarity of overall interaction formats. In our analysis, these differences concern the following two things:

Interaction work used to construct the classroom content. We have found substantial differences in how teachers and students together constitute teaching, in particular with respect to how student participation is organised, and how this in turn has consequences for how classroom content is construed. In our materials, active student participation seems to lead to a more varied distribution of classroom content, where both mathematics and issues of citizenship are present in the same lessons, whereas less active student participation seems to lead to a less varied distribution of classroom content, with a focus on mathematics only.

The character of the mathematics constituted. We have found substantial differences with respect to how mathematics is understood as relating to the outside world. In one classroom, mathematics is construed as referring to itself, and used as a tool to solve problems in the textbook, and representations are discussed in terms of their internal properties. In the other classroom, mathematics is construed as referring to phenomena outside mathematics. Representations are

318

discussed in terms of the phenomenon they represent, and consequently are used to model a world outside mathematics.

These empirical differences align to the different competencies as described in the intended national curriculum and they are present in different ways in different parts of the materials. Content-wise, emphasis is put on different things in these two classrooms. In the first one much work is done in relation to citizenship, in the other mathematical content is in focus.

Neither our research, nor this particular text have evaluative purposes, and one has to remind oneself that we have analysed only a selection of episodes from only two lessons. However, one can say that in terms of learning mathematics, SW2 seems to produce, at least at the time of recording, better ways of doing so. However, the context for doing so is quite different from SW1, in particular in terms of ability tracking of the students. In terms of learning how to become a critical and engaged citizen, with experience of a variety of shared social action, SW1 seems to produce better ways of doing so, in particular with reference to allowing for student participation and influence.

We have previously shown (Emanuelsson & Sahlström, in press) that SW1 pays 'the price of participation'. That is, interactional patterns that involve extensive participation from students tend to be washed out with respect to a clear presentation of the mathematical content. (cf. Pimm, 1994: "teaching gambit") However; in the case of SW1 extensive student participation instead seems related to an emphasis on the more general competencies of the curriculum. To continue this economical metaphor, the paid price leads to an investment in citizenship, which to us seems to be a highly relevant one.

The relevance of this finding is further highlighted by the way it is related to the social class differences between the two groups studied, where SW1 clearly is less privileged than SW2. As mentioned, with respect to mathematical content, the studied interaction also seems to favour SW2. However, the positions of favoured and unfavoured are more or less swapped when it comes to the way the students have a chance to learn about citizenship competencies. Here, SW1 clearly is the more favoured class. This finding, exploratory and based on a single case, suggests that research in and on school success could benefit from including in its interests issues of citizenship. It might be the case that the losers of today's ranking system could be the winners of tomorrow's – as are the children in the more heterogenous SW1 in this case.

NOTES

i. The transcripts in this chapter are done with a slightly simplified version of the notation developed by Gail Jefferson and her colleagues, (cf. Atkinson & Heritage, 1984). On the internet, the transcript system is found and discussed with examples at several sites, for example the transcription module found at http://www.sscnet.ucla.edu/soc/faculty/schegloff/. The translation in the transcript is as close as possible to the Swedish original, with style changes only when absolutely necessary for understanding. The original Swedish transcripts can be obtained from the chapter authors.

ACKNOWLEDGMENTS

We would like to thank sincerely the teachers and pupils who opened their classrooms for us and made this study possible. We also want to thank especially our Swedish but also our international colleagues in the Learner's Perspective Study for valuable support and comments during our work. The Bank of Sweden Tercentenary Foundation and the Swedish Research Council funded the study.

REFERENCES

Atkinson, P., & Heritage, J. (Eds.). (1984). *Structures of social action: Studies in conversation analysis.* Cambridge: Cambridge University Press.

Bauersfeld, H. (1988). Interaction, construction and knowledge: Alternative perspectives for mathematics education. In D. Grouws & T. Cooney (Eds.), *Perspectives on research on effective mathematics teaching, Vol. 1.* Reston, VA: NCTM, Lawrence Erlbaum.

Bellack, A. A., Kliebard, H. M., Hyman, R. T., & Smith F. L. (1966). *The language of the classroom.* New York: Teachers College Press.

Bentley, P-O. (2003). *Mathematics teachers and their teaching: A survey study.* Göteborg: Acta Universitatis Gothoburgensis.

Callewaert, S., & Nilsson, B. A. (1974). *Samhället, skolan och skolans inre arbete* [Society, schooling and the inner workings of the school]. Lund: Lunds bok och tidskrift AB.

Dahllöf, U. (1967). *Skoldifferentiering och undervisningsförlopp* [School differentiation and teaching cycles]. Stockholm: Almqvist och Wiksell.

Dahllöf, U. (1999). Changes within the Swedish school system and their effects. In A. Leschinsky, & K. U. Mayer (Eds.), *The comprehensive school experiment revisited: Evidence from Western Europe* (pp. 176-214). Frankfurt am Main: Verlag Peter Lang.

Eisner, E. (1991). *The enlightened eye: Qualitative inquiry and the enhancement of educational practice.* New York: McMillan

Emanuelsson, J., & Sahlström, F. (in press). The price of participation – How interaction constrains and affords classroom learning of mathematics. *Scandinavian Journal of Educational Research.*

Emanuelsson, J. (2002). *En fråga om frågor. Hur lärares frågor i klassrummet gör det möjligt att få reda på elevernas sätt att förstå det som undervisningen behandlar i matematik och naturvetenskap* [A question about questions. How teachers' questioning makes it possible to learn about the students' ways of understanding the content taught in mathematics and science. In Swedish with summary in English]. Göteborg Studies in Educational Sciences 168. Göteborg: Acta universitatis Gothoburgensis.

Evaldsson, A-C. (2005). Staging insults and mobilizing categorizations in peergroup interaction. *Discourse & Society*, 16(15).

Gustafsson, J-E. (2006). *Barns utbildningssituation. Ett bidrag till ett kommunalt barnindex.* [Childrens' educational situation. A contribution to the municipal children index] Stockholm: Save the Children Sweden.

Hiebert, J., Gallimore, R., Garnier, H., Givvin, K. B., Hollingsworth, H., Jacobs, J., Chui, A. M., Wearne, D., Smith, M., Kersting, N., Manaster, A., Tseng, E., Etterbeek, W., Manaster, C., Gonzales, P., & Stigler, J. (2003). *Teaching mathematics in seven countries: Results from the TIMSS 1999 Video Study*, NCES (2003-013), U.S. Department of Education. Washington, DC: National Center for Education Statistics.

Husén, T. (1962). *Problems of differentiation in Swedish compulsory schooling.* Stockholm: Svenska Bokförlaget/Norstedts.

Lave, J. (1993). The practice of learning. In S. Chaiklin, & J. Lave (Eds.), *Understanding practice: Perspectives on activity and context* (pp. 3-32). Cambridge: Cambridge University Press.

Liljestrand, J. (2002). *Klassrummet som diskussionsarena* [The classroom as an arena for discussions]. Örebro: Örebro Studies in Education 6.

Lindblad, S. (1994). Skolkarriär och levnadsbana. Bidrag till utredningen om den sociala snedrekryteringen till högskolan [School career and life path. Contribution to the State Commission on socially biased selection to higher education]. In R. Erikson, & J. Jonsson (Eds.), *Skola och sortering. Om snedrekrytering till utbildning och dess konsekvenser* (pp. 172-225). Stockholm: Carlssons Förlag.

Lindblad, S. & Marton, F. (2004). What is compared in comparative studies in mathematics education? In M. J. Høines and A. B. Fugelstad (Eds.), *Proceedings of the 28th Conference of the International Group for the Psychology of Mathematics Education* (Vol 1, pp 201-205). Bergen: Bergen University College.

Lindblad, S. & Sahlström, F. (1999). Ramfaktorteori och klassrumsinteraktion. Gamla mönster och nya gränser. [Frame factor theory and classroom interaction. Old patterns and new borders] *Pedagogisk Forskning i Sverige, 4*(1), 73-92.

Macbeth, D. (2004). The relevance of repair for classroom correction. *Language in Society, 33*, 703-736.

Martin, C. (2004). *From other to self: Learning as interactional change. Dissertation.* Uppsala, Uppsala University.

Marton, F., & Tsui, A. B. M. (Eds.). (2004). *Classroom discourse and the space of learning.* Mahwah, NJ: Lawrence Erlbaum.

Mehan, H. (1979). *Learning lessons. Social organization in the classroom.* Cambridge, MA: Harvard University Press.

Melander, H. M., & Sahlström F. (2006). *Learning about the size of blue whales – tracing changes in participation.* Manuscript submitted for publication.

Nuthall, G. (1997). Understanding student thinking and learning in classrooms. In B. J. Biddle, T. L. Good, & I. F. Goodson (Eds.), *International handbook of teachers and teaching. Vol. II* (pp. 681-768). Boston: Kluwer.

Pimm, D. (1994). Spoken mathematical classroom culture: Artifice and artificiality. In S. Lerman (Ed.) *Cultural perspectives on the mathematics classroom* (pp. 133-148). Dordrecht: Kluwer.

Rogoff, B (2003). *The cultural nature of human development.* Oxford: Oxford University Press.

Sahlström, F. (1999) *Up the hill backwards. On interaction constraints and affordances for equity-constitution in the classrooms of the Swedish comprehensive school.* Uppsala Studies in Education 85. Acta Universitatis Upsaliensis.

Sahlström, F. (2002) The interactional organization of hand raising in classroom interaction. *Journal of Classroom Interaction, 37*(2), 47-57.

Säljö, R. (2000) *Lärande i praktiken* [Learning in practice]. Stockholm: Prisma.

Skolverket (2003). Lusten att lära matematik [The desire to learn mathematics]. Skolverkets rapport 221. Stockholm: Skolverket.

Stigler, J. W. and Hiebert, J. (1999). *The teaching gap: Best ideas from the world's teachers for improving education in the classroom.* New York: Free Press.

Swedish Ministry of Education and Science (1996). Information on the 1994 curriculum for the compulsory school system LPO 94. Stockholm: Swedish Ministry of Education and Science.

Tholander, M. (2002). *Doing morality in school. Teasing, gossip and subteaching as collaborative action.* Linköping: Linköping Studies in Arts and Science.

Wells, G. (1993). Reevaluating the IRF sequence: A proposal for the articulation of theories of activity and discourse for the analysis of teaching and learning in the classroom. *Linguistics and Education, 5*, 1-37.

Wood, T. (1994). Patterns of interaction and the culture of mathematics classrooms. In S. Lerman (Ed.) *Cultural perspectives on the mathematics classroom* (pp. 149-168). Dordrecht: Kluwer.

JONAS EMANUELSSON AND FRITJOF SAHLSTRÖM

Jonas Emanuelsson
Department of Education
Göteborgs Universitet
Sweden

Fritjof Sahlström
Department of Education
Uppsala Universitet
Sweden

APPENDIX A: COUNTRY OVERVIEWS

CHRISTINE KEITEL

INTRODUCTION

*Overviews of the Educational Systems of Partner Countries in the
Learner's Perspective Study:
Diversity in Structure – Commonalities in Intentions*

On the following pages, twelve overviews are presented that report about the educational systems of the twelve countries whose mathematics classrooms are described in the chapters of this book. They aim at providing a comprehensive introduction to each of the national systems by focussing on the structural aspects, including their variations at the different national or federal levels and on the intentions underlying those structures. The overviews provide information not only about the specific national conditions and their materialisations within the respective educational systems and about the associated stakeholders, but also additional insights into each system arising from the authors' positions as insiders and their role within the systems, not only as mathematics educators and researchers, but also as participants in the systems. Their critical appraisal of the specific characteristics of their systems is itself maybe the most important information provided. Although these overviews mainly serve as additional means to understand the data, the analysis and the conclusions drawn by the collaborating researchers in the foregoing chapters, the fact that they are clearly written from an insider's perspective, adds substantial information by pointing to the values attributed by the researchers to these systems. This also provides the opportunity for the authors to share some of the visions they have and the struggles they undertake in their own systems and, even, to make visible their particular appreciation of their systems in comparison to those that they have encountered in other countries.

We decided that the overviews should be written from the researchers' perspectives, without the constraint of prescribed criteria, other than that they should be short, and mainly be designed to achieve the general goal of identifying the most important facets of their system and their concrete and practical realisation at the school and classroom level. As insiders' views, it is not surprising

*D. J. Clarke, C. Keitel & Y. Shimizu (Eds.), Mathematics Classrooms In Twelve Countries: The
Insider's Perspective. 323–325. © 2006 Sense Publishers. All rights reserved.*

that they are necessarily very diverse in style and perspective, and offer not only very differently framed and valued information but also focus differently on what is the necessary amount of information on the national systems. But we believe that this diversity of information selected is a value in itself; it also gives hints and a special form of contextualised connotation by combining a personal view of the authors with a professional view of the researchers as insiders. With the intention in mind to give outsiders the most condensed, most important characteristics of their systems needed to situate their chapters within the systems, the overviews can also provide a partial justification of their conclusions, together with a form of evaluative report about the system. And the overviews vary from very critical notes to very favourable remarks about each system, which serves to situate the researchers in relation to their system and can indicate their stance towards what they might see as necessary changes to that system.

One could argue that the overviews can only represent what the partners see as the most important information about their own system. Since such overviews necessarily represent a rich diversity of systems and classrooms, this has to clearly be mirrored here. Any more rigorous imposed prescription of either content or format would run the risk of silencing the insider's voice, which is such a distinctive and important feature of these overviews.

So we learn from our partner colleagues about the deep concern with the political struggle in South Africa for new approaches to social justice and inclusiveness by educational means and a new curriculum in particular, and the challenges and threats created by designing and implementing it. We learn about the diversity in a country like Israel, being confronted with a very heterogeneous population and conflicting cultural backgrounds, not necessarily visible in the structure of the system, but being played out in the classroom. The overwhelming influence of national exam procedures in many Asian countries such as China is evident in the example of Hong Kong, Shanghai and Macao, and also in Japan, the aim of which is to secure quality, economic competition and to counteract diversity in educational needs. We see structural similarities, but diversity in stratification and social justice in systems focussing on local or federal responsibility policies as apply in Australia, Germany and the USA. Similar structures in Sweden's model of welfare education are still a matter of pride, but can also provide its own problems at the same time. We look at the huge classes in the Philippines and the enormous struggle to cope with the amount and diversity of students under serious financial and organisational restrictions. While, on the other side of the world, the new developments in a former socialist country like the Czech Republic, with a rich and strong tradition and focus on mathematics in schools, might now be in danger from a new policy of economic demands and stronger stratification. And we learn from our colleagues in Singapore about the quality of the system in their country, as demonstrated by especially successful participation in international comparisons, and which they equally attribute to their centrally determined and governed educational policy and structure and to the influence of structural aspects like exams, uniform curricula and textbooks as well.

Such a summary is just a fairly value-laden snapshot. The connection that some authors have constructed implicitly between their overview and their chapter(s) is an important reflection of the situatedness of the research process. As author, one always shapes the readers' views into a wanted direction inspired by subjective feelings about the deficiencies or advantages of one's own or the foreign system. The insider's perspective is as much in evidence in the overviews as it is in the chapters. In the overviews, as in the chapters, we see the diversity of systems and classrooms mirrored, and this clearly demonstrated diversity has its own value that should not be abandoned nor changed by a prescribed uniformity or one-sidedness.

Christine Keitel
Fachbereich Erziehungswissenschaft und Psychologie
Freie Universität Berlin
Germany

GAYE WILLIAMS, CARMEL MESITI AND DAVID CLARKE

AUSTRALIA

Significant Characteristics of the School System and the
Mathematics Curriculum

In Australia, states and territories regulate their own education systems, however, national benchmarks representing minimum standards for Numeracy (in the areas of number sense, measurement and data sense, and spatial sense) help inform the individual state curricula. Australia has three school sectors: Government, Independent, and Catholic. As data collection in the Learner's Perspective Study (LPS) was restricted to Government schools, this overview focuses primarily on the types of schools from which the Learner's Perspective Study (LPS) data from Year 8 mathematics lessons was collected: Victorian government secondary schools.

In Victoria, students spend seven years at the Primary level, Prep to Year 6, and six years at the Secondary level, Year 7 to Year 12 (see Figure 1). In some other Australian states, the first year of Secondary School is Year 8. In Victoria, students are distributed across the Government, Catholic and Independent sectors as follows:
– Primary: Government, 70%, Catholic, 22%, Independent, 8%;
– Secondary: Government, 60%, Catholic, 22%, Independent, 18%.
Independent Schools each have their own governing body but need to comply with government guidelines to gain school registration and be eligible for some government funding. Catholic schools are governed by the Catholic Education Authority and are subject to the same registration constraints as independent schools. Independent and Catholic schools are autonomous in the organisation of the curriculum and to a large extent in its content. Government schools are financed by the state and the curriculum is set through the state curriculum authority. At the time of the LPS study, schools were expected to follow curriculum documents that had just been refined (Board of Studies, 1999; Board of Studies, 2000).

Research into the Middle Years (Years 5-9) of schooling in Victoria and nationally has shown student performance tends to decrease during lower secondary school, and that there is limited emphasis on higher level thinking in middle years classes (Stacey, 1997). In addition, this research has shown that many students are bored with school (Luke, Elkins, Weir, Land, Carrington, Dole, & Prendergast, 2003; Siemon & Morony, 2001) or even alienated from school (Glover, Burns, Butler, & Patton, 1998). Just prior to the LPS data collection,

D. J. Clarke, C. Keitel & Y. Shimizu (Eds.), Mathematics Classrooms In Twelve Countries: The
Insider's Perspective. 327–330. © 2006 Sense Publishers. All rights reserved.

initiatives were introduced to increase student connectedness with the school community, including reducing the number of teachers of each class by having the same teacher teach more than one subject to a given class, (for example, a class might have the same teacher for mathematics, science, and information technology).

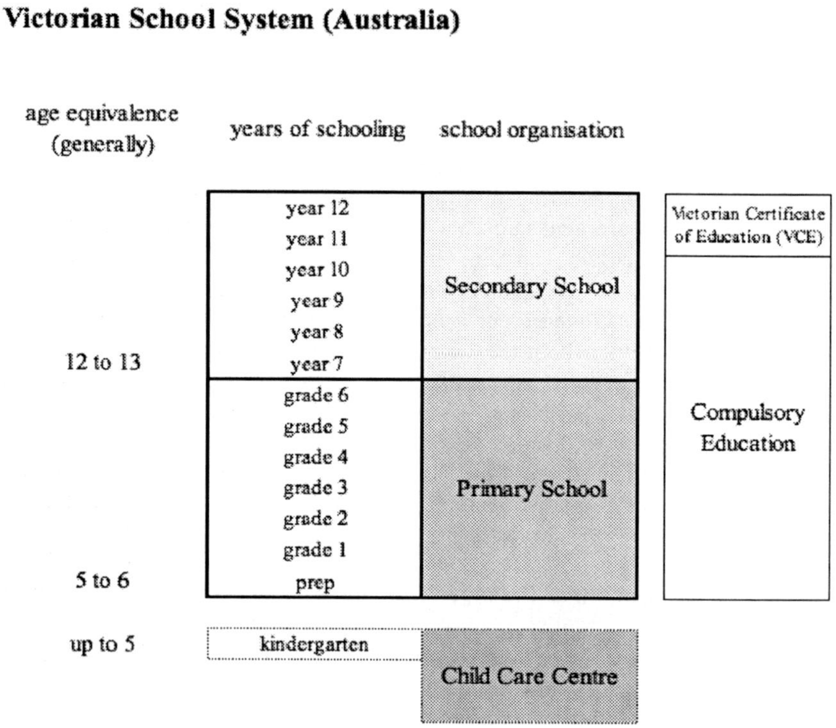

Victorian School System (Australia)

age equivalence
(generally)   years of schooling   school organisation

	year 12		Victorian Certificate of Education (VCE)
	year 11		
	year 10	Secondary School	
	year 9		
	year 8		
12 to 13	year 7		
	grade 6		Compulsory Education
	grade 5		
	grade 4		
	grade 3	Primary School	
	grade 2		
	grade 1		
5 to 6	prep		
up to 5	kindergarten	Child Care Centre	

Figure 1. Victorian School System (Australia).

The three Australian schools were videotaped in 2001 and 2002. At this time the Victorian school curriculum for the compulsory years of schooling (Preparatory to Year Ten) was guided by the *Curriculum and Standards Framework (CSF)*, a document produced by the Victorian Board of Studies. More than 2000 outcomes were expected to be covered across Years 7-10 (Board of Studies, 1994, revised 2000). The CSF framework was structured around eight key learning areas, one of which was Mathematics; the framework assumed that all students would study mathematics throughout their compulsory schooling. The major knowledge and skills within each key learning area were organised into strands. In mathematics these strands were Space, Number, Algebra, Measurement, Chance and data, and Reasoning and strategies. In addition, for each strand the CSF set six levels for student achievement over 11 years of schooling. For example, a mathematics student in Year 8 studying Measurement (substrand Using Relationships) was expected to meet the following learning outcome.

Obtaining areas by counting squares in order to develop new rules for the area of regular shapes (Board of Studies, 2000, p.145).

The framework provided indicators to help determine whether the student had achieved a particular learning outcome. For example, the following is an indicator for the aforementioned learning outcome:

The use of counting of squares to relate base, height and area of parallelograms (Board of Studies, 2000, p145).

The CSF specified that mathematics should be taught from a base of concrete experience, because the manipulation of concrete materials in the early years of schooling provided the foundation upon which children built mathematical ideas. In the CSF structure, mathematical study then progressed to focus on the patterns and relationships between numbers.

National policy developed in the late 1980s by the Australian Association of Mathematics Teachers advocated that calculators be used at all grade levels from Prep to Year 12, and curricular developments in Victoria have continued the progressive integration of calculators into instruction and assessment, including the use of graphing calculators in high-stakes Year 12 mathematics assessment.

To increase the accountability of schools, annual state testing in Years 3, 5, and 7 was intended to provide data for the measurement of state performance and for comparison against the national benchmarks. This testing was compulsory in government schools and optional in non-government schools. As students from all three school sectors were assessed by the same Year 12 high-stakes assessments that determined entry to most tertiary courses, there was commonality in the senior secondary curriculum across all three school sectors (Board of Studies, 1999).

Victoria has provided an example of how assessment can drive curricular change (Barnes, Clarke, & Stephens, 2000). The introduction of the Victorian Certificate of Education (VCE) in 1991 emphasised investigations and problem solving activity by making them a major part of Year 12 assessment. The consequences of this innovation could still be seen in the textbooks in use at the time of the LPS data collection.

REFERENCES

Barnes, M., Clarke, D., & Stephens, M. (2000). Assessment: The engine of systematic curriculum reform. *Journal of Curriculum Studies, 32*(5), 623-650.

Board of Studies. (2000). *Mathematics: Curriculum and standards framework II*. Melbourne, Victoria: Author.

Board of Studies. (1999). Victorian Certificate of Education, Mathematics Study Design. Melbourne, Victoria: Author.

Glover, S., Burns, J., Butler, H., & Patton, G. (1998). Social environments and the emotional well-being of young people. *Family Matters. Australian Institute of Family Studies,* 49, 11-16.

Luke, A., Elkins, J., Weir, K., Land, R., Carrington, V., Dole, S., & Prendergast, D. (2003). *Beyond* the *Middle: A report about literacy and numeracy development of target group students in the middle years of schooling* (Vol. 1). Canberra, ACT: Commonwealth Department of Education Science and Training.

Siemon, D., & Morony, W. (2001). Learning from experience—a framework for considering past, present and future efforts on quality teaching of mathematics for numeracy. *Proceedings of the National Conference of Quality Teaching: Improvement or Transformation* (pp. 17-19). Commonwealth Department of Employment, Training and Youth Affairs as part of Teachers for the 21st Century. Retrieved March 6, 2002 from http://www.dest.gov.au/schools/Publications/2001/teachers/Proceedings_web.pdf

Stacey, K. (1997). Teaching mathematics in Victoria: What does TIMSS show? In D. Clarke, P. Clarkson, D. Gronnet, M. Horne, M. MacKinlay, & A. McDonough (Eds.), *Mathematics - Imagine the possibilities* (pp. 362-371). Melbourne, Victoria: Mathematical Association of Victoria.

Gaye Williams
Faculty of Education
Deakin University
Australia

Carmel Mesiti
International Centre for Classroom Research
Faculty of Education
University of Melbourne
Australia

David Clarke
International Centre for Classroom Research
Faculty of Education
University of Melbourne
Australia

FREDERICK KOON SHING LEUNG, IDA AH CHEE MOK AND
RONGJIN HUANG

CHINA

Characteristics of the School System in China:
Hong Kong, Macau and Shanghai

Hong Kong and Macau are located along the south coast of China and the two cities were both under strong European influence in its history. Hong Kong had been ruled by the British for more than one and a half centuries and was returned to China in 1997. Macau had been ruled by the Portugal government and was returned to China in 1999. The two cities become two Special Administrative Regions (SAR) in the People's Republic of China (P.R.C.) after the handover. Hong Kong has an area of 1,000 km^2 with a population of 6.9 million, 97% of the population is Chinese. Macau has an area of 28km^2 and 465 thousand people, 95% of the population is Chinese. Shanghai is situated on the estuary of Yangtze River in the east coast of China. It is one of the largest cities affiliated with the central government of the People's Republic of China. It has an area of 5800 km^2 and 18.7 million people.

HONG KONG

In Hong Kong, children receive free nine years of compulsory education beginning at the age of six. The system is basically a 6+5+2 system. It includes 6 years of primary and 5 years of secondary schooling, followed by a two-year pre-university phase known as the sixth form. By the end of the secondary schooling, the students sit for a public examination, the Hong Kong Certificate of Education Examination (HKCEE). The students with good examination results can stay on for the sixth-form and take the entrance examinations into the universities in Hong Kong. Most schools use Chinese as the medium of instruction and about 100 schools use English as the medium of instruction.

MACAU

Schooling in Macau is provided by a variety of bodies from three broad categories. According to the Education Survey 2003/4, out of the total of 131 schools, only 20 schools were public/government funded, the other were private schools. Moreover, 88 private schools were integrated in the public school network. Regarding the medium of instruction, 111 schools were in Chinese, 15 schools were in English

D. J. Clarke, C. Keitel & Y. Shimizu (Eds.), Mathematics Classrooms In Twelve Countries: The Insider's Perspective. 331–333. © 2006 Sense Publishers. All rights reserved.

and 5 schools were in Portuguese. In the 1990s, Luso-Chinese schools operated a 6+5+1 system. Chinese-medium and English-medium sections of private schools either followed a 6+3+3 system (borrowed from mainland China or Taiwan) or a modified Hong Kong model of 6+5+1 system. Since there is no public examination in Macau, the schools have to set their own goals. Some cater for the General Certificate of Education examination, while others aim at entrance university examination in Mainland China or Taiwan (Bray & Koo, 2004).

SHANGHAI

In Shanghai, children start their schooling at the age of six. They receive 9 years of compulsory education. There are two different schooling systems. One is called the 5+4+3 system, which includes 5 years of primary, 4 years of junior secondary and 3 years of senior secondary. The other is the 6+3+3 system, which includes 6 years of primary, 3 years of junior secondary and 3 years of senior secondary. Since the middle of the 1980s, there has been no public examination for placing children in primary and junior secondary schools. At the end of the last year of compulsory education, students have to sit a public examination for selection into different types of senior secondary schools. After 3 years of senior secondary, the students sit for a public examination, the Shanghai Certificate of Senior Secondary and an additional public Entrance Examination for the University.

MATHEMATICS CURRICULA

Most schools in Hong Kong adopt the mathematics curriculum developed by the Curriculum Development Institute of the Education Manpower Bureau (Curriculum Development Council, 1999). Although different commercial publishers publish the textbooks in Hong Kong, the content and formats of all textbooks from the different publishers are basically the same.

The mathematics curricula in Macau are diverse. In the pro-communist schools, most of them have adopted the PRC textbooks, supplemented with Hong Kong textbooks. In the English-medium schools, many use Hong Kong textbooks supplemented with the old UK algebra textbooks. The Portuguese-medium schools just followed mathematics curriculum reform in Portugal in the early 1990s. Although the Macau government has developed provisional teaching syllabuses since 1994, the government has had great difficulty in persuading the private schools to adopt this syllabus (Tang, 2004).

Before 1988, there was only one set of mathematics textbooks published by People's Educational Press which followed the single syllabus issued by the Ministry of Education of P.R. China since its foundation in 1949. The policy of "one syllabus, many versions of textbooks" was proposed in 1988 and later developed into "many syllabuses, many versions of textbooks". Further reforms continue to take place after 1997. The reforms are under strong influence of the work by the Committee for Reforming the Curriculum and Textbooks at Primary and Secondary Schools and prominent publication such as "Action Agenda for Mathematics Education at Primary and Secondary Schools in the 21st Century"

(Gu & Zhang, 1997) and the release of "The National Mathematics Curriculum Standard at Compulsory Stage" (China Education Department, 2001).

REFERENCES

Bray, M., & Koo, R. (Eds.). (2004). *Education and society in Hong Kong and Macao: Comparative perspectives on continuity and change,* 2nd edition. Hong Kong: Comparative Education Research Centre, the University of Hong Kong.

China Education Department. (2001). *The national mathematics curriculum standard at compulsory stage* [in Chinese]. Bejing: Beijing Normal University.

Curriculum Development Council, Education Department, Hong Kong. (1999). *Syllabus for secondary schools, mathematics, secondary 1-5.* Hong Kong: CDC, Education Department.

Gu, L., & Zhang, F. (1997). *Action agenda for mathematics education at primary and secondary schools in the 21st century* [in Chinese]. Shanghai: Shanghai Academy of Educational Science.

Tang, K. C. (2004). Secondary school mathematics curricula. In M. Bray & R. Koo (Eds.), *Education and society in Hong Kong and Macao: Comparative perspectives on continuity and change,* 2nd edition (pp. 223-234). Hong Kong: Comparative Education Research Centre, the University of Hong Kong.

Frederick Koon Shing Leung
Faculty of Education,
The University of Hong Kong
Hong Kong SAR, China

Ida Ah Chee Mok
Faculty of Education
University of Hong Kong
Hong Kong SAR
China

Rongjin Huang
Faculty of Education,
University of Macau
China

JARMILA NOVOTNÁ

CZECH REPUBLIC

Czech Educational System

Education has a rich tradition and has always played an important role in Czech history. Here are some of the most important events:
- Charles University was established in 1348 (as the first European University east of Germany);
- population groups, which would not achieve any education in other countries, were often educated (e.g. Hussite women in the 15th century);
- compulsory six-year school attendance was enacted in 1774;
- general literacy in the 1930s was higher than what was common in the rest of Europe.

On the other hand, for more than 40 years there was a strong influence of the Soviet tradition, of which schools were only liberated after 1989.

Education in the Czech Republic has been undergoing development since the political change in 1989. The transition proceeded from demonopolisation to a qualitative diversification of educational opportunities. The difficulties of reforms as well as recommendations for new educational policies and structures were stated in the "Reviews of National Policies for Education" prepared by the Organisation for Economic Co-operation and Development (OECD) in 1996.

The present Czech educational system is schematically described in Figure 1. "*Basic school*" combines into one organisational unit primary and lower secondary levels of education and provides compulsory education. Its length is nine years and it is identical to the length of compulsory schooling. It is divided into a five-year long first stage and a four-year long second stage. Upon completion of the first stage, pupils who show interest and succeed in the admission procedure may transfer to a multi-year grammar school to obtain more academic education.

Upper secondary education is a multi-structured but internally co-ordinated system guaranteeing education and practical vocational training. Secondary schools are divided into four types: Secondary General School (completed with "maturita" - full secondary education), Secondary Technical School (completed with "maturita" - full secondary technical education), Secondary Vocational School (apprentice training, completed with a final apprenticeship education - secondary vocational education), Integrated Secondary technical/Vocational School (provide the same type of education as both technical and vocational schools).

A "maturita" qualification acquired at a Secondary General, Technical or Vocational School is regarded in law as equal "maturita" qualification and entitles the holder to apply to study at any university.

D. J. Clarke, C. Keitel & Y. Shimizu (Eds.), Mathematics Classrooms In Twelve Countries: The Insider's Perspective. 335–338. © *2006 Sense Publishers. All rights reserved.*

Higher vocational schools prepare pupils for demanding, skilled professions. For those secondary school graduates who passed the "maturita", they offer post-secondary vocational education ending with the absolutorium.

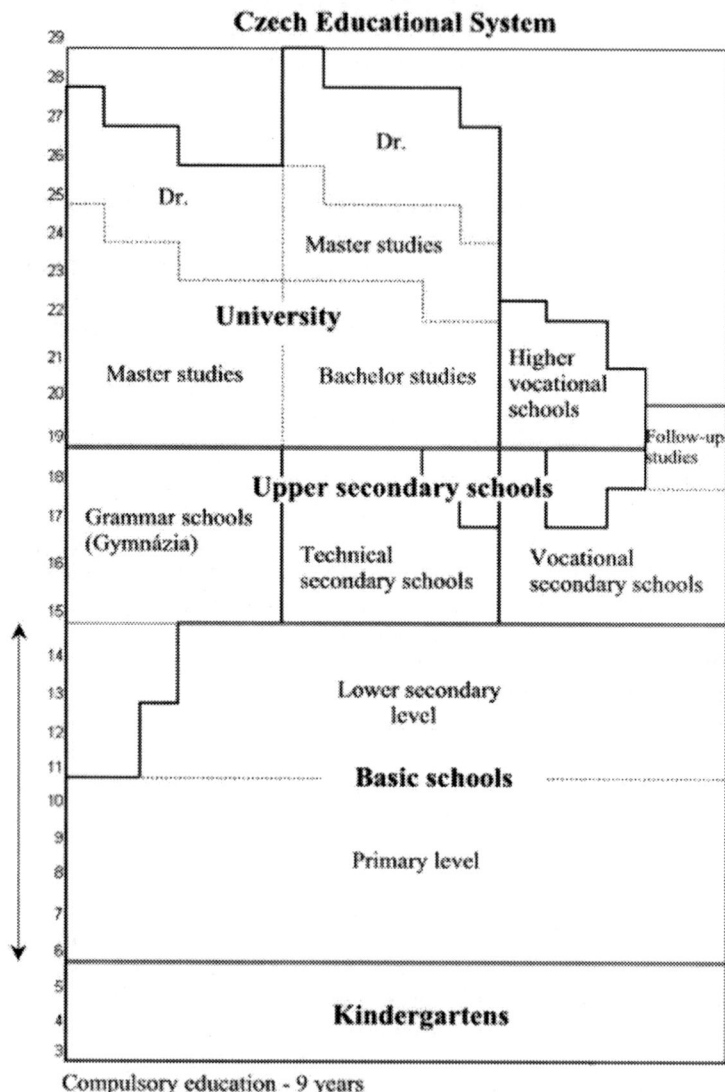

Figure 1. The present Czech educational system

Higher education institutions provide education at three levels of study programmes: Bachelor's, Master's and Doctoral. All the existing higher education institutions had university status until the end of the year 1998 despite the fact that the word "university" does not appear in the name of some technical, economic and agricultural institutions. The possibility of establishing non-university institutions of higher education appeared in 1999, and the number of these institutions is growing nowadays. In order to enter a higher education institution, a pupil must go through an admission procedure, part of which may be (and in general is) an entrance examination. A prerequisite for entering an upper secondary school was successful ending of the "primary school" education (all nine years). A prerequisite for entering tertiary education is education ending with "maturita".

Special schools are designated for children with various health (physical or mental) or social disabilities who cannot be integrated within the mainstream schools. These schools run in parallel to the mainstream schools (special primary schools, special grammar schools, special technical secondary schools, and special vocational secondary schools) - pupils achieve education equal to that achieved at ordinary schools. Children with more serious learning difficulties may go to a special support school, which provides them with basics of primary and lower secondary level education.

Basic school teachers obtain their qualification following Masters' courses (4-year for the first level teachers, 4- to 5-year for the second level teachers) usually at University Faculties of education. Teachers of general education subjects at upper secondary schools obtain their qualifications through five-year Masters' courses at University Faculties of education, philosophy, natural sciences, mathematics/physics, physical education and sport. Teachers of technical subjects in Secondary Technical and Secondary Vocational Schools prepare themselves in the course of Masters' studies at Technical Universities, Agricultural colleges, Faculties of medicine, theology, fine arts etc.

After more than 10 years of rather spontaneous and chaotic changes in the Czech educational system organisation, a new concept of educational policy was accepted in 2001. The key document is the "Rámcový vzdělávací program" (RVP), that is, Educational Programme – Frame of Reference. It stresses the multilateral cultivation of a child's personality, compact cognitive development within the cognitive domain, the domains of competences, attitudes and values. This document highlights, in accordance with the newest trends in education in EU countries, the decentralisation of education, increases the school autonomy and opens space for transformation of teaching/learning process.

An important innovation is that RVP enables different approaches to education in accordance with individual educational needs of learners and anticipates the possibility of choice and also the possible variety of educational approaches, methods and forms. It defines all that is common and imperative in the education. Foreign language teaching and information technologies are a priority.

RVP is a binding basic pedagogical document. It is the common general frame of reference for "Školní vzdělávací program" (SVP), that is, the School Educational Programme. SVP is a document used by schools to realise education. It is the responsibility of the head-teacher to prepare SVP, other teachers participate. SVP may include a specific programme for talented students as well as for students with special needs.

LPS data used in *Chapter 19: Constitution of the Classroom Environment: A Case Study* were gained in the eighth grade (students aged 14) when linear equations were solved. At the time of data collection, the school followed the official educational programme mostly used in Czech schools before the implementation of the new RVP – SVP organisation. To enable the reader to understand this background, the main characteristics of teaching pre-algebra and algebra in this programme is described:.

– Letters (possibly other symbols, e.g. *,) are commonly used in school mathematics already from the 1st grade of elementary school, in the first six grades at the intuitive level (operations with expressions containing letters are not developed; there is a strong support of acquiring algorithms for calculating with numbers). Students start to operate with letters in a systematic way from the seventh's grade.

– The impulse for introducing letters comes from the teacher and is significantly supported by textbooks and students' working materials (working books, working sheets etc.) in other subjects (pre-science, later physics, chemistry, science).

– When solving some types of problems (word problems, calculations of perimeters, areas, volumes, etc., simple constructive tasks in geometry, dependencies), students are asked to follow rules (given by the teacher, textbook etc.) for describing the solving process using letters.

REFERENCES

Education System of the Czech Republic. (1998). Prague, Institute for Information on Education.
Reviews of National Policies for Education (1996). Czech Republic. OECD.
Novotná, J. (1994). Maths teacher training in the Czech Republic. *ATEE*, Prague.
Spilková, V. (2004). RVP v kontextu vnitřní reformy české školy. *Učitelské listy*.www.eurydice.org

Jarmila Novotná
Faculty of Education
Charles University Prague
Czech Republic

CHRISTINE KEITEL AND EVA JABLONKA

GERMANY

The Educational System of Germany

The German educational system is characterised by a co-operative federalism. In the Federal Republic of Germany there are sixteen federal states ("Länder"), to each of which the "Grundgesetz" (Basic Law or Constitution) gives cultural autonomy. This means that regulations for the institutionalisation, administration and control of schools, teacher education as well as the access to schools and to teaching in schools including all decisions on resources and payment, are independently made by the federal states. This has also been applied to the new states that joined the Federal Republic of Germany in 1990. The predominant characteristic of German formal education is the *tripartite school system,* including a *unified primary school* (4 years or 6 years in some states; children enter school between 5 to 7) and the three types of secondary schools: *Hauptschule* (modern or elementary school; formerly *Volksschule*), *Realschule* (middle or technical school), and *Gymnasium* (college-bound higher secondary school). It is possible to obtain a leaving certificate from the *Hauptschule* at the end of grade 9 or 10. Students then can go on to a Part-Time Vocational School (*Berufsschule*) paralleled by an apprenticeship.

SCHOOL TIMETABLE AND THE MATHEMATICS CURRICULUM

Students have an average of 33 hours a week in the classroom. Mathematics is taught 3-4 hours per week. The historically different names given to this school subject – namely *computation and arithmetic* for the lower secondary school and *mathematics* for higher secondary – underlined the differences in quality and quantity of knowledge offered in secondary schooling. In mathematics education not only the amount but also the quality of the knowledge provided depends on the type of school. The underlying assumption is that the ordinary pupil of the lower secondary school not only is unable to acquire more than very restricted knowledge but also does not even need any more for her future vocational and private life. Consequently a highly selective function can be attributed to the school type and to mathematics education, relating the career options to the social status of pupils. This is still valid as most recent results of comparative studies show (PISA 2000, 2003) the chances of students to be admitted to higher secondary (Gymnasium) most strongly depend on social status, as those children from well–educated and socially privileged classes have a four to five times better chance to belong to the

D. J. Clarke, C. Keitel & Y. Shimizu (Eds.), Mathematics Classrooms In Twelve Countries: The Insider's Perspective. 339–343. © *2006 Sense Publishers. All rights reserved.*

20-25% of those attending the Gymnasium. The few comprehensive schools (*Gesamtschule*) introduced in some federal states in the 1970s, still struggle for acknowledgment and never have been accepted as equivalent to the Gymnasium in their quality of schooling.

According to the historical roots of schooling and teacher education in the different states of the Federal Republic of Germany, there is a variety of curricula, teacher education regulations and study programs. The "KMK" (*Kultusministerkonferenz,* Standing Commission of the Ministries of Education) serves to co-ordinate in order to gain mutual acceptance and equivalence. Syllabi of the various schools are formulated on the federal level and have been changed in the late 1970s, introducing modern mathematics, and are actually undergoing new changes to apply the newly formulated educational standards in 2004. In the 1980s teacher education also has been reformed by various, rather contradictory measures in order to improve the quality of instruction and the achievement of pupils. It changed in content and level of requirements in order to decrease the number of teacher applicants and to counteract the *surplus of educated teachers*. This policy now has lead to a remarkable lack of qualified teachers already foreseen for the years after 2015.

MATHEMATICS TEACHER EDUCATION

The strength of the teaching force within a country depends on degrees of professionalisation provided by the teacher training system and the social conditions offered to the profession, the degree of professional autonomy, freedom and self-determination, as well as the social status represented by social recognition, financial resources provided for their work, and their salary. In this respect, German teachers enjoy a rather high professionalisation and prestige. While elementary and lower secondary school teachers are considered as generalists and public servants with disciplinary tasks -- the pedagogues, the mathematics teachers at higher secondary schools (Gymnasium and Realschule) are considered as subject matter specialists and scientists. These teachers have been enjoying a rather high social status for a long time, partly given by the social appreciation of the subject of mathematics, partly by the status as prestigious public servants. Today all teachers are jointly educated at scientific institutions with university-like status with only slightly different duration of their studies, with a similar level of general quality and range of areas of study. All teachers have to pass 5-6 years of pre-service training at the university (first phase of teacher training) and in addition 2 years in the practice-bound seminaries of the second phase of teacher training. This is in comparison to most other countries a rather extensive and long pre-service preparation including two demanding state examinations. After successfully applying for a post in a school of the type she has chosen, her working place is finally determined by the state. The teacher of mathematics (for all school types) enters a profession with considerable autonomy and freedom of decision. Although she has to pay attention to a state-determined syllabus (which differs for the different school types of the tripartite system and for every federal state), these syllabi are a mixture of prescriptions (of certain topics) and recommendations (of methodical approaches, hours for themes, etc.) and function mainly as a guideline for teaching. In 2004, in reacting to the disappointing results of international comparisons and adopting international developments, the KMK has designed general "Educational Standards" for major subject areas including mathematics as recommendations for the federal states. Terms like "competencies" and "mathematics in context" have been introduced as a new terminology. These standards are ot be transformed into new syllabi by federal states.

MATHEMATICS CLASSROOM ENVIRONMENTS

Commercially produced textbooks have to pass state approval. The school teachers choose books most appropriate for their pupils from a selection of approbated textbooks (which in many states in Germany are provided by the school for pupils *on loan)*, or use teaching materials produced by the teachers. The teacher decides on her teaching style, on how to mark homework and oral or written classwork as well as on the annual promotion of the pupils. With the exception of five states that have central state examinations at the end of schooling and therefore more state control on the teaching and assessment of their teachers – the others have so far trusted the ability of their teachers to assess pupils' performance and achievement and to evaluate her teaching by school- and teacher-based assessment modes only. This specific characteristic is also questioned after the results of comparative studies that claim that federal states with state-wide exams at the end of general schooling and prior to university entrance show better achievement results in these tests.

In secondary schooling, teachers' obligations are strongly focussed on the teaching of the subject matter rather than being connected with pedagogical or general educational goals. As they are qualified in two subject areas, they regularly teach two different subject areas. The teaching load is about 24 lessons for higher secondary schools of the *Gymnasium* type up to about 28 lessons for lower secondary schools such as *Gesamtschule, Hauptschule* and primary schools. A teacher who successfully has entered her school service, looks forward to the prospect of a well-paid tenured position as a state employee; her salary is seen as one of the very best in the world and not comparable to that of most colleagues even in Western countries. A broad offering of in-service training and further education on subject matter and pedagogical innovations, for example ICT, is provided by the state for volunteers.

There is a general qualitative and quantitative difference in content matter and teaching quality in the different types of secondary schooling: Teachers of *Hauptschule* or *Gesamtschule* more often are not subject matter specialists in mathematics, but only specialists in other subjects chosen in her pre-service teacher training (German language, social studies, geography etc.). Because of a general lack of qualified mathematics teachers those teachers with a mathematics degree mostly prefer to apply for a higher paid position at the *Gymnasium.*

Although the syllabi for the different school types have been standardised in terms of general content, they differ in the depth and range of content presented, and in the demands on students intellectual capacity and mental activities. Independent and active self-initiated mathematical learning activities like general problem solving, argumentation, proof and proving are only foreseen for Gymnasium types, while for Hauptschule and comprehensive schools the focus is on applied computation and measuring in geometry, some algebra, but not on reasoning or proof.

Students' age when entering schools is between 5 and 7, the retention rate is about 5-7%, which means that they can be between 14 and 16 in grade 8. German students belong to the oldest in grade 8 compared to their counterparts in most other European or Non-European countries. The distribution of the student population in the different secondary school types is different in the various federal states: it ranges from 21% for *Gymnasium* in Bavaria to 35% in Berlin, and from 30% for lower secondary *Hauptschule* in Bavaria to only 11% in Berlin.

In the *Gesamtschule* (G3) in our study, students are placed into two mathematics courses according to their level of achievement. The course for the higher achieving is supposed to be taught according to the syllabus for the *Gymnasium* (like in our G1 and G2). The topics in grade 8 in all types of school comprise algebra (in two variables), equivalent transformations of algebraic terms and formula, linear equations, representation of functions as Cartesian graphs, calculation of percentages and interests. There is not much time devoted to geometry, which is mainly context-bound, including calculation and formula of surface areas and volumes and symmetry. Even though for the *Gesamtschule* it is recommended to start with introductory pre-proof and proving activities, proof and proving are considered as mathematical activities for students at the *Gymnasium* only, however even there it is rarely dealt with.

REFERENCES:

Damerow, P. (1977). *Reform des Mathematikunterrichts in der Bundesrepublik Deutschland. Band 1: Reform der Lehrpläne*. Stuttgart: Klett

Keitel, C. (1980). Entwicklungen im Mathematikunterricht seit 1950. In Max-Planck-Institut für Bildungsforschung und Projektpruppe Bildungsbericht (Eds.), *Bildung in der Bundesrepublik Deutschland. Daten und Analysen. Band 1* (pp. 447-449). Rowohlt und Klett, Reinbek u. Stuttgart,

Keitel, C. (1993). Notes on the unification of the two German states and the integration of two different educational systems. *The Curriculum Journal, 4*(1), *130-135*

Keitel, C. (1998). Mathematics teacher education in Germany: Facts and perspectives. In IMUK (Internationale Mathematische Unterrichtskommission) und ICM (International Congress on Mathematics) (Eds.), *Mathematics Education in Germany. Proceedings of International Congress of Mathematics (ICM 98). Special Issue of ZDM*, Karlsruhe: IMC

Jablonka, E., & Keitel, C. (2005). Funktionale Kompetenz oder mathematische Allgemeinbildung? [Functional competency or mathematical literacy?] In J. Schloemerkemper (Ed.), *Bildung und Standards. Zur Kritik der Instandardsetzung des deutschen Bildungswesens. Die Deutsche Schule*, Special Issue 8, 135-144.

Christine Keitel
Fachbereich Erziehungswissenschaft und Psychologie
Freie Universität Berlin
Germany

Eva Jablonka
Fachbereich Erziehungswissenschaft und Psychologie
Freie Universität Berlin
Germany

MIRIAM AMIT AND MICHAEL N. FRIED

ISRAEL

Mathematics Education in Israel: An Overview

THE ISRAELI EDUCATIONAL SYSTEM IN GENERAL

Israel's population comprises close to seven million citizens of which about two million are students in grades K-12. The mandatory education law requires school attendance for all children between five and fifteen years old, though more than 90% of all students continue school until the age of eighteen. The educational system is a state institution formally headed by an education minister appointed by the prime minister. The greater part of the educational system rests on professionals including ministry staff workers, inspectors, school principals, and teachers. Every school in the country is a state school and is state funded; aside from a few schools belonging to the church, there are no private schools. No tuition is required from students between four and eighteen. School books and other incidental expenses are paid for by parents.

The learning structure is as follows: preschool (ages 4 to 6)—nursery school and kindergarten; primary school (ages 6 to 12)—grades 1 to 6; middle school (ages 12 to 15)—grades 7 to 9; high school (ages 15 to 18)—grades 10 to 12. Teacher training for kindergarten, primary school, and middle school is provided by teacher colleges, which are part of the educational system; teacher training for high school teachers is provided by universities.

For every school subject, from the kindergarten through high school, there is a national curriculum. The curriculum is established by the ministry of education together with experts from academia, and it is binding upon the entire educational system. How the curriculum is implemented—including learning materials, text books, and teaching methods—is chosen by the schools or the local school authorities. A network of inspectors throughout the schools provides support for and supervision of curriculum implementation. At the head of this network, there is, for each school subject, a national superintendent at the ministry of education.

Recently, a national assessment system has been established that rigorously examines, every two years, the achievements of all students from the 4th grade up. This system supplements the previously and still existing national assessment at the conclusion of high school meant to reflect the level of knowledge achieved by students throughout their school career and to open doors to higher academic studies. The national assessment just described is carried out mostly by means of written examinations, national completion examinations known as the *Bagrut*. In practice, the *Bagrut* shapes to a very great degree the way students are taught in

D. J. Clarke, C. Keitel & Y. Shimizu (Eds.), Mathematics Classrooms In Twelve Countries: The Insider's Perspective. 345–348. © 2006 Sense Publishers. All rights reserved.

schools throughout the country, and has a decisive influence in many aspects of students' future academic and vocational life.

The language in schools serving the Jewish population is Hebrew, while that in schools serving the Moslem and Christian population is Arabic; however, the school curriculum and national completion examination in all other respects is identical.

STRUCTURE OF MATHEMATICS EDUCATION IN ISRAEL

Every student in Israel studies mathematics from kindergarten through high school according to a national curriculum set out by a professional committee appointed by the ministry of education. In the primary schools all classrooms are heterogeneous and are taught by teachers who teach mathematics along with other subjects. In recent years, more and more teachers specialising in primary school mathematics are entering the school system, resulting in a higher degree of professionalism in this area. In the middle schools, students are generally tracked into ability groups and are taught by teachers who teach mathematics alone. In the high schools, mathematics is taught at four levels, three of which are geared towards the corresponding *Bagrut* examinations. High school mathematics teachers, by in large, have a university degree in mathematics or in one of the exact sciences, Their pedagogical knowledge is attained through additional studies required for the teaching certificate.

THE ISRAELI MATHEMATICS CURRICULUM

The mathematics curriculum for primary and middle schools is structured according to the year of schooling, whereas the curriculum for the high school is structured according to topic, allowing high school teachers some flexibility as to when they teach what—so long as they arrive at the same point at the end.

Beginning this year, the primary schools will implement a new curriculum (the author was a member of the curriculum committee). The emphasis of the new curriculum is on structures both in arithmetic and in geometry; in doing so, it highlights arithmetical and geometrical insights, mastery of skills, word problem solving, investigative work, acquaintance with and correct use of mathematical terminology, as well as a positive attitude towards mathematics. The topics covered are divided between numbers, operations—including the representation and analysis of numerical data—and geometry.

The emphasis in grades 1-3 is on natural numbers and operations. Fractions and proportions are covered in grades 4-6. In geometry the emphasis is on geometrical figures and solids and the relationships between them. As for data analysis and representation, the emphasis is on projects involving the collection and analysis of real numerical data.

Middle schools (grades 7-9) teach according to a syllabus which has been used for many years; however, teaching methods, resources, and emphases are continually updated. Of the content areas, algebra and geometry are central.

Beginning with the idea of a variable, the students learn algebraic manipulations for solving equations and inequalities, exponents, graphic representations of first and second degree algebraic expressions, and, finally, linear and quadratic functions. Much weight is placed on real-life problems and on investigative skills. Every student possesses a scientific calculator and uses it frequently; graphing calculators are also used, but to a much lesser degree. In geometry a deductive approach is emphasised at levels of rigor or intuition depending on the level of the ability group. Topics studied include: parallel lines, congruence of triangles, quadrilaterals and their properties, and circles. In addition, middle school students learn basic ideas from statistics and probability.

As mentioned above, mathematics in the high school is taught at four levels—three 'academic' levels examined by the full *Bagrut* examinations and meant to prepare students for higher education and one 'basic' level presenting material in an non-rigorous way and emphasising connections with everyday life. The levels are distinguished partly by the topics taught, but chiefly by the degree of depth, the teaching methods, and the abstraction with which the topics are treated, as can be seen in the following diagram referring to the 'academic' levels:

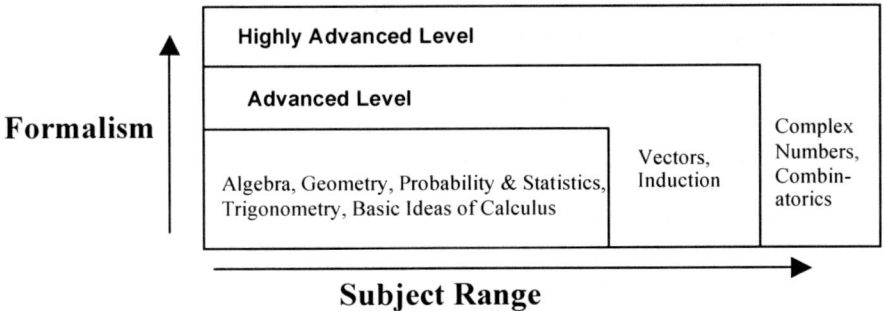

THE 8TH GRADE, UP CLOSE

Since the LPS focuses on the 8th grade, we ought to look a little more closely at this grade level. Children in the 8th grade study mathematics 4-5 hours a week. Their classrooms are in some schools heterogeneous and in others not. Class size is between 32 and 38 students. Teachers for this grade level generally also teach in the 7th and 9th grades, and sometimes 10-12th grades as well. Algebra and geometry are the main content areas in the 8th grade, though twice as much time is dedicated to algebra than to geometry; an option also exists for students to carry out projects using ideas from statistics and probability, but this is not often taken up. Yet, because of the pressure to succeed on examinations set out by the State of Israel or international examinations such as TIMSS and PISA, students generally do carry out 2-3 large investigative tasks.

Algebra topics studied include solutions of equations and inequalities in one variable, and systems of linear equations and inequalities in two variables; both algebraic and geometrical solution strategies are used. Students apply these skills in solving word problems and, to a lesser degree, non-routine problems and some simple mathematical models. Although computer technology is available in the schools, only limited use of it is made in mathematics lessons.

As noted above, a third of eighth grade mathematics lessons are dedicated to geometry. These stress Euclidean geometry and adopt a deductive approach. How formal or intuitive, abstract or concrete, the teaching style is and what degree of mathematical precision is stressed depends on the class. For more challenged students proofs are carried out in a visual and intuitive way, while for more advanced students formal deduction and justification in precise mathematical language is emphasised. Because of the number of new and central ideas in algebra and geometry introduced in the eighth grade, mathematics in this grade level is considered particularly difficult.

Miriam Amit
Program for Science and Technology Education
Ben-Gurion University of the Negev
Israel

Michael N. Fried
Program for Science and Technology Education
Ben-Gurion University of the Negev
Israel

YOSHINORI SHIMIZU

JAPAN

An Overview of the Japanese School System and Mathematics Curriculum

THE JAPANESE SCHOOL SYSTEM

The School System

The school system in Japan is essentially a single-line 6-3-3-4 system: six years of elementary school and three years of lower secondary school, followed by an optional three years of upper secondary school and four years of university or higher education. Other schooling for specific purpose such as two years colleges and special training schools, for example, are also available.

In the Japanese school system, elementary and lower secondary education is compulsory. All the students who have completed elementary school are required to study in lower secondary school for three years until the school year in which they reach the age of 15. Those students who want to learn at upper secondary schools have to take entrance examinations that is prepared and conducted at local levels. In many case, the entrance examinations to upper secondary schools play a role in determining the tracks that students will take according to their achievements. Mathematics is one of the key subjects in those entrance examinations.

National Curriculum Guidelines

The basic guidelines for school curricula to be used nationwide are prescribed in the Course of Study, which is issued by the Ministry of Education, Culture, Sports, Science and Technology (MEXT). The document includes the objectives and contents of each subject.

Specifically, the objective, goal, curricula, number of educational weeks and course subjects at each different stage of school are specified under the School Education Law, and furthermore, the objective and contents of each course are stipulated under the Courses of Study established pursuant to laws and ordinances.

In accordance with the guidelines, each school sets up and implements its own distinctive curricula, taking account the conditions of the local community and school itself, the stages of mental and physical growth and the characters of children, pupils or students, as well as other conditions for students' learning.

D. J. Clarke, C. Keitel & Y. Shimizu (Eds.), Mathematics Classrooms In Twelve Countries: The Insider's Perspective. 349–351. © 2006 Sense Publishers. All rights reserved.

The current Courses of Study were released in 1998 and implemented for elementary and lower secondary schools in April 2002 and for upper secondary schools in April 2003 (MEXT, 1998). The document emphasises the cultivation of "the zest for living," which means the ability to learn and think independently by and for oneself.

Textbooks

The School Education Law stipulates that pupils and students at all elementary, lower and upper secondary schools and schools for the blind, schools for the deaf, and schools for the other disabled children are required to use textbooks. The textbooks to be used must be either those authorised by the Minister of Education, Culture, Sports, Science and Technology, or whose copyright MEXT owns.

Textbooks are written and edited by private sector publishers, according to each author's ingenuity (1), and the Minister approves them in accordance with the Courses of Study and the Standards for Textbook Authorisation through deliberations by the Textbook Authorisation and Research Council (2). The process ensures that descriptions in proposed textbooks are both objective and impartial, and free from errors, and that proper educational considerations are paid attention. Additionally, the results of the previous year's authorisation are publicised at eight locations nationwide.

Local boards of education, the authority to set up local schools, determine which authorised textbooks to be adopted and used at each school in the locality. In order to ensure the wider realisation of the concept of free compulsory education, as stipulated in the Constitution of Japan, textbooks used at national, public and private schools at the compulsory education level are offered to pupils free of charge.

KEY FEATURES OF MATHEMATICS CURRICULUM

Objectives and Contents in Mathematics

The objectives of mathematics is to help students deepen their understanding of fundamental concepts, principles and regularities in numbers, quantities, and geometrical figures and to help them acquire the ability of representing and analysing things mathematically, thinking mathematically, as well as to help them appreciate the mathematical ways of viewing and thinking, and there by to foster their willingness of applying them (JSME, 1990).

The content in mathematics is organised in three major domains: Numbers and Algebraic Expressions, Geometrical Figures, and Mathematical Relations. In grade 8, when the Japanese data was collected, students were supposed to learn such topics like simple algebraic expressions, linear inequalities, simultaneous linear equations in Numbers and Algebraic Expressions domain. In Geometrical Figures domain, a major focus is on formal proofs in plane geometry. Key topics in this domain includes properties of parallel lines, concepts of congruence and similarity, and applying those properties and concepts to various geometrical figures. In Quantitative Relations domain, on the other hand, linear function and introductory part of statistics were the key topics.

MATHEMATICS TEACHERS

The success of formal education depends on the recruitment and placement of well-qualified teachers. To become a teacher it is necessary to acquire a teaching certificate for a particular type of school, such as an elementary school, lower secondary school or upper secondary school. In order to become a teacher at a public school, the individual who has earned a teaching certificate must then pass an examination offered by the local board of education.

After they are employed, teachers go through a broad spectrum of training implemented by their board of education. All newly employed teachers of national and public elementary, lower secondary and special schools undergo one year of induction training to cultivate practical leadership and a sense of mission as a teacher. In addition, a new system was put in place in 2003 under which training addressing topics such as course instruction and student guidance is conducted for all teachers with 10 years of experience according to their individual abilities and aptitudes.

In principle, mathematics teachers at secondary schools teach only mathematics, whereas teachers at elementary schools teach most subjects.

REFERENCES

Japan Society of Mathematical Education. (2000). *Mathematics program in Japan: Elementary, lower secondary & upper secondary schools.* Tokyo: JSME.

Ministry of Education, Culture, Sports, Science and Technology. (1998). *National curriculum standards Reform for kindergarten, elementary school, lower and upper secondary school.* Tokyo: The Ministry

Yoshinori Shimizu
Graduate School of Comprehensive Human Sciences
University of Tsukuba
Japan

KYUNGMEE PARK

KOREA

The Educational System of Korea

Korea has implemented a school system that follows a single track of 6-3-3-4 (6 years in elementary school, 3 years in each of middle school and high school, and 4 years in college or university) according to the Education Law promulgated in 1949. Education in elementary and middle school is compulsory. Korea has almost a perfect school attendance rate in the primary and secondary education levels. The enrolment rate in university and college increased dramatically in the 1980s, and the figure is now over 80%. Table 1 shows some brief statistics about Korean schools in 2005.

Table 1. Brief statistics about Korean schools in 2005

	Elementary school	Middle school	High school
Number of students per teacher	25.1	19.4	15.1
Average class size	31.8	35.3	32.8
Proportion of age group attending public/national school	98.7%	77.5%	55.2%

The minimum number of school days for the completion of one academic year for elementary, middle, and high school is 220 days. The academic year consists of two semesters, the first semester begins in March and ends in August, and the second semester spans from September to February.

CURRICULUM IN GENERAL

Korea has a uniform curriculum determined at the national level. To ensure the quality of education, the Education Law prescribes the curriculum for each school level and the criteria for the development of textbooks and instructional materials. Curricula are revised on a periodic basis to reflect the emerging needs of a changing society and the new frontiers of disciplines. Since the establishment of the Republic of Korea in 1948, there have been seven revisions of the curriculum.

D. J. Clarke, C. Keitel & Y. Shimizu (Eds.), Mathematics Classrooms In Twelve Countries: The Insider's Perspective. 353–356. © *2006 Sense Publishers. All rights reserved.*

The current 7th curriculum was announced in December, 1997. The basic direction of the revision of the 7th curriculum was to change from a 'closed educational system' to an 'open system' and from a 'producer-centred educational system' to a 'consumer- cantered one'. In order to reflect this spirit, the 7th curriculum tried to consider, as much as possible, the demands of the schools, students and parents. This effort was manifested in the form of increasing the number of hours for optional activities on the part of the school. Thus, the number of hours at the school's discretion increased from 0-1 hours a week to 2 hours a week in elementary level, and from 1-2 hours a week to 4 hours a week in middle schools. The school could use this increased time for activities deemed educationally appropriate for their students.

MATHEMATICS CURRICULUM

The 7th mathematics curriculum has brought about more drastic changes both in terms of content and format than any other previous curricula. It aims to foster students' "mathematical power". To pursue this, the 7th mathematics curriculum tends to emphasise the application of mathematics to daily life, problem solving, reasoning and communication. Also the 7th curriculum attempts to divide the mathematics contents into core and optional topics to meet the needs of students with different abilities. In this regard, the 7th mathematics curriculum is called a differentiated curriculum.

Mathematics textbooks are compiled following the curriculum framework set up by the Ministry of Education (MOE). Textbooks in elementary school are published by the MOE itself, while secondary school textbooks are published by commercial publishers authorised by the MOE.

Mathematics in grade 8 consists of 6 content domains; Numbers and Operation, Measurements, Algebra, Patterns and Functions, Probability and Statistics, and Geometric Figures. What follow is the major contents included in grade 8 mathematics.
- rational numbers, repeating decimals, terminating(finite) decimals, nonterminating (infinite) decimals
- approximate values and errors, range of true values, addition and subtraction of approximate values
- computation with monomial and polynomial, laws of exponents,
- systems of linear equations with two unknowns, simultaneous linear inequalities
- linear function and its graph
- properties of probability and computation of probability
- proposition, hypothesis, conclusion, converse, definition, theorem, simple proof regarding basic properties of triangle and rectangle,
- meaning of similarity, conditions for similarity of two triangles, area and volume of similar figures, circumcentre, incentre

TEACHER EDUCATION

Korea has a teacher education system designed and closely controlled by the government through the MOE and the 16 regional boards of education. An important characteristic of teacher education in Korea is that the government plays a critical role in pre-service education, certification, teacher selection, and in-service training. Teacher education is both prescriptive and standardised, and can only be attained at accredited institutes. Further, the government determines the admission quota of every institute on an annual basis.

All elementary school teachers in Korea are trained at one of the 11 national universities of education and one private university. Training for secondary mathematics teacher is relatively open. It is offered at comprehensive universities (both public and private) and can follow three tracks: 1) department of mathematics education in the college of education, 2) B.Sc. with parallel enrolment in teacher certification programs, and 3) graduate schools of education.

In Korea, completing a four-year pre-service teacher education at a college of education does not in itself qualify the graduates to teach in public schools. Completion of the four-year training awards the graduates with a teacher's certificate, which makes them eligible to teach in private schools only. To qualify to teach in public schools, certificate holders are required to pass a very demanding national examination, the Teachers Employment Test (TET).

Kyungmee Park
Department of Mathematics Education
Hongik University
Korea

APPENDIX: KOREA'S CURRICULUM STRUCTURE

	Elementary School						Middle School			High School		
	National Common Basic Curriculum										Elective-Courses Curriculum	
	1	2	3	4	5	6	7	8	9	10	11	12
Korean Language	Korean Language 210 (7)	238 (7)	238 (7)	204 (6)	204 (6)	204 (6)	170 (5)	136 (4)	136 (4)	136 (4)	Elective Subjects	
Moral Education	Disciplined Life 60 (2)	68 (2)	34 (1)	34 (1)	34 (1)	34 (1)	68 (2)	68 (2)	34 (1)	34 (1)		
Social Studies			102 (3)	102 (3)	102 (3)	102 (3)	102 (3)	102 (3)	136 (4)	170 (5)		
Mathematics			136 (4)	136 (4)	136 (4)	136 (4)	136 (4)	136 (4)	102 (3)	136 (4)		
Science	Intelligent Life 90 (3)	102 (3)	102 (3)	102 (3)	102 (3)	102 (3)	102 (3)	136 (4)	136 (4)	102 (3)		
Practical Arts			·	·	68 (2)	68 (2)	68 (2)	102 (3)	102 (3)	102 (3)		
Physical Education			102 (3)	102 (3)	102 (3)	102 (3)	102 (3)	102 (3)	68 (2)	68 (2)		
Music	Mathematics 120 (4)	136 (4)	68 (2)	68 (2)	68 (2)	68 (2)	68 (2)	34 (1)	34 (1)	34 (1)		
Fine Arts	Pleasant Life 180 (6)	204 (6)	68 (2)	68 (2)	68 (2)	68 (2)	34 (1)	34 (1)	68 (2)	34 (1)		
English			34 (1)	34 (1)	68 (2)	68 (2)	102 (3)	102 (3)	136 (4)	136 (4)		
School Discretion	60 (2)	68 (2)	68 (2)	68 (2)	68 (2)	68 (2)	136 (4)	136 (4)	136 (4)	204 (6)		
Extracurricular Activities	30 (1)	34 (1)	34 (1)	68 (2)	68 (2)	68 (2)	68 (2)	68 (2)	68 (2)	68 (2)	136 (4)	
Grand Total	830 (25)	850 (25)	986 (29)	986 (29)	1,088 (32)	1,088 (32)	1,156 (34)	1,156 (34)	1,156 (34)	1,224 (36)	2,448 (36)	

- The number in each cell is the minimum number of total instructional hours by subject and grade level, during a 34 week school year – except in grade 1 where the standard school year is 30 weeks.
- One instructional hour is 40 minutes in elementary school, 45 minutes in middle school and 50 minutes in high school.
- As elective subjects for mathematics, Practical Mathematics, Mathematics I, Mathematics II, Calculus, Probability and Statistics, and Discrete Mathematics are offered.

FLORENDA LOTA GALLOS AND SOLEDAD ASUNCION ULEP

THE PHILIPPINES

The Philippine School System and the
Basic Education Curriculum in Mathematics

OVERVIEW

The formal education system in the Philippines covers at least fourteen years. The Philippines consist of 7107 islands but its educational system is centrally managed from its capital, Manila. All public elementary and secondary schools (K -10) follow the prescribed national curriculum.

The Philippines are divided into three main areas: Luzon, Visayas and Mindanao. These areas are divided politically into 16 regions. Luzon is the major northern island where the National Capital Region (NCR) is found. NCR is the metropolitan district of Manila commonly referred to as Metro Manila and comprises the city of Manila, the capital of the Philippines, and its neighbouring cities and municipalities. The country has a population of around 76.5 million (National Statistical Coordination Board, 2004). There are 11 languages and 87 dialects, with Filipino and English as the official languages.

The Department of Education (DepEd) is headed by the Secretary of Education. Each region of the country is headed by a Regional Director who is responsible for all aspects of public education in that region. For the autonomous region, like the Autonomous Region in Muslim Mindanao, a Secretary heads it. Each region is further subdivided into divisions and a Schools Division Superintendent heads each division. A Principal heads each school.

Primary education is six years (ages 7-12), labelled Grade 1 to Grade 6. The secondary education is four years (ages 13-16), labelled First Year to Fourth Year. A bachelor's degree requires at least four years (ages 17 and over). Primary and secondary education is compulsory and free of fees in government schools. Mathematics, Science and English, at all levels, use the English language as the medium of instruction.

There were about 13 million pupils for the School Year 2002-2003. The time allotted to Mathematics is 80 minutes daily for Grades 1-3 and 60 minutes daily for Grades 4 - 6. The general objective of the elementary mathematics curriculum is for learners to "demonstrate understanding and skills in computing with considerable speed and accuracy, estimating, communicating, thinking analytically and critically, and in solving problems in daily life using appropriate technology" (BEC, 2002, p.2). The Grades 1 and 2 Mathematics include the study of whole numbers and fractions and its operations, basic geometry and measurements. In

D. J. Clarke, C. Keitel & Y. Shimizu (Eds.), Mathematics Classrooms In Twelve Countries: The Insider's Perspective. 357–359. © 2006 Sense Publishers. All rights reserved.

Grades 3 and 4 Mathematics, decimals and graphs are added topics. In Grades 5 and 6, the learners are expected to have skills in dealing with whole numbers, fractions and decimals. Concepts about ratio and proportion, percent, integers, simple probability, measurement, graphs and simple concepts in Algebra are also introduced.

In the School Year 2002-2003, there were about 6 million secondary students. Like in the elementary schools, the students in the public secondary schools are mandated to implement the national curriculum. This curriculum expects students at the end of Fourth Year to be "able to compute and measure accurately, come up with reasonable estimate, gather, analyse and interpret data, visualise abstract mathematical ideas, present alternative solutions to problems using technology, among others, and apply them in real-life situations" (BEC, 2002, p. 2). The mathematics curriculum for high school is compartmentalised. First Year Mathematics is Elementary Algebra. It deals with real number system, algebraic expressions, first-degree equations and inequalities and special products and factors. Second Year Mathematics is Intermediate Algebra. Included are the topics: systems of linear equations and inequalities, quadratic equations, rational algebraic equations, variation, integral exponents, radical expressions and sequences and series. Third Year Mathematics is Geometry. It deals with geometry of shapes and sizes, geometric relations, triangle congruence, properties of quadrilaterals, similarity, circles and plane coordinate geometry. The Fourth Year Mathematics contains topics on Advanced Algebra, Trigonometry and Statistics. The time allotment is 60 minutes daily.

In both levels, Mathematics textbooks and teachers' manuals used in schools are developed locally. DepEd approves drafts of books submitted by publishing companies. Schools are then given the freedom to choose which textbooks to use in their respective schools.

A typical mathematics class has a large class size, on the average about 40 children in primary grades and about 55 children in secondary levels. There are limited resources, with blackboard, chalk, textbook and the teacher's guide as the usual available resources. Learners are generally well behaved and they would just sit and apparently listen to the teacher talking. Learners' progress is usually evaluated based on the written outputs such as periodic tests and quizzes, participation and homework.

National and international assessments of students' learning outcomes revealed that there is much to be done to improve the education system in the country. Thus, the initiatives of DepEd and other sectors of the society for reforms to improve the quality of education in the country are ongoing.

REFERENCES

Bureau of Elementary Education. (2002). *Basic education curriculum. Philippine elementary learning competencies.* Pasig City, Philippines: Author.

Bureau of Secondary Education. (2002). *Operations handbook in mathematics. Basic education curriculum. Secondary level.* Pasig City, Philippines: Author

National Statistical Coordination Board. (2004). *Philippine statistical yearbook.* Makati, Philippines: Author

Florenda Lota Gallos
National Institute for Science and Mathematics Education Development
University of the Philippines
The Philippines

Soledad Asuncion Ulep
National Institute for Science and Mathematics Education Development
University of the Philippines
The Philippines

BERINDERJEET KAUR

SINGAPORE

School System and Mathematics Education in Singapore

SCHOOL SYSTEM

The education system in Singapore is a central one and the Ministry of Education (MOE) directs the formulation and implementation of educational policies. Since January 2003, primary education is compulsory for all children of schooling age. English is the official language of Singapore and the language of instruction in schools. Pupils enter primary school after the age of six years. Primary education is free and consists of a four-year foundation stage from Primary 1 to 4 and a two-year orientation stage from Primary 5 to 6. During the first four years of primary school, that is, from Primary One to Primary Four, emphasis is placed on the learning of English Language, Mathematics and Mother Tongue.

At the end of Primary Four pupils are assessed based on their performance in English Language, Mathematics and Mother Tongue and streamed according to ability. They are placed in a stream that is suited to their pace of learning. Pupils take a national examination called the Primary School Leaving Examination (PSLE) at the end of Primary Six. This examination assesses pupils' suitability for secondary education and places them in appropriate secondary school streams that suit their learning ability. There are four streams at the secondary school level. These are the Special, Express, Normal (Academic) and Normal (Technical) streams. Pupils in the Special and Express streams undergo four years of secondary education and take the General Certificate of Education (GCE) 'Ordinary' Level examination at the end of their schooling. Pupils in the Normal streams take the GCE "Normal" Level examination at the end of their fourth year and may proceed to a fifth year of study in the secondary school. At the end of their fifth year of study they take the GCE 'Ordinary' Level examination. At the end of four or five years of secondary schooling pupils may continue with post-secondary education which is offered in junior colleges (leading to the GCE 'Advanced' Level examinations), polytechnics and technical institutes or enter the workforce. There are three publicly-funded universities in Singapore. Students with GCE 'Advanced' Level qualifications, diplomas from polytechnics or other recognised qualifications may apply for admission to the universities.

D. J. Clarke, C. Keitel & Y. Shimizu (Eds.), Mathematics Classrooms In Twelve Countries: The Insider's Perspective. 361–364. © *2006 Sense Publishers. All rights reserved.*

SCHOOL MATHEMATICS CURRICULUM

Mathematics is a compulsory subject at the primary and secondary levels. The developments from 1946 to 2002 that have shaped the present School Mathematics Curricula in Singapore, are direct consequences of developments in the Education System of Singapore during the same period (Kaur, 2002a). The School mathematics curriculum, at present, emphasises a balance between mastery over basic skills and concepts in Mathematics and the application of higher order thinking skills to solve mathematical problems. It can best be described as one that offers "mathematics for all but more mathematics for some" (Kaur, 2003a) as every child in school does mathematics that is suited to his or her ability.

The conceptualisation of the mathematics syllabus is based on the framework shown in Figure 1 (MOE, 2000a; 2000b). The primary aim of the mathematics curriculum is to enable pupils to develop their ability in mathematical problem solving. Mathematical problem solving includes using and applying mathematics in practical tasks, in real life problems and within mathematics itself. The attainment of this mathematical problem solving ability is dependent on five inter-related components – Concepts, Skills, Processes, Attitudes and Metacognition.

The Ministry of Education issues the mathematics syllabuses for Primary One (Grade 1) to Secondary Two (Grade 8) levels. The Secondary Three and Four syllabuses are issued by the University of Cambridge Local Examination Board. The mathematics topics in the curriculum are arranged in a spiral manner from primary to secondary levels. The pace and scope of coverage in the curriculum is differentiated for pupils from different streams. Every ten years or so, the mathematics syllabuses undergo a periodic review to ensure that they remain relevant so as to prepare pupils for the challenges and opportunities of the future and also to keep in line with the national objectives.

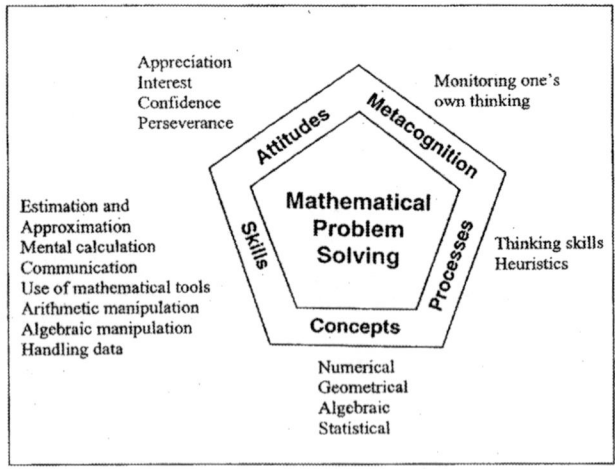

Figure 1: Framework of the School Mathematics Curriculum

Singapore's lead in the Trend in Mathematics and Science Studies (TIMSS) (Kelly, Mullis & Martin, 2000; Mullis et al., 2000; Mullis, Martin, Gonzalez & Chrostowski, 2004) has signalled to both mathematics teachers and educators in Singapore that school mathematics curricula is somewhat right but this does not mean that the process of refining it will come to a halt. A careful study of the TIMSS test data has confirmed that Singapore pupils have indeed performed very well only on items that were routine to them and tested what they had been taught at school (Kaur & Yap, 1999; Pereira-Mendoza, Kaur & Yap 1999; Kaur, 2002b; Kaur, 2003b). This suggests that pupils are not able to draw on content knowledge and explicate processes in novel situations. Hence the present revision of the school mathematics curriculum (Ministry of Education, 2000a; 2000b) and future ones hope to address some of these concerns.

REFERENCES

Kaur, B. (2002a). Singapore's school mathematics curriculum for the 21st century. In J. Abramsky (Ed.), *Reasoning, explanation and proof in school mathematics and their place in the intended curriculum – Proceedings of the QCA International Seminar* (pp. 166-177). London: Qualifications and Curriculum Authority, UK.

Kaur, B. (2002b). TIMSS & TIMSS-R: Performance of eight grade Singaporean students. *Reflections, Journal of The Mathematical Association of New South Wales, Inc. 27*(2), 7-12.

Kaur, B. (2003a). Mathematics for all but more mathematics for some – A look at Singapore's school mathematics curriculum. In B. Clark, R. Cameron, H. Forgasz & W. Seah (Eds.), *Making Mathematicians* (pp. 440-455). Australia: The Mathematical Association of Victoria.

Kaur, B. (2003b). TIMSS-R – Performance of eighth graders from Singapore. *The Mathematics Educator , 7(1)*, 62 - 79.

Kaur, B. & Yap, S. F. (1999). TIMSS – The strength and weaknesses of Singapore's lower secondary pupils performance in mathematics. In M. Waas (Ed.), *Enhancing learning: Challenge of integrating thinking and information technology into the curriculum* (pp. 436-444). Singapore: Educational Research Association.

Kelly, D. L., Mullis, I. V. S., & Martin, M. O. (2000). *Profiles of student achievement in mathematics at the TIMSS International Benchmarks: U.S. performance and standards in an international context.* MA: TIMSS International Study Center, Boston College.

Ministry of Education. (2000a). *Mathematics syllabus – Primary.* Singapore: Curriculum Planning & Development Division, Ministry of Education.

Ministry of Education. (2000b). *Mathematics syllabus – Lower secondary.* Singapore: Curriculum Planning & Development Division, Ministry of Education.

Mullis, I. V. S., Martin, M. O., Gonzalez, E. J., Gregory, K. D., Garden, R. A., O'Connor, K. M., Chrostowski, S. J. & Smith,T. A., (2000). *TIMSS 1999 International mathematics report.* MA: International Study Center, Lynch School of Education, Boston College.

Mullis, I.V. S., Martin, M. O., Gonzalez, E. J., & Chrostowski, S. J. (2004). *TIMSS 2003 International mathematics report.* MA: TIMSS & PIRLS International Study Center, Boston College.

Pereira-Mendoza, L., Kaur, B., & Yap, S. F. (1999). Some implications of the TIMSS data for primary mathematics teachers. In M. Waas (Ed.), *Enhancing learning: Challenge of integrating thinking and information technology into the curriculum* (pp. 452-457). Singapore: Educational Research Association.

Berinderjeet Kaur
National Institute of Education,
Nanyang Technological University, Singapore

JILL ADLER, BUSI GOBA, RENUKA VITHAL AND
GODFREY SETHOLE

SOUTH AFRICA

*A Summary of School Curriculum Development in South Africa and the
Implications for School Mathematics*

Curriculum development in South Africa has been an ongoing process of producing and reflecting on or critiquing documents and then reshaping or rewording them. In this summary we hope to illustrate how this process unfolded and how school mathematics has consistently been repositioned.

FROM THE FIRST CURRICULUM WAVE TO THE SECOND:

The new South African curriculum, Curriculum 2005 (C2005) was introduced as a result of, and in alignment with, the new democratic political set up in the country. "In the past," highlights the Education Policy document, "the curriculum has perpetuated race, class, gender and ethnic divisions and has emphasised separateness, rather than common citizenship and nationhood" (Department of Education (DoE), 1997, p. 1). Within a ten-year period (1995 – 2005), the new curriculum has undergone two major waves of curriculum change. It is not possible to outline, in this short space, the "course and contours of debates" (Jansen, 1999, p. 10) which shaped from one wave of curriculum change to the other. What remains consistent, though, is a commitment to a curriculum which is informed by

(1) outcomes-based education;
(2) human rights and social science and environmental justice;
(3) a high level of skills and knowledge for all;
(4) balance of progression and intention; as well as
(5) clarity and understanding (DoE, 2001, p. 8).

The aim of this curriculum is thus, in sum, to produce a learner who
– possesses the skills and knowledge relevant for each subject (as highlighted by principles 3 and 4); and
– is able to apply knowledge within and across each subject.
Whilst there is a clear indication over the type of learners the curriculum aims to develop, there is no specification on the teaching method to be used for this purpose. Leaving teaching methodologies to teachers for a new education system most are unfamiliar with has led some critiques to view C2005 as a "destination

D. J. Clarke, C. Keitel & Y. Shimizu (Eds.), Mathematics Classrooms In Twelve Countries: The Insider's Perspective. 365–367. © 2006 Sense Publishers. All rights reserved.

without a map" (Potenza & Monyokolo, 1999). It is within this context of fluidity that C2005 was to be implemented.

CHANGING THE SCHOOL MATHEMATICS CURRICULUM

Changes resulting from the new curriculum

C2005 has, in two ways, increased the scope of what school mathematics entails. It particularly recognises that:
- Mathematical ideas and concepts build on one another to create a coherent whole. The use of the term Learning Area instead of subject is intended to convey the envisaged interrelatedness between different aspects of mathematics (e.g. Mathematical Literacy, Statistics etc.);
- The teaching of mathematics should incorporate social contexts relevant for learners.

At a pedagogical level, C2005 is also non-prescriptive in relation to how mathematics should be taught. Instead, there is an implied suggestion that whilst competency in subject matter is essential, teachers should be 'open to opinions held by the learners which may differ from his/her own" (DoE, 2001, p. 6). Because the underlying philosophy of the new curriculum is outcomes-based, there is an underspecification of content but an explication of the product (the kind of learner envisaged).

Increasing the scope of what school mathematics entails can be seen as a response to concerns raised long before 1994 elections over the "very formal and highly abstract" and "decontextualised" (Adler cited in Christie, 1991, p. 287) school mathematics curriculum. Such a school mathematics programme concealed both the relationship between mathematics and other subjects and mathematics and everyday contexts. Yet, there are others who contend that increasing the scope of what school mathematics entails will produce a "generally educated man" and achieve a condition of false equality between different domains and different participants (Muller & Taylor, 1985, p. 235).

Changes resulting from debates

Seemingly, in an attempt to accommodate and create space for the formal aspects of mathematics and its use within the everyday, school mathematics has been split into two Learning areas. One Learning area, Mathematics is meant is more formal and is meant for learners who plan to proceed with mathematics up to university level. Another Learning Area, Mathematical Literacy, has been introduced (for the last three years of schooling) for learners who choose not to take mathematics. Mathematical Literacy is described as "a subject driven by life-related applications of mathematics". (DoE, 2005, p. 7). This, perhaps, is an acknowledgement on the difficulties of having one mathematics programme which satisfactorily speaks to the abstract and specialised nature of mathematics and the use of mathematics in real life contexts.

REFERENCES

Christie, P. (1991). *The right to learn: The struggle for education in South Africa.* Cape Town: Ravan Press

Department of Education. (1997). *Senior Phase (Grades 7 to 9) policy document.* South Africa: Author

Department of Education. (2001). *Revised National Curriculum Statement: Learning Area Statement for Mathematics.* South Africa: Author

Department of Education. (2005). *National Curriculum Statement: Grades 10 – 12, Mathematical Literacy.* South Africa: Author

Jansen, J. (1999). Why outcomes-based education will fail: An elaboration. In J. Jansen & P. Christie (Eds.), *Changing Curriculum – Studies on Outcomes-based Education in South Africa* (pp. 145-154). Kenwyn: Juta & Co.

Muller, J., & Taylor, N. (1985). Schooling and everyday life: Knowledges sacred and profane. *Social Epistemology, 9* (3), 257 – 275.

Potenza, E., & Monyokolo, M., (1999). A destination without a map: Premature implementation of curriculum 2005. In J. Jansen & P. Christie (Eds.), *Changing curriculum – Studies on outcomes-based education in South Africa* (pp. 231-246). Kenwyn: Juta & Co.

Jill Adler
Faculty of Humanities
University of the Witwatersrand
South Africa

Busi Goba
Faculty of Education
University of KwaZulu-Natal
South Africa

Renuka Vithal
Faculty of Education
University of KwaZulu-Natal
South Africa

Godfrey Sethole
Faculty of Education
Tshwane University of Technology
South Africa

367

SVERKER LINDBLAD

SWEDEN

The Education System and Mathematics Education in Sweden

THE SWEDISH MODEL OF WELFARE STATE EDUCATION

A way to introduce the Swedish case is to relate it to predominant discourses on a welfare State in transition. In such discourses from the 1940s to the 1970s, Swedish welfare State politics is characterised by a combination of equity strivings and centralised State governance. A breaking up from the earlier forms of political governance has taken place from the 1980s, in terms of decentralisation and deregulation. Privatisation and market mechanisms have been introduced, at the same time as the State has parted with key economic governance tools in relation to detailed control of the municipalities. The 1990s is considered to be a period of extensive discursive change in Sweden. The continuous restructuring of Swedish economy, characterised by an expansion of knowledge intensive production, a further growth of the service sector and a rise of the general education level, was intensified in this period. The economic crisis, reaching its climax in 1992-1994, resulted in very high reductions in the number of jobs and in an historical rise of unemployment. Additionally, during the 1990s Sweden witnessed substantial demographic changes, above all increasing numbers of immigrants and refugees seeking asylum, a further concentration of the population to a few areas while a big part of the municipalities were depopulated, and an ageing population.

In order to understand the changing meaning of education in Sweden we need to consider how basic aspects and education ideas have transformed from the time when the classic model of progressive welfare state education was at its peak (1970) to the present restructured education system (2005). We can consider that in relation to social exclusion in Table 1.

D. J. Clarke, C. Keitel & Y. Shimizu (Eds.), Mathematics Classrooms In Twelve Countries: The Insider's Perspective. 369–372. © *2006 Sense Publishers. All rights reserved.*

Table 1: *Swedish welfare state education in 1970 and 2005: basic aspects and ideas in relation to social exclusion.*

Aspects and ideas	1970	2005
Expanding:	Credentialising opportunity	Normalising necessity
Education for all	*Exclusion by means of missing access*	*Exclusion by means of no attainment*
Governing:	Government	Governance
Planning and agency	*Exclusion by means of policy-made restrictions*	*Exclusion by means of students' lack of agency*
Teaching process:	Teachers teaching and students listening	Students' studenting and teachers helping
Weaker framing and classification	*Exclusion by means of teachers work or content offers*	*Exclusion by means of students' self-selection*

In sum we are seeing a progressive education system in transition. Education expanded in the name of Democracy and Education for All. We saw the birth of the comprehensive school and the opening up of secondary education and later on higher education. Still in 1970 secondary education and higher education was still a privilege in the Nordic countries. To pass the Gymnasium exam meant that the world was more or less open to you. In a way we in the Nordic countries had implemented the idea of education for all when we came to the turn of the century, since, for example, in Sweden, 90 percent of a cohort applied for upper secondary education and 98 (!) percent did actually enter this education. There is a dark side here, however: If you did not pass the secondary education exam at the end of the millennium the risk for social exclusion turned out to be very high. Education turned to a normalising system.

Considering governing: Dewey and Childs (1933) talk about "...a society which is not so much planned as planning". I guess this is at the core of current restructuring discourses on education. A main point is that you are leaving a governing from behind, by means of directives, to a governing by means of goals and results. That is you are going from a planned system of education towards a planning system. Education restructuring can be regarded as such a change from government to (new) governance. Such a shift in governing implies changes in school management and steering. This includes greater use of private sector management practices and explicit and measurable standards of performance.

In this text curriculum codes and teaching are of vital importance: Bernstein (1996) deals with issues of power and knowledge in a way that strives to integrate different layers in a 'micro-macro dimension'. His theory on framing and classification and curriculum codes is a well known example of a way to conceptualise changes and alternatives in pedagogical discourses. Classification refers to relations between categories. The concept attempts to measure the degree of insulation between different discourses, specialisation, academic disciplines and so on. Framing refers to control of communication in pedagogical discourses, that is selection, pace and sequence as well as criteria for performances. In Sweden and in many other pats of 'Little Europe', there has been strong education policy ambitions to change the curriculum code, to weaken classification as well as framing. In relation to Bernstein's concept of framing we argue that in Sweden framing is weaker in 2005 compared to 1970, since classroom interaction is not that dominated by the teachers. The students work to a much larger extent individually or in groups. The 2005 students control to a higher degree the pacing and sequencing of their work compared to lessons in 1970. In sum: our studies point to a transition in classroom interaction from teaching to studenting, where students to a larger extent govern their own work with an always present possibility of self-exclusion from the teaching project and with frequent crashing of the teaching project from the students' side.

MATHEMATICS EDUCATION IN THE COMPREHENSIVE SCHOOL

There is little of organisational differentiation of students in mathematics education. A main idea is that "a school for everybody" should include all kinds of students in the same classroom. Differences between students should be dealt with by means of individualisation of students learning. Governing by goals and results in Mathematics education is combined with an increased emphasis on problem solving and perceived meaningfulness of mathematics from the students perspective and less of conceptual exercises. The directives in the National Curriculum are more general compared to earlier and there are no recommendations concerning topics and methods in the current curriculum. The goals in the NC deal with students achievements in grade 5 and grade 9, respectively. Below are two selected goals of relevance here to be attained by the end of the ninth year in school:
– Pupils should have acquired the knowledge in mathematics needed to be able to describe and manage situations, as well as solve problems that occur regularly in the home and society, which is needed as a foundation for further education.
– Pupils should be able to interpret and use simple formulae, solve simple equations, as well as be able to interpret and use graphs for functions describing real relationships and events.
Looking at international comparisons we find an interesting trend in the fact that students' self-confidence have increased in combination with lower test performances. Teacher directed lessons have become less frequent and individualised work has increased. The teachers are more like mentors for the individual students and less like lecturers (Petterson & Kjellström, 2005).

CONCLUSIONS

The Swedish welfare state education can be regarded as an ambitious attempt to implement progressive ideas in education. We can compare this to the US. Here,

Lagemann (2000) stated "I have often argued that one cannot understand the history of education in the United States during the twentieth century, unless one understands that Edward L. Thorndike won and John Dewey lost." Dewey had a strong impact on education reforms and curriculum construction in Sweden. Considering this I would conclude: If there is one thing you should know about Northern European Welfare state education it is: Dewey did win – but it was a Pyrrhic victory! To my understanding we can include mathematics education in that general conclusion.

REFERENCES

Bernstein, B. (1996). *Pedagogy, symbolic control and identity: Theory, research, critique.* London: Taylor & Francis.

Dewey, J., & Childs, J. L. (1933). The social-economic situation and education. In W. H. Kilpatrick (Ed.), *The Educational Frontier* (pp. 32-72). New York: D.Appleton-Century.

Lagemann, E. C. (2000) *An elusive science. The troubling history of educational research.* Chicago: The University of Chicago Press

Lindblad, S., & Popkewitz, T. (Eds). (2001). Education, governance, social integration and exclusion: Studies in the powers of reason and the reasons of power. *Uppsala Reports on Education, number 39.*

Lindblad, S., & Sahlström, F. (1999). Gamla mönster och nya gränser. Om ramfaktorer och klassrumsinteraktion. *Pedagogisk Forskning i Sverige, årgång 4*(1) 73-92.

Pettersson, A., & Kjellström, K. (2005). *Matematiken och ungdomskulturen – är det förenligt? I Grundskolans ämnen i ljuset av den nationella utvärderingen 2003.* Stockholm: Skolverket.

Sverker Lindblad
Department of Education
Göteborgs Universitet
Sweden

TERRY WOOD

THE UNITED STATES OF AMERICA

Mathematics Education Reform and the United States Educational System

The United States is a large country covering 9, 372,610 square kilometres and has a population of approximately 294 million (The Economist, 2006). The country is divided into 50 states, 48 states on the mainland and the Hawaiian Islands and Alaska. The population is one of the most diverse with most inhabitants coming to the U.S. as immigrants at various time periods beginning in 1600. Currently, approximately 83% of the population is white with Hispanics as the largest minority population, 12% African American, 3.3% Asian/Pacific and 0.8% Native Americans. The US spends 5.7% of the GNP on education and the people are staunchly adamant about the 'public' nature of schools as is evidenced by the lack of support of various schemes to support private schools system and the support of political candidates who promise increased federal spending for education. Fundamentally, the people of the US believe education is the key to their economic success. Education in the US is relatively consistent in structure. In Figure 1 the various levels of education are presented. Basically, compulsory education is 13 years (5-16 years) and consists of three levels; elementary, middle/secondary and tertiary (post-secondary). Elementary and middle/secondary are part of the K-12 publicly funded school system and tertiary education is supported by some combination of public and private funding. In terms of age enrolment, 98% of the population is enrolled in elementary, 94% secondary and 81% tertiary education (The Economist, 2006).

Nevertheless, there is considerable variation among the states in terms of educational spending. For example in 1998 Utah spent US$3,632 per K-12 student while New Jersey spent US$10,140 (Alexander, 2000). In addition, the system is decentralised in terms of curriculum goals and school policies. As Alexander (2000) correctly states:

> The [US] federal government does devise national education policies and does invest in education,. . .Yet state laws and [school] district decisions have a far greater impact on the day-to-day work of schools than pronouncements from Washington (p. 103).

This decentralisation creates challenges for attempts to reform curriculum as evidenced by the 15 year effort to reform mathematics education in K-12. In the US school mathematics is viewed as consisting predominantly of computational procedures which are mastered through drill and rote learning. Two attempts to

D. J. Clarke, C. Keitel & Y. Shimizu (Eds.), Mathematics Classrooms In Twelve Countries: The Insider's Perspective. 373–375. © *2006 Sense Publishers. All rights reserved.*

reform mathematics education have occurred; the first - in the 1960's - was led by mathematicians who reconceptualised the curriculum to fit their view of the order and sequence of the content. Although the curriculum was conceptual in nature, the reform did not consider issues of learning nor the need for teacher education. The end result was a strong 'back to basics' movement that eliminated this curriculum.

The next reform in mathematics education was instigated by the 1983 report, *A Nation at Risk*, from which the first President Bush's administration established the *National Education Goals for 2000* with overarching broadly delineated goals, and extended by the second President Bush to include a national accountability system under the program No Child Left Behind. The mathematics education reform was the forerunner to all subject matter standards-based reform in the US. The National Research Council (NRC) and the National Council of Teachers of Mathematics (NCTM) and the National Science Foundation (NSF) formed a coalition to support substantial reform in school mathematics. NCR led the way with the document *Everybody Counts* which provided the rational, evidence and direction for reform in mathematics education. This was followed the 1989 NCTM *Curriculum and Evaluation Standards* a document that delineated a visionary plan for reform of curriculum, learning and teaching. This was followed by two other documents which defined changes in teaching and assessment to accommodate the curriculum changes. NSF backed the reform with a commitment of 15 years of funding to support necessary research, development and dissemination. The result was several influential research projects redefining the nature of teaching and learning mathematics in schools along with a reform of curriculum at K-12 grade levels. However, the reform was not without detractors; these were mainly from the conservative population and strangely a few mathematicians. This attempt to discredit the reform movement was known as the 'Math Wars' that lasted a decade. In an attempt to settle the disagreements, NCR formed a task force of mathematics educators, mathematicians and others to resolve differences. This culminated in the 2001 report, *Adding it Up,* and a revision in the original 1989 standards document, renamed the NCTM *Principles and Standards for School Mathematics* (2000). At this time, it is still the case that textbooks dominate instruction at all levels of education; thus determine what is actually taught in schools. These textbooks can be categorised as either 'traditional' or 'reform'. Although the reform curricula are gaining in influence, "for the most part, with the reform movement well into its second decade, its goals have yet to be realized in the large majority of school districts in North America" (Van de Walle, 2004, p. 9).

REFERENCES

Alexander, R. (2000). *Culture and pedagogy.* Oxford, UK: Blackwell Publishers.
National Commission of Excellence in Education. (1983). *A nation at risk: The imperative for educational reform.* Washington, DC: U.S. Government Printing Office.
National Council of Teachers of Mathematics. (1989). *Curriculum and evaluation standards for school mathematics.* Reston, VA: Author.
National Council of Teachers of Mathematics. (2000). *Principles and standards for school mathematics.* Reston, VA: Author.

National Research Council (1989). *Everybody counts: A report to the nation on the future of mathematics education.* Washington, DC: National Academy Press.

National Research Council (2001). *Adding it up: Helping children learn mathematics.* J. Kilpatrick, J. Swafford, & B. Findell (Eds.). Washington, DC: National Academy Press.

The Economist. (2005). *Pocket world in figures 2006 Edition,* London: The Economist Newspaper Ltd.

Van de Walle, J. (2004). Elementary and middle school mathematics: Teaching developmentally. Boston: Allyn and Bacon.

Terry Wood
College of Education
Purdue University
USA

AUTHOR INDEX

SUBJECT INDEX

Further Reading Book 2:

Making Connections
Comparing Mathematics Classrooms Around The World

Editors: **David Clarke**, University of Melbourne, Australia
Jonas Emanuelsson, Götenborgs Universitet, Sweden
Eva Jablonka Freie Universität Berlin, Germany
Ida Ah Chee Mok, The University of Hong Kong, Hong Kong

Paperback ISBN 90-77874-79-8
Hardback ISBN 90-77874-90-9

In this book, comparisons are made between the practices of classrooms in a variety of different school systems around the world. The abiding challenge for classroom research is the realization of structure in diversity. The structure in this case takes the form of patterns of participation: regularities in the social practices of mathematics classrooms. The expansion of our field of view to include international rather than just local classrooms increases the diversity and heightens the challenge of the search for structure, while increasing the significance of any structures, once found. In particular, this book reports on the use of 'lesson events' as an entry point for the analysis of lesson structure.

International research offers opportunities to study settings and characteristics untenable in the researcher's local situation. Importantly, international comparative studies can reveal possibilities for practice that would go unrecognized within the established norms of educational practice of one country or one culture. Our capacity to conceive of alternatives to our current practice is constrained by deep-rooted assumptions, reflecting cultural and societal values that we lack the perspective to question. The comparisons made possible by international research facilitate our identification and interrogation of these assumptions. Such interrogation opens up possibilities for innovation that might not otherwise be identified, expanding the repertoire of mathematics teachers internationally, and providing the basis for theory development.

Other Series of Interest:

New Directions in Mathematics and Science Education

Rationale:
Mathematics and science education are in a state of change. Received models of teaching, curriculum, and researching in the two fields are adopting and developing new ways of thinking about how people of all ages know, learn, and develop. The recent literature in both fields includes articles focusing on issues and using theoretical frames that were unthinkable a decade ago. For example, we see an increase in the use of semiotics as a theoretical tool to understand how students learn, how textbooks are written, and how different forms of knowledge are interconnected. Science and mathematics educators also have turned to issues such as identity and emotion as salient to the way in which people of all ages display and develop knowledge ability. And they use dialectical or phenomenological approaches to answer ever arising questions about learning and development in science and mathematics.

The purpose of this series is to invite and encourage the publication of books that are close to the cutting edge of both fields. The series will be a leader in contributing cutting edge work—rather than out-of-date reproductions of past states of the art—shaping (producing) both fields as much as reproducing them, thereby closing the traditional gap that exists between journal articles and books in terms of their salience about what is new. The series is intended not only to foster books concerned with knowing, learning, and teaching in schools but also with learning in the two fields across the lifespan (e.g., science in kindergarten; mathematics at work); and it is to be a vehicle for publishing books that fall between the two domains—such as when scientists learn about graphs and graphing as part of their work.

Other Titles in Mathematics Education:

Theorems in School
From History, Epistomology and Cognition to Classroom Practice
Paolo Boero, Università di Genova, Italy (ed.)
Paperback ISBN 9077874-21-6 Hardback ISBN 90-77874-22-4

Handbook of Research on the Psychology of Mathematics Education
Past, Present and Future
Angel Gutiérrez, Universidad de Valencia, Spain
Paolo Boero, Università di Genova, Italy (eds.)
Paperback ISBN 9077874-19-4 Hardback ISBN 90-77874-66-6

Traveling Through Education
Uncertainty, Mathematics, Responsibility
Ole Skovsmose, Aalborg University, Denmark
Paperback ISBN 9077874-03-8 Hardback ISBN 90-77874-67-4

Lightning Source UK Ltd.
Milton Keynes UK
19 February 2010